DYNAMIC SYSTEM IDENTIFICATION
Experiment Design and Data Analysis

This is Volume 136 in
MATHEMATICS IN SCIENCE AND ENGINEERING
A Series of Monographs and Textbooks
Edited by RICHARD BELLMAN, *University of Southern California*

The complete listing of books in this series is available from the Publisher
upon request.

DYNAMIC SYSTEM IDENTIFICATION
Experiment Design and Data Analysis

GRAHAM C. GOODWIN   AND   ROBERT L. PAYNE

Department of Electrical Engineering
The University of Newcastle
New South Wales, Australia

ACADEMIC PRESS   *New York   San Francisco   London*   1977

A Subsidiary of Harcourt Brace Jovanovich, Publishers

ACADEMIC PRESS, INC.
111 Fifth Avenue, New York, New York 10003

United Kingdom Edition published by
ACADEMIC PRESS, INC. (LONDON) LTD.
24/28 Oval Road, London NW1

Library of Congress Cataloging in Publication Data

Goodwin, Graham Clifford.
Dynamic system identification.

(Mathematics in science and engineering ; v.
Includes bibliographical references.
1. System analysis. 2. Mathematical models.
3. Experimental design. I. Payne, Robert L.,
joint author. II. Title. III. Series.
QA402.G66 1977 511'.8 76-50396
ISBN 0–12–289750–1

PRINTED IN THE UNITED STATES OF AMERICA

Contents

Chapter 6 Experiment Design

Chapter 7 Recursive Algorithms

Appendix A Summary of Results from Distribution Theory

Appendix B Limit Theorems

Preface

This book is devoted to the theory of mathematical model building, using experimental data. There are many situations where accurate mathematical models of real systems are desirable. For example, engineers often require models for prediction or control purposes. Other areas of application include economics, biology, and sociology.

The material covered is suitable for a postgraduate course on estimation techniques for dynamic systems. It has been presented to higher degree students at Imperial College, the University of New South Wales and at the University of Newcastle. The material can be covered adequately in a one-semester course of 40 hours duration. The book is also intended as a reference for researchers and practitioners in the field.

Each chapter concludes with a set of problems. These problems range from straightforward exercises to those which are intended to challenge the reader's understanding of the subject. In all cases, outline solutions have been provided in the back of the book. Thus, even with the most difficult problem, the reader can gain an appreciation of the solution and proceed to the rest of the book with confidence.

A prerequisite is a knowledge of probability, statistics, stochastic processes, and linear system theory. Suitable background material is contained in Papoulis [10], Silvey [11], and Chen [5]. Suggested concurrent reading is contained in Eykhoff [12], Åström [14], Deutsch [25], and Dhrymes [17]. Additional background material may be found in [16, 21, 23, 24, 29, 32, 33] among others. There are also many papers on identification available in the technical literature. Useful starting points are the four IFAC Symposia on Identification and System Parameter Estimation held in Prague, 1967, 1970; The Hague/Delft, 1973; and Tbilisi, 1976. Also, the December 1974 issue of the IEEE Transactions on Automatic Control has been devoted to System Identification and Time Series Analysis.

During the writing of the text, a number of points of controversy arose. The authors found the following quotation from a paper by Tukey [147] was useful in resolving these conflicts: "Far better an approximate answer to the right question, which is often vague, than an exact answer to the wrong question, which can always be made precise."

The authors gratefully acknowledge many stimulating discussions with colleagues and students. Unfortunately, we cannot name everyone who has assisted, but the following deserve special mention: Professor Brian Anderson, Dr. Reg Brown, Dr. Tony Cantoni, Paul Kabaila, Professor David Mayne, Professor John Moore, Tung Sang Ng, Professor Neville Rees, Professor Lennart Ljung, Professor Tom Kailath, and Marting Zarrop.

The manuscript was adroitly prepared by Lorraine Pogonoski and Dianne Piefke whose skill virtually eliminated the need for proofreading, save for the detection of errors of the authors' making. The bound copies for review were expertly prepared by Kristene Sanson.

The diagrams were prepared by Martin Ooms.

Finally, the authors wish to thank their wives, Rosslyn and Kären, without whose understanding and encouragement the book would not have been written.

CHAPTER

1

Introduction
and Statistical Background

1.1. INTRODUCTION

Fundamental to most physical sciences is the concept of a mathematical model. For example, models are essential for prediction and control purposes. The type and accuracy of the model depends upon the application in mind. For example, models for aerospace applications usually need to be very precise, whereas models for industrial processes, such as blast furnaces, can often be very crude. Models can be obtained from physical reasoning or by analyzing experimental data from the system. In the latter case, our ability to obtain an accurate model is limited by the presence of random fluctuations such as unmeasured disturbances and measurement errors.

The problem of obtaining mathematical models of physical systems from noisy observations is the subject of this book. In particular, we study the problem of estimation of the parameters within models of dynamic systems. We also investigate the effects of various experimental conditions upon model accuracy.

Our approach to the problem is basically statistical in nature, i.e., we treat the disturbances and measurement errors as realizations of stochastic processes. We draw upon many results from probability theory and statistics and, to a lesser extent, linear systems theory. We assume that the reader will have met many of these results previously, but, for completeness, we summarize them in the appendices. Our primary aim has been to give insight. We have, therefore, resisted the temptation to stray either into unnecessary mathematical technicalities, or into ad hoc practical side issues. We therefore hope that the book will be of use, both as a starting point for researchers and as a guide for persons intending to use identification methods in practice.

We have attempted to present a state of the art description of design and analysis of experiments for dynamic systems. We have emphasized those aspects of the subject that are of current interest, and have included many results not previously published in book form. We now briefly review the contents of the book.

In Chapter 2 we discuss the simplest estimation procedure, viz., least squares. We study the properties of the estimator both with and without the assumption of Gaussian modeling errors. Least squares methods are important because of their inherent simplicity and because they form the basis of many more sophisticated procedures. Because of their importance, we also briefly look at a numerically robust procedure for solving least square problems.

In Chapter 3 we introduce the important maximum likelihood approach to estimation. Properties of the estimator are investigated for nondynamic models.

Chapter 4 is concerned with the development of mathematical models for dynamic systems. We discuss various canonical forms and their relevance to estimation of parameters in linear dynamic systems. We also develop a general model structure that is capable of describing a wide class of stochastic systems.

Estimation procedures that are relevant for dynamic systems are described in Chapter 5. In particular we mention least squares, maximum likelihood, instrumental variables, and generalized least squares. We study, in some detail, multiple-input, multiple-output systems, and optimization algorithms. We also treat systems operating under closed loop conditions, i.e., in the presence of feedback. This latter problem is frequently met in practice since it is often not feasible to operate systems in open loop.

In Chapter 6, the problem of accuracy of estimation is considered. It is shown that the choice of experimental conditions (inputs, sampling times, presampling filters, etc.) has a significant effect upon the achievable accuracy. Experiment design procedures in both time and frequency domains are developed.

Finally, in Chapter 7, we describe recursive estimation and experiment design techniques. These are shown to be particularly relevant in situations where the dynamics may be time varying. We also show the application of these ideas to stochastic control.

In the remainder of this chapter we shall review a number of results from the theory of statistical inference. Our treatment of the results will be somewhat brief as we anticipate that many readers will have met the concepts previously. However, we summarize the key results for revision purposes and to establish notation. A reader with a good background in statistical inference could proceed immediately to Chapter 2.

1.2 PROBABILITY THEORY[†]

1.2.1 *Probability spaces* Fundamental to a study of probability theory is the concept of a probability space $(\Omega, \mathscr{A}, P)$, Ω is called the *sample space* and elements ω, of Ω are called *outcomes*. $\mathscr{A}$ is a class of sets of Ω, closed under all countable set operations. $\mathscr{A}$ is called a *σ-algebra* of sets of Ω and elements of $\mathscr{A}$ are called *events*. P is a function defined on $\mathscr{A}$ satisfying the following axioms.

(I) $0 \leq P(A) \leq 1$ for all $A \in \mathscr{A}$

(II) $P(\Omega) = 1$

(III) $P(\cup A_n) = \sum P(A_n)$ for all countable sequences $\{A_n\}$ of sets in $\mathscr{A}$ satisfying $A_i \cap A_j = \varnothing$ for all $i \neq j$

A function satisfying I, II, and III is called a *probability measure*.

1.2.2 *Random variables and distribution functions* A real valued function $X(\cdot)$ defined on the space $(\Omega, \mathscr{A}, P)$ is called a *random variable* if the set $\{\omega : X(\omega) < x\} \in \mathscr{A}$. The probability of the event $\{\omega : X(\omega) < x\}$ is called the *distribution function* and is denoted by $F_X(x)$. We can always write $F_X(x)$ as the sum of an absolutely continuous component and a component that is piecewise constant with a countable number of discontinuities (Lebesgue decomposition lemma [37]). If $F_X(x)$ is absolutely continuous, then it can be expressed in terms of a *probability density function* $p_X(x)$ as follows:

$$F_X(x) = \int_{-\infty}^{x_1} \cdots \int_{-\infty}^{x_n} P_X(y) \, dy_1 \cdots dy_n$$

where x_i and y_i are the ith components of x and y, respectively. Commonly occurring density functions are described in Appendix A.

† See [37, 38, 104–106, 40].

1.2.3 *Conditional probabilities* Consider sets A, $C \in \mathcal{A}$ such that $P(A) \neq 0$. Then the *conditional probability* of A given C (denoted by $P(A|C)$) is defined by

$$P(A|C) = P(A \cap C)/P(A)$$

We say that two events A and C are *statistically independent* if $P(A|C) = P(A)$.

For random variables, we can define a *conditional distribution function* as

$$F_{Y|X}(y|X < x) = P[(Y < y) \cap (X < x)]/P(X < x)$$

Similarly, we can define a *conditional probability density* (if it exists) by

$$p_{Y|X}(y|x) = p_{Y,X}(y, x)/p_X(x)$$

where $p_X(x)$ is the *marginal density function* for x given by

$$p_X(x) = \int_{-\infty}^{\infty} \cdots \int_{-\infty}^{\infty} p_{Y,X}(y, x)\, dy_1 \cdots dy_n$$

1.2.4 *Expectations*   Let X be a random variable. Then we define the *expected value* of X as

$$E_X(X) = \int_{-\infty}^{\infty} \cdots \int_{-\infty}^{\infty} x\, dF_X(x)$$

where the integral is of the Stieltjes type.

If $F_X(x)$ is absolutely continuous, then we can express $E_X(X)$ in terms of the density function as

$$E_X(X) = \int_{-\infty}^{\infty} \cdots \int_{-\infty}^{\infty} xp_X(x)\, dx$$

Similarly, for conditional expectations

$$E_{X|Y}(X) = \int_{-\infty}^{\infty} \cdots \int_{-\infty}^{\infty} xp_{X|Y}(x|y)\, dx$$

1.3 POINT ESTIMATION THEORY[†]

Consider a random variable Y defined on a probability space $(\Omega, \mathcal{A}, P)$, where P is a member of a parametric family $\mathcal{P}$. We denote a general member of $\mathcal{P}$ by P_θ where $\theta \in \Theta$ and $\Theta \subset R^p$.

† See [11, 15, 30, 27].

We are given a realization of Y which we shall call the *data y*. We define an *estimator* as a function $g(Y)$ of the random variable Y. For given data, we call $g(y)$ an *estimate*.

Clearly we would like $g(y)$ to be a "good" estimate of the particular value of θ which corresponds to the "true state of nature" (i.e., $P_\theta = P$). However, we must define exactly what we mean by the term "good." Properties that are commonly used to describe estimators are defined below. However, it should be clear that possession of one or more of these properties does not necessarily imply that the estimator is "best" for a given purpose.

Definition 1.3.1 An estimator $g(Y)$ for θ is said to be *unbiased* if $g(Y)$ has expected value θ under P_θ for all $\theta \in \Theta$, i.e.,

$$E_{Y|\theta}\, g(Y) = \theta \qquad \text{for all} \quad \theta \in \Theta \qquad \nabla \qquad (1.3.1)$$

Definition 1.3.2 An estimator $g(Y)$ for θ is said to be *uniformly minimum mean square error* if

$$E_{Y|\theta}\{(g(Y) - \theta)(g(Y) - \theta)^{\mathrm{T}}\} \le E_{Y|\theta}\{(\tilde{g}(Y) - \theta)(\tilde{g}(Y) - \theta)^{\mathrm{T}}\} \quad (1.3.2)$$

for all $\theta \in \Theta$ and every other estimator $\tilde{g}$. ∇

Comment The requirement of minimum mean square error uniformly in θ is too stringent in general since the right-hand side of (1.3.2) is zero for $\tilde{g} = \theta$. The requirement becomes more meaningful if we restrict the class of estimators.

Definition 1.3.3 An estimator $g(Y)$ is said to be a *minimum variance unbiased estimator* (MVUE) if it has minimum mean square error uniformly in θ among the class of unbiased estimators. ∇

Definition 1.3.4 An estimator $g(Y)$ is a *best linear unbiased estimator* (BLUE) if it has minimum mean square error uniformly in θ among the class of unbiased estimators that are linear functions of the data. ∇

Among the class of unbiased estimators, it is often not feasible to establish the existence of suitable MVUE or BLUE estimators. From a practical point of view it often suffices to show that the estimator being used approaches a lower bound on the variance of all unbiased estimators. A suitable lower bound is described in the following theorem:

Theorem 1.3.1 (*The Cramer–Rao inequality*) Let $\{P_\theta : \theta \in \Theta\}$ be a family of distributions on a sample space Ω, $\Theta \subset R^p$, and suppose that, for each θ, P_θ is defined by a density $p_{Y|\theta}(\cdot|\theta)$. Then subject to certain regularity conditions, the covariance of any unbiased estimator $g(Y)$ of θ satisfies the inequality

$$\text{cov } g \geq M_\theta^{-1} \tag{1.3.3}$$

where

$$\text{cov } g = E_{Y|\theta}\{(g(Y) - \theta)(g(Y) - \theta)^{\mathrm{T}}\} \tag{1.3.4}$$

and where M_θ (*known as Fisher's information matrix*) is defined by

$$M_\theta = E_{Y|\theta}\{[\partial \log p(Y|\theta)/\partial\theta]^{\mathrm{T}}[\partial \log p(Y|\theta)/\partial\theta]\} \tag{1.3.5}$$

Proof Since $g(Y)$ is an unbiased estimator of θ, we have

$$E_{Y|\theta}\{g(Y)\} = \theta \tag{1.3.6}$$

i.e.,

$$\int_\Omega g(y)p(y|\theta)\,dy = \theta, \quad \text{so} \quad \partial/\partial\theta \int_\Omega g(y)p(y|\theta)\,dy = I$$

We now assume sufficient regularity to allow differentiation under the integral sign and thus obtain

$$\int_\Omega g(y) \frac{\partial p(y|\theta)}{\partial\theta}\,dy = I$$

or

$$\int_\Omega g(y) \frac{\partial \log p(y|\theta)}{\partial\theta} p(y|\theta)\,dy = I$$

i.e.,

$$E_{Y|\theta}\left\{g(Y) \frac{\partial \log p(Y|\theta)}{\partial\theta}\right\} = I \tag{1.3.7}$$

Now we also have

$$E_{Y|\theta} \frac{\partial \log p(Y|\theta)}{\partial\theta} = \int_\Omega \frac{\partial \log p(y|\theta)}{\partial\theta} p(y|\theta)\,dy = \int_\Omega \frac{\partial p(y|\theta)}{\partial\theta}\,dy$$

$$= \frac{\partial}{\partial\theta} \int_\Omega p(y|\theta)\,dy = \frac{\partial}{\partial\theta}(1) = \mathbf{0}^{\mathrm{T}} \tag{1.3.8}$$

Thus, using (1.3.6) and (1.3.8), the covariance of $\partial \log p(Y|\theta)/\partial\theta$ and $g(Y)$ is

$$E_{Y|\theta}\left\{\left[\begin{array}{c}(g(Y)-\theta) \\ \left(\dfrac{\partial \log p(Y|\theta)}{\partial\theta}\right)^{\mathrm{T}}\end{array}\right]\left[(g(Y)-\theta)^{\mathrm{T}} \quad \left(\dfrac{\partial \log p(Y|\theta)}{\partial\theta}\right)\right]\right\}$$

$$= \begin{bmatrix} \operatorname{cov} g & I \\ I & M_\theta \end{bmatrix} \tag{1.3.9}$$

where (1.3.4), (1.3.5), (1.3.7), and (1.3.8) have been used.

The matrix in Eq. (1.3.9) is clearly nonnegative definite since it is a covariance matrix. Hence we have

$$[I : -M_\theta^{-1}]\begin{bmatrix} \operatorname{cov} g & I \\ I & M_\theta \end{bmatrix}\begin{bmatrix} I \\ -M_\theta^{-1} \end{bmatrix} \geq 0$$

i.e.,

$$\operatorname{cov} g - M_\theta^{-1} \geq 0 \qquad \triangledown$$

Following Theorem 1.3.1 we are led to the following definition.

Definition 1.3.5 An unbiased estimator is said to be *efficient* if its covariance is equal to the *Cramer–Rao lower bound* (i.e., the inverse of Fisher's information matrix). $\triangledown$

Necessary and sufficient conditions for the existence of efficient unbiased estimators are given in the following theorem.

Theorem 1.3.2 Subject to regularity conditions, there exists an efficient unbiased estimator for θ if and only if we can express $\partial \log p(y|\theta)/\partial\theta$ in the form

$$[\partial \log p(y|\theta)/\partial\theta]^{\mathrm{T}} = A(\theta)[g(y) - \theta] \tag{1.3.10}$$

where $A(\theta)$ is a matrix not depending upon y.

Proof Sufficiency Assume (1.3.10) holds, then Eq. (1.3.9) becomes

$$E_{Y|\theta}\left\{\begin{bmatrix} g(Y)-\theta \\ A(\theta)[g(Y)-\theta] \end{bmatrix}[(g(Y)-\theta)^{\mathrm{T}} \quad (g(Y)-\theta)^{\mathrm{T}}A(\theta)^{\mathrm{T}}]\right\}$$

$$= \begin{bmatrix} \operatorname{cov} g & \operatorname{cov} g\, A^{\mathrm{T}}(\theta) \\ A(\theta)\operatorname{cov} g & A(\theta)\operatorname{cov} g\, A^{\mathrm{T}}(\theta) \end{bmatrix} = \begin{bmatrix} \operatorname{cov} g & I \\ I & M_\theta \end{bmatrix} \qquad \text{(from (1.3.9))}$$

which gives

$$A(\theta) \operatorname{cov} g = I \qquad \text{and} \qquad A(\theta) \operatorname{cov} g \ A^{\mathrm{T}}(\theta) = M_\theta$$

Hence

$$\operatorname{cov} g = M_\theta^{-1}$$

Necessity: Assume $\operatorname{cov} g = M_\theta^{-1}$: then from (1.3.9),

$$E_{Y|\theta}\left\{\left[\begin{array}{c}(g(Y) - \theta) \\ \left(\dfrac{\partial \log p(Y|\theta)}{\partial \theta}\right)^{\mathrm{T}}\end{array}\right]\left[(g(Y) - \theta)^{\mathrm{T}} \quad \dfrac{\partial \log p(Y|\theta)}{\partial \theta}\right]\right\} = \left[\begin{array}{cc} M_\theta^{-1} & I \\ I & M_\theta \end{array}\right]$$

Hence, premultiplying by $[M_\theta, \ -I]$ and postmultiplying by $[M_\theta, \ -I]^{\mathrm{T}}$, gives

$$E_{Y|\theta}\left[M_\theta(g(Y) - \theta) - \left(\dfrac{\partial \log p(Y|\theta)}{\partial \theta}\right)^{\mathrm{T}}\right]\left[M_\theta(g(Y) - \theta) - \left(\dfrac{\partial \log p(Y|\theta)}{\partial \theta}\right)^{\mathrm{T}}\right]$$

$$= 0$$

This implies that $(\partial \log p(Y|\theta)/\partial \theta)^{\mathrm{T}} = M_\theta(g(Y) - \theta)$ and the theorem is proved. ∇

Corollary 1.3.1   We can see from the proof of the above theorem that if (1.3.10) is satisfied, then the matrix $A(\theta)$ is Fisher's information matrix. ∇

We now proceed to consider the asymptotic properties of estimators which become valid for large sample sizes.

Definition 1.3.6   Let g_N be the estimator based on N samples. The sequence $\{g_N, N = 1, \ldots, \infty\}$ is said to be a *consistent* sequence of estimators of θ if the limit of g_N as $N \to \infty$ is θ almost surely (see Appendix B), i.e.,

$$g_N \xrightarrow{\text{a.s.}} \theta \qquad \nabla \qquad (1.3.11)$$

Definition 1.3.7   The sequence $\{g_N, N = 1, \ldots, \infty\}$ is said to be a *weakly consistent* sequence of estimators of θ if the limit of g_N as $N \to \infty$ tends to θ in probability (see Appendix B), i.e.,

$$g_N \xrightarrow{\text{prob}} \theta \qquad \nabla \qquad (1.3.12)$$

1.4. SUFFICIENT STATISTICS[†]

We first define a *statistic*.

Definition 1.4.1   A statistic $s(y)$ is a (measurable) function of the data. ∇

An estimator is, of course, a statistic. Any statistic will, in general, be less "informative" than the original data, i.e., less accurate estimates can be obtained from a statistic than from the data itself.

Definition 1.4.2   A *sufficient statistic* $s(y)$ is a statistic such that the conditional density $p_Y(\cdot|s(y))$ does not depend on θ. ∇

The search for sufficient statistics is facilitated by the following theorem.

Theorem 1.4.1   (*Factorization theorem*)   A statistic $s(y)$ is sufficient for the class $\mathscr{P} = \{p(\cdot|\theta) : \theta \in \Theta\}$ if and only if we can factorize $p(y|\theta)$ (the likelihood function) as

$$p(y|\theta) = f(s(y), \theta)g(y) \qquad (1.4.1)$$

where $g(y)$ is not dependent upon $\theta \in \Theta$. ∇

Proof   See [68].

1.5. HYPOTHESIS TESTING[‡]

The theory of hypothesis testing is concerned with the question: "Is a given set of observations consistent with some hypothesis about the true state of nature?" The key ideas of the classical approach to hypothesis testing were developed by Neyman and Pearson [63].

In the Neyman and Pearson theory, two hypotheses are involved; the so-called *null hypothesis* denoted by H_0 which is the hypothesis of prime interest, and the so-called *alternative hypothesis* denoted by H_A which is the complement of the null hypothesis. A statistical test of the null hypothesis amounts to a partitioning of the sample space into a set which is consistent with the null hypothesis, and the complement of this set. The Neyman–Pearson theory recognizes the following two possible errors:

(a)   Type I   Reject the null hypothesis when it is actually true.
(b)   Type II   Accept the null hypothesis when it is actually false.

† See [15, 27].
‡ See [31, 15].

We denote the partitions of the sample space by: A (the set consistent with the null hypothesis in which we accept the null hypothesis) and $\bar{A}$ (the complement of A, called the critical region in which we reject the null hypothesis). The probability of a Type I error is denoted by α and the probability of a Type II error is denoted by β. We have

$$\alpha = P(\bar{A}|H_0), \qquad \beta = P(A|H_A) = 1 - P(\bar{A}|H_A)$$

The quantity $1 - \beta$ is called the *power function* of the test and α is called the *significance* level of the test.

The basic idea of the Neyman–Pearson theory is to fix α (the significance level) and then to find a test for which the power function $(1 - \beta)$ is maximized uniformly over all alternative hypotheses. Such a test is said to be *uniformly most powerful* (UMP).

The easiest hypothesis testing situation arises when one wishes to test a simple hypothesis θ_0 against a simple alternative θ_1. (A simple hypothesis is one that corresponds to a single state of nature.) In this case we do need the concept of uniformly most powerful since the alternative is a single state. Here the critical region, in which we reject the null hypothesis, is given by the *likelihood ratio test*. This result is known as the Neyman–Pearson fundamental lemma and is stated below:

Theorem 1.5.1   The region $\bar{A}$ in the sample space for which

$$\alpha = P(\bar{A}|\theta_0) = \int_{\bar{A}} p(y|\theta) \, dy \tag{1.5.1}$$

is *fixed* and

$$1 - \beta = P(\bar{A}|\theta_1) = \int_{\bar{A}} p(y|\theta) \, dy \tag{1.5.2}$$

is *maximized* is given by

$$\bar{A} = \left\{ y : \frac{p(y|\theta_1)}{p(y|\theta_0)} \geq k_\alpha \right\} \tag{1.5.3}$$

where k_α is a positive constant chosen so that (1.5.1) is satisfied for the given value of α.

Proof   Choose any other region A^* satisfying (1.5.1), i.e.,

$$\alpha = \int_{\bar{A}} p(y|\theta_0) \, dy = \int_{\bar{A}^*} p(y|\theta_0) \, dy \tag{1.5.4}$$

then

$$\int_{\overline{B}} p(y|\theta_0)\, dy = \int_{\overline{B}^*} p(y|\theta_0)\, dy \qquad (1.5.5)$$

where $\overline{B} = \overline{A} - \overline{A} \cap \overline{A}^*$ and $\overline{B}^* = \overline{A}^* - \overline{A} \cap \overline{A}^*$. Also

$$\int_{\overline{A}} p(y|\theta_1)\, dy - \int_{\overline{A}^*} p(y|\theta_1)\, dy = \int_{\overline{B}} p(y|\theta_1)\, dy - \int_{\overline{B}^*} p(y|\theta_1)\, dy \quad (1.5.6)$$

Now $\overline{B} \subset \overline{A}$ and since $\overline{A}$ satisfies (1.5.3), we have

$$\int_{\overline{B}} p(y|\theta_1)\, dy \geq \int_{\overline{B}} k_\alpha p(y|\theta_0)\, dy \qquad (1.5.7)$$

Also $\overline{B}^* \subset \Omega - \overline{A}$ and hence,

$$\int_{\overline{B}^*} p(y|\theta_1)\, dy \leq \int_{\overline{B}^*} k_\alpha p(y|\theta_0)\, dy \qquad (1.5.8)$$

Substituting (1.5.7) and (1.5.8) into the right-hand side of (1.5.6) gives

$$\int_{\overline{A}} p(y|\theta_1)\, dy - \int_{\overline{A}^*} p(y|\theta_1)\, dy \geq k_\alpha \left\{ \int_{\overline{B}} p(y|\theta_0)\, dy - \int_{\overline{B}^*} p(y|\theta_0)\, dy \right\} (1.5.9)$$

Finally, substituting (1.5.5) into the right-hand side of (1.5.9) gives

$$\int_{\overline{A}} p(y|\theta_1)\, dy - \int_{\overline{A}^*} p(y|\theta_1)\, dy \geq 0 \qquad (1.5.10)$$

So, of all regions satisfying (1.5.1), the region $\overline{A}$ specified in (1.5.3) maximizes the power at θ_1. Hence we have the most powerful test of significance level α. ∇

The above result can be used to construct tests in simple situations. Detailed discussion of the result together with extensions and applications is given in [69]. Unfortunately, however, we are seldom interested in tests for simple hypotheses. Our main purpose in presenting the result here is because it provides motivation for the likelihood ratio test. This test is not UMP in general but it does possess a number of interesting properties. These properties are studied in Chapter 3 of this book.

1.6. THE BAYESIAN DECISION THEORY APPROACH[†]

In the Bayesian approach, in addition to a family of probability distributions $P = \{P_\theta : \theta \in \Theta\}$, we allow a probability distribution $p(\cdot)$ defined on Θ

[†] See [18, 35, 34, 36, 71, 19].

which expresses our degrees of belief that θ lies in various subsets of Θ prior to performing the experiment. We call $p(\theta)$ the prior probability distribution (density).

Bayesian inference is simply the process of transforming prior probability statements to posterior probability statements via Bayes' rule:

$$p(\theta|y) = \frac{p(y|\theta)p(\theta)}{p(y)} \tag{1.6.1}$$

We refer to $p(\cdot|y)$ as the *posterior distribution* defined on Θ. The function $p(\cdot|y)$ summarizes all our knowledge about θ gained both from the prior distribution and from the data. The marginal distribution of y, $p(y)$ is a scaling factor which ensures that $p(\theta|y)$ integrates unity. It is given by

$$p(y) = \int_{\Theta} p(y|\theta)p(\theta) \, d\theta \tag{1.6.2}$$

Presumably we had some purpose in mind when we performed the experiment. If we can quantify our purpose, we are led to the *Bayesian decision theory*, the main ingredients of which are now summarized.

We assume the existence of a *decision space* $\mathscr{D}$ of possible decisions d, a sample space Ω of data y, a parameter space Θ of possible true states of nature and a family $\mathscr{E}$ of possible experiments ε. We further assume that we can define a *loss function* J on $\mathscr{D} \times \Omega \times \Theta \times \mathscr{E}$. We take $J(d, y, \theta, \varepsilon)$ to be the cost associated with performing the experiment ε and making the decision d when the data are y and the true state of nature is θ.

We distinguish two problems: *optimal decision making* and *optimal experiment design*.

Definition 1.6.1   The optimal decision rule is a function $d^*: \Omega \times \mathscr{E} \to \mathscr{D}$ which minimizes the *posterior loss* or *risk* $\bar{J}$, i.e.,

$$\bar{J}(d^*(y, \varepsilon), y, \varepsilon) \le \bar{J}(\tilde{d}(y, \varepsilon), y, \varepsilon)$$

for all other functions $\tilde{d}: \Omega \times \mathscr{E} \to \mathscr{D}$, where

$$\bar{J}(d, y, \varepsilon) = E_{\theta|y, \varepsilon}\{J(d, y, \theta, \varepsilon)\} \qquad \triangledown$$

Definition 1.6.2   The optimal experiment design is a function $\varepsilon^*: \mathscr{D} \to \mathscr{E}$ which minimizes the *expected posterior loss*, i.e.,

$$\hat{J}(\varepsilon^*(d), d) \le \hat{J}(\tilde{\varepsilon}(d), d)$$

for all other $\tilde{\varepsilon}: \mathscr{D} \to \mathscr{E}$, where

$$\hat{J}(\varepsilon, d) = E_{Y|\varepsilon}\{\bar{J}(d, y, \varepsilon)\} \qquad \triangledown$$

It is clear from Definitions 1.6.1 and 1.6.2 that the *joint optimal decision rule and experiment design* d^{**} and ε^{**} may be defined as follows. ε^{**} minimizes

$$E_{Y|\varepsilon}\bar{J}(d^*(y, \varepsilon), y, \varepsilon)$$

where $d^*(y, \varepsilon)$ is specified in Definition 1.6.1, d^{**} is then given by

$$d^{**}(y) = d^*(y, \varepsilon^{**}) \qquad \nabla$$

The Bayesian approach outlined above gives a conceptually pleasing framework within which to state the problems of experiment design and optimal statistical decision making. Obvious difficulties with the approach are

(1)   complexity,
(2)   difficulty of obtaining a suitable loss function, and
(3)   the subjective nature of prior probabilities.

It could, however, be argued that these difficulties are fundamental and the Bayesian approach has merely focused attention on them. Even if not used explicitly, the Bayesian approach provides a useful standard against which other methods can be compared.

An example of Bayesian decision theory is given in Problem 1.9.

1.7.   INFORMATION THEORY APPROACH[†]

Motivated by the problem of describing the information content of signals in communication systems, Shannon [139] introduced the mathematical concept of information. This concept is of a technical nature and cannot be directly equated to the usual emotive meaning of information. However, Shannon's information measure has certain additive properties which link it with the intuitive concept of information. Information theory provides an alternative approach to statistical inference [141, 142]. Here we shall briefly review some of the key concepts.

Definition 1.7.1   The *entropy* of a random variable X having probability density function $p(X)$ is defined to be

$$H_x = -E_X[\log p(X)]$$

[†] See [141].

Result 1.7.1   For the Gaussian case, the entropy is given by

$$H_x = \text{const} + \tfrac{1}{2}\log\det\Sigma$$

where Σ is the covariance matrix of the m-vector random variable X.

Proof

$$p(x) = [(2\pi)^m \det\Sigma]^{-1/2} \exp\{-\tfrac{1}{2}(x-\mu)^{\mathrm{T}}\Sigma^{-1}(x-\mu)\}$$

$$
\begin{aligned}
H_x &= -E_X[\log p(X)] \\
&= -E_X[-\tfrac{1}{2}m\log 2\pi - \tfrac{1}{2}\log\det\Sigma - \tfrac{1}{2}(X-\mu)^{\mathrm{T}}\Sigma^{-1}(X-\mu)] \\
&= \tfrac{1}{2}m\log 2\pi + \tfrac{1}{2}\log\det\Sigma + \tfrac{1}{2}\operatorname{trace}\Sigma^{-1}E_X(X-\mu)(X-\mu)^{\mathrm{T}} \\
&= \tfrac{1}{2}m(\log 2\pi + 1) + \tfrac{1}{2}\log\det\Sigma \qquad \nabla
\end{aligned}
$$

Result 1.7.2   A vector random variable X of m components and covariance Σ has its maximum entropy when its distribution is Gaussian.

Proof   See [139]. ∇

Result 1.7.3   Consider the class of vector random variables having covariance matrices with the same trace. Those random vectors with independently, identically distributed Gaussian components have maximum entropy.

Proof   See Problem 1.10. ∇

Entropy can be very loosely thought of as a measure of disorder or lack of information. We now define a measure of the amount of information provided by an experiment based on the concept of entropy.

Definition 1.7.2 [140]   Lindley's measure of the *average amount of information* provided by an experiment ε with data y and parameters θ is defined to be

$$\mathscr{I}(\varepsilon) = H_\theta - E_y[H_{\theta|y}]$$

where

$$H_\theta = -E_\theta[\log p(\theta)], \qquad H_{\theta|y} = -E_{\theta|y}[\log p(\theta|y)]$$

and $p(\theta)$ and $p(\theta|y)$ are the prior and posterior density functions for θ. ∇

Lindley's information measure is closely related to Shannon's concept of channel capacity [139]. It has also been called the mean information in y

about θ [141], and mutual information [143]. The concept can be given a heuristic interpretation as follows: We note from Result 1.7.1 that entropy is related to the dispersion (or covariance) of the density function and hence the uncertainty. Thus $\mathscr{I}(\varepsilon)$ can be seen to be the difference between the prior uncertainty and the expected posterior uncertainty. Of course this should not be interpreted as a justification of the measure. It is a mathematical rather than a physical concept.

Result 1.7.4   The average information provided by an experiment is nonnegative, being zero if and only if $p(y, \theta) = p(y)p(\theta)$, i.e., if and only if the data are statistically independent of the parameters.

Proof   See [141, p. 15]. ∇

In the next section we describe some commonly used estimators. These estimators are motivated by the concepts introduced in this and in previous sections.

1.8.   COMMONLY USED ESTIMATORS

Definition 1.8.1   Given a function $f: R^N \times R^p \to R^N$, then a least squares estimator $g^*(y)$ is defined as follows:

$$\|f(y, g^*)\|_Q \le \|f(y, \tilde{g})\|_Q$$

for any other function $\tilde{g}(y)$, where $y \in R^N$ are the given data and $\|x\|_Q = x^T Q x$, Q positive definite. ∇

The least squares estimator minimizes the weighted sum of squares of the vector $f(y, g)$. The least squares approach is independent of any underlying probabilistic description of the data-generating mechanism. The least squares estimator is studied in Chapter 2 for the case where $f(y, g)$ is of the form $(y - Xg)$ where X is an $N \times p$ matrix.

Definition 1.8.2   Given a parametric family, $\mathscr{P} = \{p(\cdot|\theta) : \theta \in \Theta\}$, of probability densities on a sample space Ω, then the *maximum likelihood estimator* (MLE), $g^*: \Omega \to \Theta$ is defined by

$$p(y|g^*) \le p(y|\tilde{g})$$

where $\tilde{g}$ is any other function. ∇

The function $p(y|\cdot)$ is called the *likelihood function*. The heuristic justification of the maximum likelihood estimator is that the likelihood function can be interpreted as giving a measure of the plausability of the data under different parameters. Thus we pick the value of θ which makes the data most plausible as measured by the likelihood function.

Result 1.8.1 If $p(y|\theta)$ is Gaussian with mean $\tau(\theta)$ and known non-singular covariance Σ, then the maximum likelihood estimate for θ is identical to the least squares estimate obtained with $f(y, g) = y - \tau(g)$ and $Q = \Sigma^{-1}$.

Proof

$$p(y|\theta) = [(2\pi)^N |\Sigma|]^{-1/2} \exp\{-\tfrac{1}{2}(y - \tau(\theta))^T \Sigma^{-1}(y - \tau(\theta))\}$$

Since log is a monotonic function and Σ is known, maximizing $p(y|\theta)$ is equivalent to minimizing

$$J = (y - \tau(\theta))^T \Sigma^{-1}(y - \tau(\theta)) = \|y - \tau(\theta)\|_{\Sigma^{-1}} \qquad \triangledown$$

Definition 1.8.3 A *maximum a posteriori estimate* (MAP) is the mode of the posterior distribution $p(\theta|y)$. $\triangledown$

Result 1.8.2 For a uniform prior distribution (often called a *non-informative prior*), the MAP estimator coincides with the maximum likelihood estimator.

Proof

$$p(\theta|y) = p(y|\theta)p(\theta)/p(y) = kp(y|\theta)$$

where k does not depend upon θ. The result follows. $\triangledown$

Definition 1.8.4 Consider a Bayesian decision problem in which the decision space $\mathscr{D}$ is isomorphic to the parameter space Θ. If the loss function is $J(d, y, \theta, \varepsilon)$, then we define the *minimum risk estimator* d^* by

$$\bar{J}(d^*(y, \varepsilon), y, \varepsilon) \le \bar{J}(\bar{d}(y, \varepsilon), y, \varepsilon)$$

for all other functions $\bar{d}: \Omega \times \mathscr{E} \to \mathscr{D}$, where $\bar{J}$ is the risk defined by

$$\bar{J}(d, y, \varepsilon) = E_{\theta|y, \varepsilon}\{J(d, y, \theta, \varepsilon)\} \qquad \triangledown$$

Result 1.8.3 For a quadratic loss function of the form

$$J(d, y, \theta, \varepsilon) = (d - \theta)^T W(d - \theta)$$

where W is any symmetric positive definite matrix function of y and ε, the minimum risk estimator, is the posterior mean.

Proof

$$\bar{J}(d, y, \varepsilon) = E_{\theta|y, \varepsilon} J(d, y, \theta, \varepsilon) = E_{\theta|y, \varepsilon}(d - \theta)^{\mathrm{T}} W (d - \theta)$$

$$= E_{\theta|y, \varepsilon} \, \mathrm{trace}\{W(d - \theta)(d - \theta)^{\mathrm{T}}\} = \mathrm{trace}\, WP + (d - \bar{\theta})^{\mathrm{T}} W (d - \bar{\theta})$$

where $\bar{\theta} = E_{\theta|y, \varepsilon} \theta$, the posterior mean, and $P = E_{\theta|y, \varepsilon}\{(\theta - \bar{\theta})(\theta - \bar{\theta})^{\mathrm{T}}\}$, the posterior covariance. It follows that $\bar{J}$ is minimized when d is chosen as $\bar{\theta}$. $\triangledown$

Corollary 1.8.1   If the posterior distribution is symmetric, then the minimum risk estimator under quadratic loss corresponds to the MAP estimator.

Proof   For a symmetric distribution, the mode and the mean coincide. $\triangledown$

The estimators defined above have, with the exception of the least squares case, relied upon a particular probabilistic description of the data. However, in some cases, we may be reluctant to specify a probability distribution. Jaynes [142] has argued that we should be as noncommittal as possible regarding the things we do not know. He used the concept of entropy (cf. Section 1.7) to develop the principle of minimum prejudice which roughly states that the minimally prejudiced assignment of probabilities is that which maximizes the entropy subject to the given information about the situation. For parameter estimation problems the principle leads to the following minimax entropy estimator.

Definition 1.8.5   The *minimax entropy* estimator is constructed so that the parameters in a model which determines the value of the maximum entropy are assigned values which minimize the maximum entropy. $\triangledown$

The minimax entropy estimator requires careful interpretation in practice since it is often unclear how to choose the entropy function. For a particular assignment of the entropy function, the minimax entropy estimator can be related to the maximum likelihood estimator with Gaussian errors, i.e.,

Result 1.8.4   If the entropy function is defined in terms of the modeling errors, then minimizing the maximum entropy function (subject to the constraint that the error covariance is fixed) is equivalent to maximizing the log likelihood function obtained under a Gaussian assumption.

Proof [The reader may care to leave this proof until after Chapter 3 where maximum likelihood will be studied in more detail. See Problem 3.9.]

Let $(y_1, \ldots, y_N)$ denote a set of data. Assume that the data can be (approximately) modeled by a known function $f_t(\theta)$, where θ is a parameter vector. The sample covariance of the modeling errors with parameter value $\hat{\theta}$ is

$$D(\hat{\theta}) = \frac{1}{N} \sum_{t=1}^{N} \omega_t \omega_t^{\mathrm{T}} \qquad \text{where} \quad \omega_t(\hat{\theta}) = y_t - f_t(\hat{\theta})$$

From Results 1.7.2 and 1.7.3 we know that if the covariance of $\{\omega_t\}$ is fixed at $D(\hat{\theta})$, then the entropy H_ω of $\omega_1, \ldots, \omega_N$ is maximized when these random variables are independent and Gaussian. Hence using Result 1.7.1 we have, subject to $\mathrm{cov}(\omega_t) = D(\hat{\theta})$ that maximum $H_\omega = N[\frac{1}{2}m(\log 2\pi + 1) + \frac{1}{2} \log \det D(\hat{\theta})]$.

On the other hand, if we assume that the data are Gaussian and independently distributed with mean $\{f_t(\theta)\}$ and common covariance Σ, then the likelihood function becomes

$$p(y_1, \ldots, y_N | \theta, \Sigma) = [(2\pi)^m \det \Sigma]^{-N/2} \exp\left\{ -\frac{1}{2} \sum_{t=1}^{N} [y_t - f_t(\theta)]^{\mathrm{T}} \Sigma^{-1} [y_t - f_t(\theta)] \right\}$$

Since log is monotonic, the maximum likelihood estimator is obtained by maximizing

$$\log p(y_1, \ldots, y_N | \theta, \Sigma) = \frac{-Nm}{2} \log 2\pi - \frac{N}{2} \log \det \Sigma$$

$$-\frac{1}{2} \sum_{t=1}^{N} [y_t - f_t(\theta)]^{\mathrm{T}} \Sigma^{-1} [y_t - f_t(\theta)]$$

For any value of θ, say $\hat{\theta}$, the above expression for $\log p(y_1, \ldots, y_N | \theta)$ can be minimized with respect to Σ giving

$$\hat{\Sigma} = \frac{1}{N} \sum_{t=1}^{N} [y_t - f_t(\hat{\theta})][y_t - f_t(\hat{\theta})]^{\mathrm{T}} = D(\hat{\theta})$$

We can use the above expression to eliminate Σ from the likelihood function yielding

$$\log p(y_1, \ldots, y_N | \theta, \Sigma = D(\theta)) = -\tfrac{1}{2}Nm(\log 2\pi + 1) - \tfrac{1}{2}N \log \det D(\theta)$$

Comparing the above expression with the previous expression for max H_ω establishes the equivalence of the maximum likelihood and mini-max entropy estimators in this case. $\triangledown$

1.9. CONCLUSIONS

This chapter has surveyed some of the basic concepts from the theory of statistical inference. We shall frequently call upon the results of this chapter in our later work. In the next chapter, we begin our detailed study of estimation procedures with the least squares approach.

PROBLEMS

1.1. Show that $E_{X,Y}(X + Y) = E_X(X) + E_Y(Y)$.

1.2. Show that

$$E_\theta\{g(\theta)E_{Y|\theta}\{f(Y)\}\} = E_Y\{f(Y)E_{\theta|Y}\{g(\theta)\}\}$$

1.3. Consider a random variable Y with probability density function $p(y|\mu, \sigma^2) = (1/\sigma\sqrt{2\pi}) \exp\{-(1/2\sigma^2)(y - \mu)^2\}$. Find the mean and variance of Y.

1.4. Consider N identically distributed random variables $X_1, \ldots, X_N$ with common mean μ. Show that the sample mean $\bar{x} = (1/N)\sum_{i=1}^N x_i$ is an unbiased estimator of μ. Show that $\frac{1}{2}(x_1 + x_N)$ is also an unbiased estimator of μ. Comment on the results.

1.5. Consider N independent samples, each with associated density

$$p(y_i|\mu, \sigma^2) = (1/\sigma\sqrt{2\pi}) \exp\{-(1/2\sigma^2)(y_i - \mu)^2\}$$

Show that a sufficient statistic for μ and σ^2 is

$$S(y_1, \ldots, y_N) = \begin{bmatrix} \dfrac{1}{N} \sum_{i=1}^N y_i \\ \dfrac{1}{N} \sum_{i=1}^N y_i^2 \end{bmatrix}$$

1.6. Evaluate the information matrix for μ and σ^2 in Problem 1.5.

1.7. Evaluate the covariance of the estimator

$$g = \frac{1}{N} \sum_{i=1}^N y_i$$

in Problem 1.5. Show that cov g achieves the Cramer–Rao lower bound.

1.8. Consider three independent normal random variables x_K, ω_K, and v_K where $x_K \sim N(\hat{x}_K, P_K)$, $\omega_K \sim N(0, Q_K)$, $v_K \sim N(0, R_K)$ (see Section A.2). We define two further random variables x_{K+1} and y_K by the following linear equations:

$$x_{K+1} = A_K x_K + \omega_K, \qquad y_K = C_K x_K + v_K$$

(a)   Find the joint distribution of x_{K+1} and y_K (Hint: See Result A.2.4.)

(b)   Show that the conditional distribution for x_{K+1} given y_K is normal with mean $\hat{x}_{K+1}$ and covariance P_{K+1} where

$$\hat{x}_{K+1} = A_K \hat{x}_K + A_K K_K(y_K - C_K \hat{x}_K)$$

$$P_{K+1} = A_K(P_K - K_K C_K P_K)A_K^T + Q_K$$

where

$$K_K = P_K C_K^T (C_K P_K C_K^T + R_K)^{-1}$$

(c)   By assuming $x_0 \sim N(\bar{x}_0, P_0)$, use induction to show that the conditional distribution of x_i given $y_{i-1}, y_{i-2}, \ldots, y_1$ is normal and that the conditional mean and covariances can be obtained recursively from the equations found in part (b). Comment: This question gives a derivation of the well-known Kalman–Bucy filter (see [14] for further details).

1.9.   Consider a two-party political system. We denote the parties by W and H. An opinion poll just prior to an election indicates that the probability of the outcome θ being W, $p(\theta = W)$ is 0.6 and that the probability of the outcome θ being H, $p(\theta = H)$ is 0.4.

We wish to place a $100 bet on the outcome of the election. The betting agency will pay $200 ($100 profit plus our $100 stake) if we correctly back W and $300 ($200 profit plus and $100 stake) if we correctly back H.

Before placing our bet (which we assume will be either on W or H but not split), we have time to consult either a man in the street A (at no cost), or B, a respected political commentator (at a cost of $100). (Thus we have described two allowable experiments ε_A, ε_B.)

The probability that $\varepsilon = \varepsilon_A$ or ε_B will predict the outcome $y = W$ or H when the real outcome will actually be $\theta = W$ or H is summarized as follows:

$$p(y = W | \varepsilon = \varepsilon_A, \theta = W) = 0.9 \qquad p(y = W | \varepsilon = \varepsilon_B, \theta = W) = 0.9$$

$$p(y = H | \varepsilon = \varepsilon_A, \theta = W) = 0.1 \qquad p(y = H | \varepsilon = \varepsilon_B, \theta = W) = 0.1$$

$$p(y = W | \varepsilon = \varepsilon_A, \theta = H) = 0.8 \qquad p(y = W | \varepsilon = \varepsilon_B, \theta = H) = 0.1$$

$$p(y = H | \varepsilon = \varepsilon_A, \theta = H) = 0.2 \qquad p(y = H | \varepsilon = \varepsilon_B, \theta = H) = 0.9$$

Note that A has bias for the W party.

(a)   Show that the gain (negative loss) associated with the two decisions d_W (back W) and d_H (back H) when the real outcome is $\theta = W$ or H and we perform experiments ε_A or ε_B is given by the following table:

	ε_A		ε_B	
	d_W	d_H	d_W	d_H
$\theta = W$	+$100	−$100	$0	−$200
$\theta = H$	−$100	+$200	−$200	+$100

Hence define a suitable loss function $J(d, y, \theta, \varepsilon)$.

(b)   Use Bayes' rule to determine the posterior distribution for θ under the two possible experiments $\varepsilon = \varepsilon_A$ and $\varepsilon = \varepsilon_B$ and the two possible predictions $y = W$ or $y = H$. Show that the results are

$$p(\theta = W \,|\, \varepsilon = \varepsilon_A, y = W) = 0.54/0.86$$

$$p(\theta = W \,|\, \varepsilon = \varepsilon_A, y = H) = 0.06/0.14$$

$$p(\theta = H \,|\, \varepsilon = \varepsilon_A, y = W) = 0.32/0.86$$

$$p(\theta = H \,|\, \varepsilon = \varepsilon_A, y = H) = 0.08/0.14$$

$$p(\theta = W \,|\, \varepsilon = \varepsilon_B, y = W) = 0.54/0.58$$

$$p(\theta = W \,|\, \varepsilon = \varepsilon_B, y = H) = 0.06/0.42$$

$$p(\theta = H \,|\, \varepsilon = \varepsilon_B, y = W) = 0.04/0.58$$

$$p(\theta = H \,|\, \varepsilon = \varepsilon_B, y = H) = 0.36/0.42$$

(c)   For each possible experiment, prediction, and decision, show that the expected gain (negative risk)

$$\bar{J}(d, y, \varepsilon) = E_{\theta|y, \varepsilon} J(d, y, \theta, \varepsilon)$$

is given by

	ε_A		ε_B	
	$y = W$	$y = H$	$y = W$	$y = H$
d_W	$22/0.86	−$2/0.14	−$8/0.58	−$72/0.42
d_H	$10/0.86	$10/0.14	−$104/0.58	$24/0.42

(d)   Show that the joint optimal decision rule and optimal experiment is to ask the man in the street for his prediction, and then to back the party he predicts. Show that the expected return before performing the optimal experiment is $32.

1.10.   Use Result 1.7.2 to establish Result 1.7.3.

CHAPTER

2

Linear Least Squares
and Normal Theory

2.1. INTRODUCTION

In this chapter we shall study the relatively simple linear least squares approach to identification. The basic notion of least squares is that parameter estimates are chosen so that the model output agrees with the data as closely as possible as measured by the sum of the squares of the differences between them. The least squares technique was originally developed independently by Gauss and Legendre in the early 19th century. The concept of least squares analysis is explained by Gauss in Theoria Motus Corporum Coelestium [67] (Theory of the Motion of Heavenly Bodies):

> If the astronomical observations and other quantities on which the computation of orbits is based were absolutely correct, the elements also, whether deduced from three or four observations would be strictly accurate (so far indeed as the motion is supposed to take place exactly according to the laws of Kepler) and, therefore, if other observations were used, they might be confirmed but not corrected. But since all our measurements and observations are nothing more than approximations to the truth, the same must be true of all calculations

resting upon them, and the highest aim of all computations concerning concrete phenomena must be to approximate, as nearly as practicable to the truth. But this can be accomplished in no other way than by a suitable combination of more observations than the number absolutely requisite for the determination of the unknown quantities.

Gauss goes on to say that

... the most probable value of the unknown quantities will be that in which the sum of the squares of the differences between the actually observed and the computed values multiplied by numbers that measure the degree of precision is a minimum.

The linear least squares problem may be motivated as follows: Suppose we have a stochastic process $\{Y_t, t \in (1, \ldots, N)\}$ for which the mean value is a linear function of a parameter vector θ; i.e.,

$$E[Y_t] = x_t^T\theta \tag{2.1.1}$$

where x_t is a known vector for $t = 1, \ldots, N$. Our aim is to obtain a good estimate (good in the least squares sense) of θ from a realization of the stochastic process $\{y_t, t \in (1, \ldots, N)\}$.

We shall investigate various properties of the least squares estimator. Also of interest are the distributions of the estimator and of the functions of residuals (differences between observed data and model output) when the process $\{Y_t, t \in (1, \ldots, N)\}$ is Gaussian. We shall illustrate by means of examples how these distributions may be used to test hypotheses regarding the model and to obtain confidence intervals for the parameters.

2.2.   THE LEAST SQUARES SOLUTION

We begin by using an intuitive approach. We seek the value of $\hat{\theta}$ which minimizes the sum of the squares of the deviations between y_t and the mean predicted by the model, namely, $x_t^T\theta$, i.e., we attempt to minimize

$$S = \sum_{t=1}^{N} (y_t - x_t^T\theta)^2 \tag{2.2.1}$$

Equation (2.2.1) can be written in vector form

$$S = (Y - X\theta)^T(Y - X\theta) \tag{2.2.2}$$

where

$$Y = \begin{bmatrix} y_1 \\ \vdots \\ y_N \end{bmatrix}, \quad X = \begin{bmatrix} x_1^T \\ \vdots \\ x_N^T \end{bmatrix} \tag{2.2.3}$$

Differentiating Eq. (2.2.2) with respect to θ shows that the value $\hat{\theta}$ which minimizes S satisfies the equation (often called the *normal equation*)

$$(X^TX)\hat{\theta} = X^TY \tag{2.2.4}$$

If X^TX is invertible, then there is a unique solution which can be expressed as

$$\hat{\theta} = [X^TX]^{-1}X^TY \tag{2.2.5}$$

Equation (2.2.5) is often called the *least squares estimator* since it minimizes the sum of the squares given in (2.2.1). For the case where X^TX is singular, the normal equation does not have a unique solution and there is a family of least squares estimates which may be determined in any particular case by the usual methods for solving linear equations [84, 86].

Example 2.2.1 Consider the following (moving average) model of a stochastic process

$$E[Y_t] = \sum_{k=0}^{p} h_k u_{t-k}, \qquad t = 1, \ldots, N$$

where the u_t, $t = 1 - p, \ldots, N$ are known and h_k, $k = 0, \ldots, p$ are unknown parameters.

Thus, using the notation already introduced

$$x_t^T = (u_t, u_{t-1}, \ldots, u_{t-p})$$

and

$$X^TX = \sum_{t=1}^{N} \begin{bmatrix} u_t^2 & u_{t-1}u_t & \cdots & u_{t-p}u_t \\ u_t u_{t-1} & u_{t-1}^2 & \cdots & \\ \vdots & \vdots & & \vdots \\ u_t u_{t-p} & & \cdots & u_{t-p}^2 \end{bmatrix}, \qquad X^TY = \sum_{t=1}^{N} \begin{bmatrix} u_t y_t \\ u_{t-1}y_t \\ \vdots \\ u_{t-p}y_t \end{bmatrix}$$

The least squares estimate is obtained from Eq. (2.2.5). ∇

The least squares problem can be given a geometric interpretation by looking at Eq. (2.2.2). We observe that $\hat{Y} = X\hat{\theta}$ defines a subspace of R^N spanned by the columns of X. The quantity S in Eq. (2.2.2) thus represents the square of the distance from some point $\hat{Y}$ in this subspace to the given vector Y. For any $\hat{\theta}$ satisfying (2.2.4), the error vector $Y - \hat{Y}$ is perpendicular (or normal) to the subspace spanned by the columns of X. This follows from (2.2.4) since

$$X^TX\hat{\theta} - X^TY = 0$$

i.e.,

$$X^T(\hat{Y} - Y) = 0 \tag{2.2.6}$$

(In fact, from (2.2.6), $\hat{Y} - Y$ is in the nullspace of X^{T}, i.e., $\hat{Y} - Y$ is in the orthogonal complement of the range space of X [39].) This property explains the use of the term normal equation for (2.2.4).

Example 2.2.2 Consider

$$E[Y_t] = (3 - t)\theta, \qquad t = 1, 2$$

Thus, $X = (2 \quad 1)^{\mathrm{T}}$. Say we observe

$$y_1 = 4, \qquad y_2 = 1$$

For this example, the geometric interpretation of the least squares solution is depicted in Fig. 2.2.1.

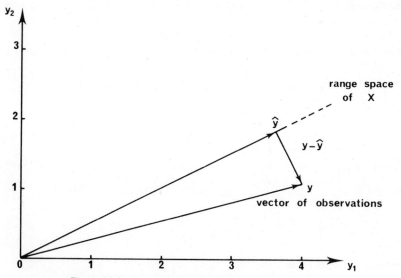

Figure 2.2.1. Geometric interpretation of least squares.

In subsequent sections we investigate further properties of the least squares and related estimators.

2.3. BEST LINEAR UNBIASED ESTIMATORS

A best linear unbiased estimator (BLUE) is defined as the estimator that has minimum variance among the class of all linear unbiased estimators (see Definition 1.3.4). For our current problem, we thus require $\hat{\theta}$ to be a linear function of Y, i.e.,

$$\hat{\theta} = CY \tag{2.3.1}$$

We restrict attention to stochastic processes having mean given by (2.1.1) with θ the true parameter value, and having covariance given by

$$E[(y_t - x_t^T\theta)(y_s - x_s^T\theta)] = \sigma^2 \quad \text{if} \quad t = s \tag{2.3.2}$$
$$= 0 \quad \text{otherwise} \tag{2.3.3}$$

where E denotes expectation over the distribution of y given θ. Combining (2.3.2) and (2.3.3),

$$E[(Y - X\theta)(Y - X\theta)^T] = \sigma^2 I \tag{2.3.4}$$

Properties of the least squares estimator under the above assumptions are discussed in the following theorem:

Theorem 2.3.1   Consider a second-order stochastic process Y, having mean $X\theta$ and covariance $\sigma^2 I$. Then the least squares estimator of θ given by

$$\hat{\theta} = [X^T X]^{-1} X^T Y \tag{2.3.5}$$

has the following properties

(i)   It is a linear function of the data
(ii) $E_Y[\hat{\theta}] = \theta$ (i.e., $\hat{\theta}$ is unbiased—see Definition 1.3.1)   (2.3.6)
(iii) $E_Y[(\hat{\theta} - \theta)(\hat{\theta} - \theta)^T] = (X^T X)^{-1}\sigma^2$ (2.3.7)
(iv) $E_Y[(\hat{\theta} - \theta)(\hat{\theta} - \theta)^T] \leq E_Y[(\tilde{\theta} - \theta)(\tilde{\theta} - \theta)^T]$ (2.3.8)

where $\tilde{\theta}$ is any other linear unbiased estimator (i.e., $\hat{\theta}$ is BLUE).

Proof   (i)   From (2.3.5),

$$\hat{\theta} = LY \tag{2.3.9}$$

where $L = (X^T X)^{-1} X^T$.
(ii)

$$E_Y[\hat{\theta}] = E_Y[(X^T X)^{-1} X^T Y] = (X^T X)^{-1} X^T E_Y Y = (X^T X)^{-1} X^T X\theta = \theta$$

(iii)

$$\hat{\theta} - \theta = (X^T X)^{-1} X^T Y - \theta = (X^T X)^{-1} X^T (Y - X\theta)$$
$$E_Y[(\hat{\theta} - \theta)(\hat{\theta} - \theta)^T] = (X^T X)^{-1} X^T E_Y (Y - X\theta)(Y - X\theta)^T X (X^T X)^{-1}$$
$$= (X^T X)^{-1} X^T I X (X^T X)^{-1}\sigma^2 = (X^T X)^{-1}\sigma^2$$

(iv)   Consider any linear estimator $\tilde{\theta}$ where $\tilde{\theta} = CY$. If $\tilde{\theta}$ is an unbiased estimator, then

$$E_Y(\tilde{\theta}) = \theta$$

i.e.,

$$E_Y(CY) = \theta, \qquad CX\theta = \theta \tag{2.3.10}$$

For (2.3.10) to be true for any θ, we require

$$CX = I \tag{2.3.11}$$

The covariance of $\tilde{\theta}$ is given by

$$\begin{aligned} E_Y[(\tilde{\theta} - \theta)(\tilde{\theta} - \theta)^T] &= E_Y[(CY - \theta)(CY - \theta)^T] \\ &= E_Y[(CY - CX\theta + CX\theta - \theta)(CY - CX\theta + CX\theta - \theta)^T] \\ &= E_Y[(CY - CX\theta)(CY - CX\theta)^T] \end{aligned}$$

(using (2.3.11))

$$= CE_Y[(Y - X\theta)(Y - X\theta)^T]C^T = CC^T\sigma^2 \tag{2.3.12}$$

We now define D in terms of C as

$$D = C - (X^TX)^{-1}X^T \tag{2.3.13}$$

Now we form DD^T as

$$\begin{aligned} DD^T &= [C - (X^TX)^{-1}X^T][C - (X^TX)^{-1}X^T]^T \\ &= CC^T - (X^TX)^{-1}X^TC^T - CX(X^TX)^{-1} + (X^TX)^{-1} \\ &= CC^T - (X^TX)^{-1} \qquad \text{(using (2.3.11))} \\ &\geq 0 \end{aligned} \tag{2.3.14}$$

Thus

$$CC^T\sigma^2 \geq (X^TX)^{-1}\sigma^2$$

This establishes part (iv) of the theorem. ∇

Corollary 2.3.1   Consider the stochastic process Y having mean $X\theta$ and covariance Σ (where Σ is known positive definite matrix). Then the BLUE for θ is

$$\hat{\theta} = (X^T\Sigma^{-1}X)^{-1}X^T\Sigma^{-1}Y \tag{2.3.15}$$

Proof   We can write Σ in the form PP^T where P is nonsingular. Hence defining

$$\overline{Y} = P^{-1}Y \tag{2.3.16}$$

Then

$$E[\overline{Y}] = E[P^{-1}Y] = P^{-1}X\theta \tag{2.3.17}$$

We can thus write

$$E[\overline{Y}] = \overline{X}\theta, \qquad \overline{X} = P^{-1}X \qquad (2.3.18)$$

Also

$$E[(\overline{Y} - \overline{X}\theta)(\overline{Y} - \overline{X}\theta)^{\mathrm{T}}] = E[P^{-1}(Y - X\theta)(Y - X\theta)^{\mathrm{T}}P^{-\mathrm{T}}]$$
$$= P^{-1}\Sigma P^{-\mathrm{T}} = P^{-1}PP^{\mathrm{T}}P^{-\mathrm{T}} = I \qquad (2.3.19)$$

Hence, from Theorem 2.3.1 the BLUE is

$$\hat{\theta} = (\overline{X}^{\mathrm{T}}\overline{X})^{-1}\overline{X}^{\mathrm{T}}\overline{Y} = (X^{\mathrm{T}}P^{-\mathrm{T}}P^{-1}X)^{-1}X^{\mathrm{T}}P^{-\mathrm{T}}P^{-1}Y$$
$$= (X^{\mathrm{T}}\Sigma^{-1}X)^{-1}X^{\mathrm{T}}\Sigma^{-1}Y \qquad (2.3.20)$$

This establishes the Corollary.   ▽

Note   The BLUE estimator described in Corollary 2.3.1 is also the estimator which minimizes the following weighted sum of squares

$$S = (Y - X\theta)^{\mathrm{T}}\Sigma^{-1}(Y - X\theta) \qquad (2.3.21)$$

We can thus give Gauss's interpretation to Σ^{-1} as expressing the relative precision of our measurements (see Section 2.1).

The covariance of θ described in Eq. (2.3.20) is readily shown to be

$$E[(\hat{\theta} - \theta)(\hat{\theta} - \theta)^{\mathrm{T}}] = (X^{\mathrm{T}}\Sigma^{-1}X)^{-1} \qquad (2.3.22)$$

A possible interpretation of Corollary 2.3.1 is that, if Σ is known (at least up to a scalar multiplier), then by means of a suitable linear transformation of the data (i.e., $\overline{Y} = P^{-1}Y$), the problem may be reduced to the ordinary least squares situation of Theorem 2.3.1. The transformation P^{-1} is sometimes called a *prewhitening filter*. The case where Σ is unknown is also amenable to solution but we defer discussion on the most general problem until a later chapter (cf. Sections 3.3 and 5.3). Here we shall concentrate on the case where Σ takes the special form $\sigma^2 D$ with σ^2 unknown but D known. The known matrix D can be eliminated by use of a prewhitening filter and thus we assume $D = I$ without loss of generality.

2.4.   UNBIASED ESTIMATION OF BLUE COVARIANCE

It was shown in Theorem 2.3.1 that the covariance of $\hat{\theta}$ was given by

$$E_Y[(\hat{\theta} - \theta)(\hat{\theta} - \theta)^{\mathrm{T}}] = (X^{\mathrm{T}}X)^{-1}\sigma^2 \qquad (2.4.1)$$

Of course, σ^2 is a parameter of the distribution of Y and hence will also need to be estimated in general. A suitable unbiased estimator of σ^2 is described in the following theorem.

Theorem 2.4.1 An unbiased estimator of σ^2 is

$$\hat{V} = S(\hat{\theta})/(N - p) \tag{2.4.2}$$

where N is the number of observations, p the number of parameters, and S the sum of the squares of the residuals

$$S(\hat{\theta}) = (Y - X\hat{\theta})^{\mathsf{T}}(Y - X\hat{\theta}) \tag{2.4.3}$$

Proof Substituting (2.3.5) into (2.4.3) yields

$$\begin{aligned}
S(\hat{\theta}) &= (Y - X(X^{\mathsf{T}}X)^{-1}X^{\mathsf{T}}Y)^{\mathsf{T}}(Y - X(X^{\mathsf{T}}X)^{-1}X^{\mathsf{T}}Y) \\
&= Y^{\mathsf{T}}(I_N - X(X^{\mathsf{T}}X)^{-1}X^{\mathsf{T}})(I_N - X(X^{\mathsf{T}}X)^{-1}X^{\mathsf{T}})Y \\
&= Y^{\mathsf{T}}(I_N - X(X^{\mathsf{T}}X)^{-1}X^{\mathsf{T}})Y = \operatorname{trace}(I_N - X(X^{\mathsf{T}}X)^{-1}X^{\mathsf{T}})Y Y^{\mathsf{T}}
\end{aligned}$$

Therefore,

$$\begin{aligned}
E[S] &= \operatorname{trace}(I_N - X(X^{\mathsf{T}}X)^{-1}X^{\mathsf{T}})[\sigma^2 I_N + X\theta\theta^{\mathsf{T}}X^{\mathsf{T}}] \\
&= \sigma^2 \operatorname{trace}(I_N - X(X^{\mathsf{T}}X)^{-1}X^{\mathsf{T}}) \\
&= \sigma^2 \operatorname{trace} I_N - \sigma^2 \operatorname{trace} X(X^{\mathsf{T}}X)^{-1}X^{\mathsf{T}} \\
&= \sigma^2 \operatorname{trace} I_N - \sigma^2 \operatorname{trace}(X^{\mathsf{T}}X)^{-1}X^{\mathsf{T}}X \\
&\quad (\text{since } \operatorname{trace} AB = \operatorname{trace} BA) \\
&= \sigma^2 \operatorname{trace} I_N - \sigma^2 \operatorname{trace} I_p = \sigma^2(N - p)
\end{aligned} \tag{2.4.4}$$

i.e., $E[\hat{V}] = \sigma^2$ as required. $\triangledown$

2.5. NORMAL THEORY

The theory developed above made no requirements on the form of distribution of Y save that the mean be $X\theta$ and variance $\sigma^2 I$. We now make the further assumption that Y is distributed normally, i.e., $Y \sim N(X\theta, \sigma^2 I)$ (see Section A.2). We now have the following stronger result on the optimality of the least squares estimator.

Theorem 2.5.1 Consider a normal stationary stochastic process having mean $X\theta$ and covariance $\sigma^2 I$. The minimum variance unbiased estimator (MVUE) (see Definition 1.3.3) of θ is the least squares estimator

$$\hat{\theta} = (X^{\mathsf{T}}X)^{-1}X^{\mathsf{T}}Y \tag{2.5.1}$$

Proof From Eq. (1.3.5), Fisher's information matrix is

$$M_\beta = E_{Y|\beta}\left\{ \left[\frac{\partial \log p(Y|\beta)}{\partial \beta}\right]^{\mathsf{T}} \left[\frac{\partial \log p(Y|\beta)}{\partial \beta}\right] \right\} \tag{2.5.2}$$

where

$$\beta = (\theta^{\mathrm{T}}, \sigma^2)^{\mathrm{T}} \tag{2.5.3}$$

Now using the assumed normal distribution for the data, we have

$$p(Y|\beta) = (2\pi\sigma^2)^{-N/2} \exp\{-(1/2\sigma^2)(Y - X\theta)^{\mathrm{T}}(Y - X\theta)\} \tag{2.5.4}$$

$$\partial \log p(Y|\beta)/\partial\theta = (1/\sigma^2)(Y - X\theta)^{\mathrm{T}}X = v_1^{\mathrm{T}} \tag{2.5.5}$$

$$\partial \log p(Y|\beta)/\partial\sigma^2 = -(N/2\sigma^2) + (1/2\sigma^4)(Y - X\theta)^{\mathrm{T}}(Y - X\theta) = v_2 \tag{2.5.6}$$

Hence

$$M_\beta = E_{Y|\beta} \begin{bmatrix} v_1 \\ v_2 \end{bmatrix} [v_1^{\mathrm{T}} \quad v_2] \tag{2.5.7}$$

$$= E_{Y|\beta} \begin{vmatrix} v_1 v_1^{\mathrm{T}} & v_1 v_2 \\ v_2 v_1^{\mathrm{T}} & v_2^2 \end{vmatrix} \tag{2.5.8}$$

Now

$$E_{Y|\beta}(v_1 v_2) = 0 \qquad \text{(since first and third central moments of a normal distribution are zero)} \tag{2.5.9}$$

$$E_{Y|\beta}(v_1 v_1^{\mathrm{T}}) = E_{Y|\beta}(1/\sigma^4)X^{\mathrm{T}}(Y - X\theta)(Y - X\theta)^{\mathrm{T}}X$$
$$= (1/\sigma^4)X^{\mathrm{T}}[\sigma^2 I]X = (1/\sigma^2)X^{\mathrm{T}}X \tag{2.5.10}$$

Hence from the Cramer–Rao inequality (Theorem 1.3.1), the covariance of any unbiased estimator $\hat{g}(Y)$ of β satisfies the inequality

$$\operatorname{cov} \hat{g} \geq M_\beta^{-1} = \begin{bmatrix} \dfrac{1}{\sigma^2}X^{\mathrm{T}}X & 0 \\ 0 & E_{Y|\beta} v_2^2 \end{bmatrix}^{-1} = \begin{bmatrix} (X^{\mathrm{T}}X)^{-1}\sigma^2 & 0 \\ 0 & (E_{Y|\beta} v_2^2)^{-1} \end{bmatrix} \tag{2.5.11}$$

However, it was shown in Theorem 2.3.1 that

$$\operatorname{cov} \hat{\theta} = (X^{\mathrm{T}}X)^{-1}\sigma^2 \tag{2.5.12}$$

where $\hat{\theta}$ is the least squares estimator of θ.

Thus $\hat{\theta}$ achieves the Cramer–Rao lower bound. Since $\hat{\theta}$ is also unbiased (Theorem 2.3.1), it follows that $\hat{\theta}$ is a MVUE of θ. ▽

Theorem 2.5.2 The estimator $\hat{V}$ given in Eq. (2.4.2) is a MVUE for σ^2.

Proof See Zacks [27, Chapter 3]. ▽

We now establish some properties of the estimator $\hat{\theta}$ and the residual vector $R = (Y - X\hat{\theta})$:

Theorem 2.5.3 Under the same conditions as Theorem 2.5.1, the estimator $\hat{\theta} = (X^TX)^{-1}X^TY$ is normally distributed with mean θ and covariance $(X^TX)^{-1}\sigma^2$.

Proof $\hat{\theta}$ is a linear function of Y and thus has a normal distribution since Y has a normal distribution. We have previously established the mean and covariance and thus the theorem is proved. ▽

The distribution of the estimator $\hat{\theta}_i$ is depicted diagrammatically in Fig. 2.5.1. Note that the results are clustered around the true parameter value θ_i. The standard deviation for $\hat{\theta}_i$ is $(\sigma^2 C_{ii})^{1/2}$ where C_{ii} is the ith diagonal element of $[X^TX]^{-1}$.

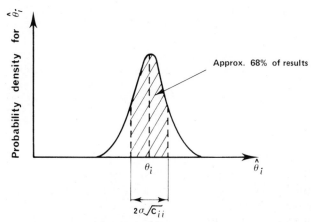

Figure 2.5.1. Normal distribution for $\hat{\theta}_i$ with mean θ_i and variance $\sigma^2 C_{ii}$.

Theorem 2.5.4 Under the same conditions as Theorem 2.5.1, the residual vector $R = Y - X\hat{\theta}$ is normally distributed and is statistically independent of $\hat{\theta}$.

Proof The normality of R follows from the normality of Y and $\hat{\theta}$.

We now consider the following covariance:

$$\begin{aligned}
E_Y[(\hat{\theta} - \theta)R^T] &= E_Y[((X^TX)^{-1}X^TY - \theta)(Y - X(X^TX)^{-1}X^TY)^T] \\
&= E_Y[(X^TX)^{-1}X^TY(Y - X(X^TX)^{-1}X^TY)^T] \\
&= E_Y[(X^TX)^{-1}X^TYY^T(I - X(X^TX)^{-1}X^T)] \\
&= (X^TX)^{-1}X^T[\sigma^2 I + X\theta\theta^TX^T][I - X(X^TX)^{-1}X^T] \\
&= \sigma^2(X^TX)^{-1}X^T[I - X(X^TX)^{-1}X^T] = 0 \qquad (2.5.13)
\end{aligned}$$

Therefore the independence of $\hat{\theta}$ and R follows from the normality of $\hat{\theta}$ and R.

Theorem 2.5.5   The normalized sum of squares of residuals $s(\hat\theta)/\sigma^2 = R^{\mathrm{T}}R/\sigma^2$ has a $\chi^2(N - p)$ distribution.

Proof

$$
\begin{aligned}
R^{\mathrm{T}}R/\sigma^2 &= (1/\sigma^2)[(Y - X\hat\theta)^{\mathrm{T}}(Y - X\hat\theta)] \\
&= (1/\sigma^2)[(Y - X(X^{\mathrm{T}}X)^{-1}X^{\mathrm{T}}Y)^{\mathrm{T}}(Y - X(X^{\mathrm{T}}X)^{-1}X^{\mathrm{T}}Y)] \\
&= (1/\sigma^2)Y^{\mathrm{T}}(I_N - X(X^{\mathrm{T}}X)^{-1}X^{\mathrm{T}})Y \\
&= (1/\sigma^2)(Y - X\theta)^{\mathrm{T}}(I_N - X(X^{\mathrm{T}}X)^{-1}X^{\mathrm{T}})(Y - X\theta) \qquad (2.5.14)
\end{aligned}
$$

Now we define

$$Z = (1/\sigma)(Y - X\theta) \qquad (2.5.15)$$

$$A = I_N - X(X^{\mathrm{T}}X)^{-1}X^{\mathrm{T}} \qquad (2.5.16)$$

Note A is idempotent $(A^2 = A)$ and Z is normally distributed $\sim N(0, I)$. Hence from Corollary A.6.2, $R^{\mathrm{T}}R/\sigma^2$ has a $\chi^2(r)$ distribution where

$$r = \text{trace } A = \text{trace}(I_N - X(X^{\mathrm{T}}X)^{-1}X^{\mathrm{T}}) = N - p \qquad (2.5.17)$$

This completes the proof of the theorem. ∇

The general shape of the distribution of the normalized sum of squares of residuals is shown in Fig. 2.5.2. (Note that for $N - p$ large the $\chi^2(N - p)$ distribution can be approximated by a normal distribution with mean $N - p$ and variance $2(N - p)$, see Sections 16.3–16.6 of reference [15]). Notice that the values of $R^{\mathrm{T}}R/(N - p)$ cluster around σ^2 for $N - p$ large.

In Theorem 2.5.3, we showed that $\hat\theta \sim N(\theta, (X^{\mathrm{T}}X)^{-1}\sigma^2)$. However, in practice we shall only have an estimate of σ^2, i.e., Eq. (2.4.2). We therefore establish the following result.

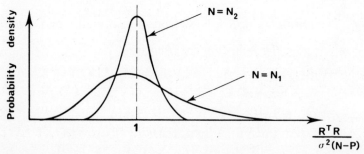

Figure 2.5.2. Approximate distribution for normalized sum of squares of residuals for $N = N_1$ and $N = N_2$ with $N_2 > N_1$.

Theorem 2.5.6 The distribution of $(\hat{\theta}_i - \theta_i)/\{C_{ii} S(\hat{\theta})/(N - p)\}^{1/2}$ is the student t distribution on $(N - p)$ degrees of freedom where $\hat{\theta}_i$ and θ_i are the ith components of $\hat{\theta}$ and θ, respectively, C_{ii} is the ith diagonal element of $(X^T X)^{-1}$, and

$$S(\hat{\theta}) = (Y - X\hat{\theta})^T(Y - X\hat{\theta})$$

Proof From Theorem 2.5.3, $(\hat{\theta} - \theta) \sim N(0, (X^T X)^{-1}\sigma^2)$. Hence the marginal distribution for the ith component is

$$\hat{\theta}_i - \theta_i \sim N(0, C_{ii}\sigma^2)$$

or

$$v_i = (\hat{\theta}_i - \theta_i)/\sigma\sqrt{C_{ii}} \sim N(0, 1)$$

Also

$$z = (1/\sigma^2)S(\hat{\theta}) \sim \chi^2(N - p)$$

from Theorem 2.5.5. Theorem 2.5.4 implies that v_i and z are independent. Hence from Result A.5.1 it follows that

$$v_i/[z/(N - p)]^{1/2} \sim t(N - p)$$

i.e.,

$$(\hat{\theta}_i - \theta_i)/[C_{ii} S(\hat{\theta})/(N - p)]^{1/2} \sim t(N - p)$$

This completes the proof of the theorem. ∇

The above result can be used to construct a $(1 - \alpha)\%$ confidence interval for θ_i since

$$P_r\{-t_\alpha < (\hat{\theta}_i - \theta_i)/[C_{ii} S(\hat{\theta})/(N - p)]^{1/2} < t_\alpha\} = 1 - \alpha$$

where t_α is the upper α percentile of the t distribution on $(N - p)$ degrees of freedom.

The $(1 - \alpha)\%$ confidence interval for θ_i is

$$\hat{\theta}_i - t_\alpha[C_{ii} S(\hat{\theta})/(N - p)]^{1/2} < \theta_i < \hat{\theta}_i + t_\alpha[C_{ii} S(\hat{\theta})/(N - p)]^{1/2} \qquad \nabla$$

As might be expected from Figs. 2.5.1 and 2.5.2, for $(N - p)$ large the $t(N - p)$ distribution can be approximated by an $N(0, 1)$ distribution. Thus Theorem 2.5.6 reduces to Theorem 2.5.3 with σ^2 replaced by $S(\hat{\theta})/(N - p)$ for large N.

Example 2.5.1 A process $\{y_t\}$ is known to be either constant or to increase linearly with time. A suitable model for $\{y_t\}$ is

$$y_t = h_1 + h_2 t + \varepsilon_t, \qquad t = 0, \ldots, N$$

where $\{\varepsilon_t\}$ is an i.i.d. Gaussian process with zero mean and variance σ^2.

With $N = 11$, the following results were obtained for the minimum variance unbiased estimates of (h_1, h_2, σ^2):

$$\hat{h}_1 = 0.53, \qquad \hat{h}_2 = 0.05, \qquad \hat{\sigma}^2 = 0.132$$

We shall use the t distribution to obtain a 95 % confidence interval for h_2. From Theorem 2.5.6 the distribution of $(\hat{h}_2 - h_2)/\{C_{22} S(\hat{\theta})/10\}^{1/2}$ has a student t distribution on 10 degrees of freedom, where

$$\hat{\theta} = (\hat{h}_1, \hat{h}_2)^{\mathsf{T}}$$

$$S(\hat{\theta}) = (N - p)\,\hat{\sigma}^2 = 1.32$$

$$C_{22} = (2, 2)\text{th element of } (X^{\mathsf{T}}X)^{-1}$$

$$= (2, 2)\text{th element of } \left\{\sum_{t=0}^{11} \begin{bmatrix} 1 & t \\ t & t^2 \end{bmatrix}\right\}^{-1}$$

$$= 1/110$$

From tables of the t distribution the upper 97.5 percentile is 2.228.

Application of Theorem 2.5.6 leads to the following result for the 95 % confidence interval:

$$0.05 - 2.228 \left|\frac{1}{110} \cdot \frac{1.32}{10}\right|^{1/2} < h_2 < 0.05 + 2.228 \left|\frac{1}{110} \cdot \frac{1.32}{10}\right|^{1/2}$$

i.e., $-0.027 < h_2 < 0.127$. $\triangledown$

We now turn to consider the situation where there may be linear relationships between the parameters. This may occur, for example, when too many parameters have been included in the parameter vector. We wish to establish a test for the presence of these constraints. The essential results are contained in the following theorem.

Theorem 2.5.7   Consider the normal linear model with the side constraint $L\theta = C$ where L is an $s \times p$ matrix and is assumed to have rank s. Then the minimum variance unbiased estimator $\tilde{\theta}$ subject to the side constraint $L\tilde{\theta} = C$ is given by

$$\tilde{\theta} = \hat{\theta} - (X^{\mathsf{T}}X)^{-1}L^{\mathsf{T}}(L(X^{\mathsf{T}}X)^{-1}L^{\mathsf{T}})^{-1}(L\hat{\theta} - C)$$

where $\hat{\theta}$ is the unconstrained least squares estimate of θ.

We also define the following quantities:

$$S(\hat{\theta}) = (Y - X\hat{\theta})^{\mathsf{T}}(Y - X\hat{\theta}) \tag{2.5.18}$$

$$S(\tilde{\theta}) = (Y - X\tilde{\theta})^{\mathsf{T}}(Y - X\tilde{\theta}) \tag{2.5.19}$$

$$S_H = S(\tilde{\theta}) - S(\hat{\theta}) \tag{2.5.20}$$

Then under the null hypothesis that $L\theta = C$ we have

(i) $(1/\sigma^2)S_H \sim \chi^2(s)$

(ii) S_H and $S(\hat{\theta})$ are independent

(iii) $\dfrac{S_H/s}{S(\hat{\theta})/(N-p)} \sim F(s, N-p)$

Proof Consider

$$J = S(\tilde{\theta}) - S(\hat{\theta}) \tag{2.5.21}$$

where $\hat{\theta} = (X^TX)^{-1}X^TY$. Hence

$$\begin{aligned} J &= (Y - X\tilde{\theta})^T(Y - X\tilde{\theta}) - (Y - X\hat{\theta})^T(Y - X\hat{\theta}) \\ &= (Y - X\tilde{\theta})^T(Y - X\tilde{\theta}) - Y^TY + \hat{\theta}^TX^TX\hat{\theta} \\ &= -2\tilde{\theta}^TX^TY + \tilde{\theta}^TX^TX\tilde{\theta} + \hat{\theta}^TX^TX\hat{\theta} = (\tilde{\theta} - \hat{\theta})^TX^TX(\tilde{\theta} - \hat{\theta}) \end{aligned} \tag{2.5.22}$$

We now introduce the constraint via a Lagrangian multiplier to form

$$J_c = (\tilde{\theta} - \hat{\theta})^TX^TX(\tilde{\theta} - \hat{\theta}) + 2\lambda^T(L\tilde{\theta} - C)$$

Differentiating w.r.t. $\tilde{\theta}$ leads to

$$L^T\lambda = X^TX(\hat{\theta} - \tilde{\theta})$$

or defining $A = X^TX$,

$$L^T\lambda = A(\hat{\theta} - \tilde{\theta}) \tag{2.5.23}$$

Multiplying on the left by LA^{-1} gives

$$LA^{-1}L^T\lambda = L(\hat{\theta} - \tilde{\theta}) = L\hat{\theta} - C \qquad \text{since} \quad L\tilde{\theta} = C$$

Also, $LA^{-1}L^T$ is positive definite since L has rank s. So

$$\lambda = (LA^{-1}L^T)^{-1}(L\hat{\theta} - C) \tag{2.5.24}$$

From (2.5.23),

$$\tilde{\theta} = \hat{\theta} - A^{-1}L^T\lambda$$

Using (2.5.24),

$$\tilde{\theta} = \hat{\theta} - A^{-1}L^T(LA^{-1}L^T)^{-1}(L\hat{\theta} - C) \tag{2.5.25}$$

Thus establishing the first part of the theorem.

Substituting into (2.5.22), gives

$$J = (L\hat{\theta} - C)^T(LA^{-1}L^T)^{-1}LA^{-1}L^T(LA^{-1}L^T)^{-1}(L\hat{\theta} - C) = S_H \tag{2.5.26}$$

If the null hypothesis is true, then $C = L\theta$ where θ is the true parameters.

Therefore

$$S_H = (\hat{\theta} - \theta)^{\mathrm{T}} L^{\mathrm{T}} (LA^{-1}L^{\mathrm{T}})^{-1} LA^{-1} L^{\mathrm{T}} (LA^{-1}L^{\mathrm{T}})^{-1} L(\hat{\theta} - \theta)$$
$$= (\hat{\theta} - \theta)^{\mathrm{T}} L^{\mathrm{T}} (LA^{-1}L^{\mathrm{T}})^{-1} L(\hat{\theta} - \theta)$$

But

$$\hat{\theta} - \theta = (X^{\mathrm{T}}X)^{-1} X^{\mathrm{T}} (Y - X\theta) = A^{-1} X^{\mathrm{T}} (Y - X\theta)$$
$$S_H = (Y - X\theta)^{\mathrm{T}} X A^{-1} L^{\mathrm{T}} (LA^{-1}L^{\mathrm{T}})^{-1} LA^{-1} X^{\mathrm{T}} (Y - X\theta)$$
$$= (Y - X\theta)^{\mathrm{T}} A_H (Y - X\theta) \qquad\qquad (2.5.27)$$

where

$$A_H = X A^{-1} L^{\mathrm{T}} (LA^{-1}L^{\mathrm{T}})^{-1} LA^{-1} X^{\mathrm{T}}$$

Clearly A_H is idempotent (i.e., $A_H{}^2 = A_H$).

Now, from (2.5.14)

$$S(\hat{\theta})/\sigma^2 = (1/\sigma)(Y - X\theta)^{\mathrm{T}} (I_N - X(X^{\mathrm{T}}X)^{-1}X^{\mathrm{T}})(Y - X\theta)(1/\sigma)$$

From (2.5.27),

$$(S_H/\sigma^2) = (1/\sigma)(Y - X\theta)^{\mathrm{T}} A_H (Y - X\theta)(1/\sigma)$$

Now we form

$$A_1 = X^{\mathrm{T}} (X^{\mathrm{T}}X)^{-1} X^{\mathrm{T}} - A_H$$

Clearly

$$A_1 + A_H + I_N - X(X^{\mathrm{T}}X)^{-1}X^{\mathrm{T}} = I_N$$

Also, rank $A_H = s$

$$\mathrm{rank}(I_N - X(X^{\mathrm{T}}X)^{-1}X^{\mathrm{T}}) = N - p$$

and

$$\mathrm{rank}\, A_1 = p - s \qquad \text{since} \quad A_1 \text{ is also idempotent}$$

Therefore

$$\mathrm{rank}\, A_H + \mathrm{rank}(I_N - X(X^{\mathrm{T}}X)^{-1}X^{\mathrm{T}}) + \mathrm{rank}\, A_1 = N$$

Furthermore,

$$(1/\sigma)(Y - X\theta) \sim N(0, I)$$

Hence, from the Fisher–Cochrane theorem (Theorem A.6.1), $S(\hat{\theta})/\sigma^2$ and S_H/σ^2 are independent χ^2 random variables with $(N - p)$ and s degrees of freedom, respectively.

This completes the proof of parts (i) and (ii) of the theorem. Part (iii) now immediately follows from Result A.4.1. ▽

The general shape of the F distribution is shown in Fig. 2.5.3.

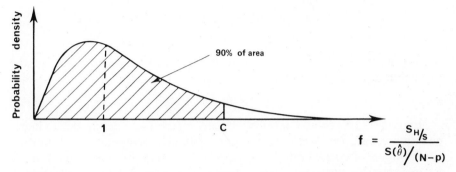

Figure 2.5.3. General form of the F distribution. Note: C decreases with increasing N.

We illustrate the application of Theorem 2.5.6 in the following example:

Example 2.5.2 With the same data as in Example 2.5.1, the following results were obtained for the MVUE of h_1, σ^2 when h_2 was constrained to be zero

$$\hat{h}_1 = 0.54, \qquad \hat{\sigma}^2 = 0.204$$

We shall use the F distribution to test the hypothesis that $h_2 = 0$ at a 95% risk level. From Theorem 2.5.6, the following variable has an $F(1, 10)$ distribution under the null hypothesis $(h_2 = 0)$

$$f = S_H(S(\hat{\theta})/10)$$

where

$$S(\hat{\theta}) = 1.32, \qquad S_H = 2.04 - 1.32 = 0.72$$

therefore

$$f = 0.72/0.132 = 5.4$$

Now, from tables, $F_{0.95}(1, 10) = 4.96$. Since $f > 4.96$, we conclude that there is good evidence to reject the null hypothesis at a 5% risk level (cf. the confidence interval found for h_2 in Example 2.5.1). ▽

Example 2.5.2 is a form of decision problem in which a choice is desired between two competing model structures. We have adopted a classical approach by fixing the probability that the null hypothesis is rejected when it is actually true. Classical methods for testing statistical hypotheses are treated in many texts, for example [69].

2.6. NUMERICAL ASPECTS

The least squares estimates have been shown to be solutions of the normal equations (2.2.4), i.e.,

$$(X^\mathsf{T}X)\hat{\theta} = X^\mathsf{T}Y \tag{2.6.1}$$

Equation (2.6.1) could be solved by any method for solving sets of linear equations, e.g., Guassian elimination [84]. Unfortunately, the normal equations are often ill-conditioned, especially when the number of parameters is large. In these circumstances special numerical procedures are required. We briefly describe one numerically robust algorithm which exploits the structure of the least squares problem [86].

Consider a Householder transformation matrix defined as follows:

$$Q = I - 2uu^\mathsf{T} \tag{2.6.2}$$

where u is any unit length vector.

Lemma 2.6.1 The transformation matrix Q defined in (2.6.2) has the following properties:

(i) Q is symmetric, i.e.,

$$Q^\mathsf{T} = Q \tag{2.6.3}$$

(ii) Q is is orthonormal

$$Q^\mathsf{T}Q = I \tag{2.6.4}$$

Proof Left as an exercise. ∇

Lemma 2.6.2 The product of Householder transformations is orthonormal, i.e., if $\Psi = Q_p Q_{p-1} \cdots Q_1$ where $Q_1, \ldots, Q_p$ are Householder transformations, then

$$\Psi^\mathsf{T}\Psi = I \tag{2.6.5}$$

Proof Left as an exercise. ∇

Lemma 2.6.3   Given any vector x, there exists a Householder transformation Q such that

$$Qx = \begin{bmatrix} \lambda \\ 0 \\ \vdots \\ 0 \end{bmatrix} \qquad (2.6.6)$$

where

$$\lambda = (x^{\mathsf{T}}x)^{1/2} \qquad (2.6.7)$$

Proof   Using (2.6.3) and (2.6.4) in (2.6.6) gives

$$x = Q \begin{bmatrix} \lambda \\ 0 \\ \vdots \\ 0 \end{bmatrix} \qquad (2.6.8)$$

Using (2.6.2),

$$\begin{bmatrix} x_1 \\ \vdots \\ \vdots \\ x_N \end{bmatrix} = \begin{bmatrix} (1 - 2u_1{}^2)\lambda \\ -2u_1 u_2 \lambda \\ \vdots \\ -2u_1 u_N \lambda \end{bmatrix} \qquad (2.6.9)$$

Thus

$$u_1 = [\tfrac{1}{2}(1 - x_1/\lambda)]^{1/2} \qquad (2.6.10)$$

$$u_i = -x_i/(2u_1\lambda), \qquad i = 2, \ldots, N \qquad (2.6.11)$$

Finally, from (2.6.6)

$$x^{\mathsf{T}}Q^{\mathsf{T}}Qx = \lambda^2 \qquad (2.6.12)$$

i.e.,

$$\lambda = (x^{\mathsf{T}}x)^{1/2} \qquad \triangledown \qquad (2.6.13)$$

Lemma 2.6.4   Given any $N \times p$ matrix X ($N > p$), there exists an orthonormal matrix Ψ such that

$$\Psi X = \begin{vmatrix} R \\ 0 \end{vmatrix} \qquad (2.6.14)$$

where R is a $p \times p$ upper triangular matrix and Ψ is the product of p Householder transformations.

Proof From Lemma 2.6.3 it is clear that a transformation Q_1 can be found such that

$$Q_1 X = \begin{bmatrix} \lambda_1 & x'_{12} & \cdots & x'_{1p} \\ 0 & & & \vdots \\ \vdots & & & \vdots \\ 0 & x'_{N2} & \cdots & x'_{Np} \end{bmatrix} \qquad (2.6.15)$$

Similarly, if Q_2 is defined by $I - 2u'u'^{\mathrm{T}}$ where the first element of u' is zero, then Q_2 can be chosen so that

$$Q_2 Q_1 X = \begin{bmatrix} \lambda_1 & x'_{12} & x'_{13} & \cdots & x''_{1p} \\ 0 & \lambda_2 & x''_{23} & \cdots & x''_{2p} \\ \vdots & & 0 & & \vdots \\ 0 & 0 & x''_{N3} & \cdots & x''_{Np} \end{bmatrix} \qquad (2.6.16)$$

Continuing this process yields Eq. (2.6.14) with

$$\Psi = Q_p Q_{p-1} \cdots Q_1 \qquad (2.6.17)$$

which is orthonormal from Lemma 2.6.2.   ▽

Theorem 2.6.1   Consider the least squares problem with sum of squares

$$S(\theta) = (Y - X\theta)^{\mathrm{T}}(Y - X\theta) \qquad (2.6.18)$$

The least squares estimate satisfies

$$R\hat{\theta} = \eta_1 \qquad (2.6.19)$$

where R is upper triangular and is given by

$$\Psi X = \begin{vmatrix} R \\ 0 \end{vmatrix} \qquad (2.6.20)$$

where Ψ is given by (2.6.17) and the p-vector η_1 in Eq. (2.6.19) is given by

$$\begin{vmatrix} \eta_1 \\ \eta_2 \end{vmatrix} = \Psi Y \qquad (2.6.21)$$

The sum of squares of residuals is given by

$$S(\hat{\theta}) = \eta_2{}^{\mathrm{T}}\eta_2 \qquad (2.6.22)$$

Proof

$$S(\theta) = (Y - X\theta)^{\mathrm{T}}(Y - X\theta) \qquad (2.6.23)$$

Using Lemma 2.6.2,

$$S(\theta) = (Y - X\theta)^{\mathrm{T}}\Psi^{\mathrm{T}}\Psi(Y - X\theta) \qquad (2.6.24)$$

Using Lemma 2.6.4, the matrix Ψ can be found such that

$$\Psi X = \begin{vmatrix} R \\ 0 \end{vmatrix} \tag{2.6.25}$$

where R is upper triangular. Let

$$\Psi Y = \begin{vmatrix} \eta_1 \\ \eta_2 \end{vmatrix} \tag{2.6.26}$$

where η_1 is a p-vector and η_2 is an $(N - p)$-vector. Substituting (2.6.25) and (2.6.26) into (2.6.24) gives

$$S(\theta) = (\eta_1 - R\theta)^{\mathrm{T}}(\eta_1 - R\theta) + \eta_2{}^{\mathrm{T}}\eta_2 \tag{2.6.27}$$

Since R is square, $S(\theta)$ is minimized by $\hat{\theta}$ satisfying

$$R\hat{\theta} = \eta_1$$

and the sum of squares of residuals is

$$S(\hat{\theta}) = \eta_2{}^{\mathrm{T}}\eta_2 \qquad \triangledown$$

Equation (2.6.19) can be readily solved by back-substitution since R is triangular. The algorithm outlined above for determining $\hat{\theta}$ has superior numerical properties compared with naïve algorithms based on direct solution of the normal equations (2.6.1) [86].

2.7. CONCLUSIONS

This chapter has considered the problem of estimation when the stochastic process has a mean value which is expressible as a known linear function of a parameter vector, i.e., $E\{Y\} = X\theta$ where X is known. The results are fundamental to any treatment of identification and we shall often have reason to refer back to this chapter. In the next chapter we consider more general estimation problems via the maximum likelihood approach.

PROBLEMS

2.1. In an agricultural experiment, the crop yields (y) were obtained from a sequence of experiments with different amounts of fertilizer per acre (u).

u:	1	2	3	4	5	6	7	8	9	10	11	12
y:	1.1	0.9	1.1	1.4	1.3	1.1	1.2	1.6	2.1	2.0	1.6	1.7

(i) Assuming that the 12 experiments are independent (i.e., the errors are uncorrelated) fit a model to the data of the following form

$$y_k = \theta_0 + \theta_1 u_k + \varepsilon_k$$

where $\{\varepsilon_k\}$ is assumed to be an i.i.d. sequence of *normal random variables* each with zero mean and variance σ^2.

(ii) Using the results of part (i), construct 95% confidence intervals for the parameters θ_0 and θ_1 (use t distribution).

(iii) It is hypothesized that the fertilizer has no effect on the yield per acre (i.e., it is hypothesized that θ_1 is zero).

Test, at the 5% risk level, that the data give evidence against the hypothesis (use F distribution).

2.2. Use the least squares technique to fit a quadratic function of the form

$$E[y_t] = at^2 + bt + c$$

to the data given below.

t:	1	2	3	4	5	6	7	8	9	10
y_t:	9.6	4.1	1.3	0.4	0.05	0.1	0.7	1.8	3.8	9.0

2.3. Consider a linear continuous time dynamic system having transfer function

$$H(s) = B(s)/A(s) \qquad (s \text{ is the Laplace transform variable})$$

where

$$B(s) = b_0 + b_1 s + \cdots + b_{n-1} s^{n-1}$$
$$A(s) = 1 + a_1 s + \cdots + a_n s^n$$

By applying sinewaves of frequency $\omega_1, \omega_2, \ldots, \omega_N$ to the system, we are able to measure $H(j\omega_i)$, $i = 1, \ldots, N$. Of course, there will be unavoidable errors in these measurements, so we denote the results by $\hat{H}(j\omega_i)$, $i = 1, \ldots, N$.

Since $H(j\omega) = B(j\omega)/A(j\omega)$, we have

$$A(j\omega)H(j\omega) = B(j\omega)$$

We therefore propose that the coefficients in A and B be estimated by minimizing the sum of squares of moduli of the differences:

$$J = \sum_{i=1}^{N} e_i^* e_i$$

where $e_i = (A(j\omega_i)\hat{H}(j\omega_i) - B(j\omega_i))$ and * denotes complex conjugate transpose.

(a) Show that J can be expressed as

$$J = (Y - X\theta)^*(Y - X\theta)$$

where

$$\theta = (a_1, a_2, \ldots, a_n, b_0, b_1, \ldots, b_{n-1})^{\mathsf{T}}$$

$$X = \begin{bmatrix} -j\omega_1 \hat{H}(j\omega_1), \ldots, (-j\omega_1)^n \hat{H}(j\omega_1), \ 1, \ j\omega_1, \ldots, (j\omega_1)^{n-1} \\ \vdots \qquad\qquad\qquad\qquad\qquad\qquad \vdots \\ -j\omega_N \hat{H}(j\omega_N), \ldots, (-j\omega_N)^n \hat{H}(j\omega_N), \ 1, \ j\omega_N, \ldots, (j\omega_N)^{n-1} \end{bmatrix},$$

$$Y = \begin{bmatrix} \hat{H}(j\omega_1) \\ \vdots \\ \hat{H}(j\omega_N) \end{bmatrix}$$

(b) Show that the value of θ minimizing J is

$$\hat{\theta} = [\mathrm{Re}(X^*X)]^{-1} \, \mathrm{Re}[X^*Y]$$

where Re denotes real part. (The algorithm described in this question is often called Levy's method.)

2.4. Is the estimator found in Problem 2.3b a linear function of the data $\hat{H}(j\omega_1), \ldots, \hat{H}(j\omega_N)$? Comment on the applicability of the theory of Section 2.3 to this problem.

2.5. Establish Eq. (2.3.22).

2.6. Let X be a k-variate normal random variable with mean μ and covariance Σ. Show that the characteristic function for X is

$$\phi_X(t) = \exp\{jt^{\mathsf{T}}\mu - \tfrac{1}{2}t^{\mathsf{T}}\Sigma t\}$$

2.7. Let X_m denote the mth component of the vector X. Show that

$$E_X\{X_{i_1} X_{i_2} \cdots X_{i_n}\} = \frac{1}{(j)^n} \frac{\partial^n \phi_X(t)}{\partial t_{i_1} \, \partial t_{i_2} \cdots \partial t_{i_n}}\bigg|_{t=0}$$

where $i_1, i_2, \ldots, i_n$ are integers and $\phi_X(t)$ is the characteristic function.

2.8. Let X be a k-variate normal random variable with zero mean and covariance matrix Σ. Show that the general 4th central moment of the distribution can be expressed as follows:

$$E_X\{X_l X_m X_n X_p\} = \Sigma_{mn}\Sigma_{lp} + \Sigma_{lm}\Sigma_{np} + \Sigma_{ln}\Sigma_{mp}$$

where Σ_{ij} denotes the ijth element of Σ. Hint: Use the answer to Problem 2.7.

2.9. Let Y denote an N-variate normal random variable having mean $X\theta$ and covariance matrix $\sigma^2 I$ where σ^2 is a scalar and I is an $N \times N$ identity matrix. The regression matrix X is known but θ and σ^2 are unknown parameters (the dimension of θ is p and hence X is $N \times p$).

Determine the information matrix M_β for the parameter vector

$$\beta = \begin{bmatrix} \theta \\ \sigma^2 \end{bmatrix} \qquad (\beta \text{ is a } (p+1)\text{-vector})$$

Hence show that the lower bounds on the covariance of any unbiased estimator of θ and σ^2 are given, respectively, by

$$\operatorname{cov}(\hat{\theta}) \geq (X^T X)^{-1} \sigma^2 \qquad \text{and} \qquad \operatorname{cov}(\hat{\sigma}^2) \geq 2\sigma^4/N$$

Hint: See Theorem 2.5.1.

2.10. We know that the estimator $\hat{V} = S/(N-p)$ given in Eq. (2.4.2) is an unbiased estimator for σ^2. Show that the variance of $\hat{V}$ is given by

$$\operatorname{var}(\hat{V}) = 2\sigma^4/(N-p)$$

Comment: $\hat{V}$ clearly does not achieve the Cramer–Rao lower bound (cf. problem 2.9) but $\hat{V}$ is MVUE (see Theorem 2.5.2).

2.11. Show that $S(\tilde{\theta})/\sigma^2$, where $S(\tilde{\theta})$ is defined in Eq. (2.5.19), has a $\chi^2(N-p+s)$ distribution.

2.12. Show that the estimators

$$\hat{\theta} = (X^T X)^{-1} X^T Y \qquad \text{and} \qquad \hat{V} = [1/(N-p)](Y - X\hat{\theta})^T (Y - X\hat{\theta})$$

are sufficient statistics for θ and σ^2 if Y is assumed to have Gaussian distribution with mean $X\theta$ and covariance $\sigma^2 I$.

CHAPTER

3

Maximum Likelihood Estimators

3.1. INTRODUCTION

The theory of Chapter 2 was concerned with models which were linear in the parameters. The resulting least squares estimator was shown to have a number of optimal properties. The estimator was analyzed in the case of Gaussian disturbances. In this chapter we consider a more general class of problems.

For an arbitrary family of probability distributions we can think of the data as giving information about the family, since it is usually the case that some data are more probable under some members of the family than under others. In maximum likelihood we choose the member which makes the given data most probable in the sense that the likelihood function $p(y|\theta)$ is maximized.

We devote this chapter to the study of maximum likelihood estimators and we shall show that they have a number of important properties for large sample lengths. We shall also show that for the linear model with Gaussian disturbances, the maximum likelihood estimator reduces to the least squares estimator of Chapter 2.

3.2. THE LIKELIHOOD FUNCTION
AND THE ML ESTIMATOR

We assume that the data-generating mechanism may be represented by a member of a parametric family of probability distributions $\mathscr{P} = \{p(\cdot|\theta): \theta \in \Theta\}$ where θ is a parameter and $\Theta \subset R^P$.

For given data y we can regard $p(y|\theta)$ as a function of θ on Θ. We then refer to $p(y|\cdot)$ as the likelihood function on Θ.

We are now in a position to define the maximum likelihood estimator (MLE).

Definition 3.2.1   The MLE is a function $\hat{\theta}(y)$ such that

$$p(y|\hat{\theta}) = \sup_{\theta \in \Theta} p(y|\theta) \tag{3.2.1}$$

where $p(y|\theta)$ is the likelihood function for the parameters (cf. Definition 1.8.2). ∇

Example 3.2.1   Consider a process $\{x_t\}$ where $\{x_t\}$ is a sequence of i.i.d. random variables having probability density function $\theta^2 x e^{-\theta x}$, $\theta > 0$. The likelihood function for N observations of x is

$$p(y|\theta) = \prod_{k=1}^{N} p(x_k|\theta) = \theta^{2N} \left| \exp\left(-\theta \sum_{k=1}^{N} x_k \right) \right| \prod_{k=1}^{N} x_k \tag{3.2.2}$$

It can be shown by differentiation that the maximum likelihood estimate of θ is

$$\hat{\theta} = 2N \left/ \sum_{k=1}^{N} x_k \right. \quad \nabla \tag{3.2.3}$$

It can be shown that $\hat{\theta}$ given in (3.2.3) is a biased estimator of θ (see Problem 3.1). We thus conclude that it is not generally the case that MLE's are unbiased. The estimators do however have optimal properties asymptotically. These properties are discussed in a later section.

3.3. MAXIMUM LIKELIHOOD FOR THE NORMAL
LINEAR MODEL

We shall now develop expressions for the maximum likelihood estimators of the parameters θ and σ^2 of the normal linear model considered in Chapter 2.

Result 3.3.1 Consider an m variate random vector Y, having a joint Gaussian distribution with mean $X\theta$ and covariance Σ (assumed known). Then the maximum likelihood estimator for θ is equivalent to the weighted least squares estimator

$$\hat{\theta} = (X^T\Sigma^{-1}X)^{-1}X^T\Sigma^{-1}Y \tag{3.3.1}$$

Proof The likelihood function is

$$p(y|\theta) = [(2\pi)^m \det \Sigma]^{-1/2} \exp\{-\tfrac{1}{2}(y - X\theta)^T\Sigma^{-1}(y - X\theta)\} \tag{3.3.2}$$

It is equivalent to minimize $\log p(y|\theta)$ since log is monotonic. Differentiating w.r.t. θ and setting the result to zero gives (3.3.1). The second derivative matrix is $-(X^T\Sigma^{-1}X)$ showing that this corresponds to the global maximum of the likelihood function. ▽

Result 3.2.2 Consider the same system as considered in Result 3.3.1 save that Σ has the special form $\sigma^2 I$ where σ^2 is to be estimated. The maximum likelihood estimators for θ and σ^2 are

$$\hat{\theta} = (X^TX)^{-1}X^TY \tag{3.3.3}$$

$$\hat{\sigma}^2 = (1/m)(Y - X\hat{\theta})^T(Y - X\hat{\theta}) \tag{3.3.4}$$

Proof Left as an exercise (Problem 3.3). ▽

Result 3.3.3 Consider N independent observations of an m-variate random vector $\{Y_t; t \in (1, 2, \ldots, N)\}$ such that each Y_t has a Gaussian distribution with mean $X_t\theta$ and common covariance Σ. Then a necessary condition for $\hat{\theta}$ and $\hat{\Sigma}$ to be maximum likelihood estimators is that they simultaneously satisfy

$$\hat{\theta} = \sum_{t=1}^{N} [X_t^T\hat{\Sigma}^{-1}X_t]^{-1} \sum_{t=1}^{N} X_t^T\hat{\Sigma}^{-1}Y_t \tag{3.3.5}$$

$$\hat{\Sigma} = \frac{1}{N} \sum_{t=1}^{N} (Y_t - X_t\hat{\theta})(Y_t - X_t\hat{\theta})^T \tag{3.3.6}$$

Proof The likelihood function is

$$p(y|\theta, \Sigma) = [(2\pi)^m \det \Sigma]^{-N/2} \exp\left\{-\frac{1}{2} \sum_{t=1}^{N} (y_t - X_t\theta)^T \Sigma^{-1}(y_t - X_t\theta)\right\}$$

$$\tag{3.3.7}$$

Taking logarithms yields

$$\log p(y|\theta, \Sigma) = \frac{-Nm}{2} \log 2\pi - \frac{N}{2} \log \det \Sigma$$

$$- \frac{1}{2} \sum_{t=1}^{N} (y_t - X_t \theta)^T \Sigma^{-1} (y_t - X_t \theta) \quad (3.3.8)$$

Differentiating w.r.t. θ and Σ yields

$$- \sum_{t=1}^{N} X_t^T \Sigma^{-1} (y_t - X_t \theta) = 0 \qquad (3.3.9)$$

$$- \frac{N}{2} \Sigma^{-1} + \tfrac{1}{2} \Sigma^{-1} \left(\sum_{t=1}^{N} (y_t - X_t \theta)(y_t - X_t \theta)^T \right) \Sigma^{-1} = 0 \qquad (3.3.10)$$

where we have used the following matrix differentiation results (see Appendix E)

$$(\partial/\partial\Sigma) \log \det \Sigma = (\Sigma^{-1})^T \qquad (3.3.11)$$

$$(\partial/\partial\Sigma) \text{ trace } A\Sigma^{-1}B = -(\Sigma^{-1}BA\Sigma^{-1})^T \qquad (3.3.12)$$

Rearrangement of (3.3.9) and (3.3.10) gives the desired result. ∇

Result 3.3.4   (a)   For fixed $\hat{\Sigma}$, then $\hat{\theta}$ given in (3.3.5) yields the unique maximum of the likelihood function.

(b)   For fixed $\hat{\theta}$, then $\hat{\Sigma}$ given in (3.3.6) yields the unique maximum of the likelihood function.

Proof   Part (a) is immediate, since the second derivative matrix is

$$\frac{\partial^2 J}{\partial \theta^2} = - \sum_{t=1}^{N} [X_t^T \hat{\Sigma}^{-1} X_t] \qquad (3.3.13)$$

which is negative definite for all θ.

Part (b) can be established by considering the second variation of the log likelihood function. We proceed as follows:

The first derivative of log likelihood function J with respect to Σ is

$$\frac{\partial J}{\partial \Sigma} = - \frac{N}{2} \Sigma^{-1} + \frac{N}{2} \sum_{t=1}^{N} \Sigma^{-1} \omega_t \omega_t^T \Sigma^{-1} \qquad (3.3.14)$$

where

$$\omega_t = Y_t - X_t \theta \qquad (3.3.15)$$

From Eq. (3.3.14) there is a unique stationary point at the point $\Sigma = \hat{\Sigma}$ where $\hat{\Sigma}$ is given by Eq. (3.3.6). To show that this is a maximum we consider the second derivative, viz.,

$$\frac{\partial^2 J}{\partial \Sigma_{rs}\, \partial \Sigma_{ij}} = \frac{N}{2}\, [\Sigma^{-1}]_{ri}[\Sigma^{-1}]_{js} - \frac{1}{2} \sum_{t=1}^{N} [\Sigma^{-1}]_{ri}[\Sigma^{-1}]_{j\cdot}\, \omega_t\, \omega_t^{\mathrm{T}} [\Sigma^{-1}]_{\cdot s}$$

$$- \frac{1}{2} \sum_{t=1}^{N} [\Sigma^{-1}]_{r\cdot}\, \omega_t\, \omega_t^{\mathrm{T}} [\Sigma^{-1}]_{\cdot i} [\Sigma^{-1}]_{js} \qquad (3.3.16)$$

where $[\Sigma^{-1}]_{ij}$, $[\Sigma^{-1}]_{j\cdot}$ and $[\Sigma^{-1}]_{\cdot i}$ denote the ijth element, jth row, and ith columns of Σ^{-1}, respectively. Evaluating (3.3.16) at $\Sigma = \hat{\Sigma}$ yields

$$\frac{\partial^2 J(\hat{\Sigma})}{\partial \Sigma_{rs}\, \partial \Sigma_{ij}} = -\frac{N}{2}\, [\hat{\Sigma}^{-1}]_{ri}[\hat{\Sigma}^{-1}]_{js} \qquad (3.3.17)$$

For J to be a maximum at $\Sigma = \hat{\Sigma}$, we require

$$\delta^2 J = \Sigma_{ijrs} \frac{\partial^2 J(\hat{\Sigma})}{\partial \Sigma_{rs}\, \partial \Sigma_{ij}} \Gamma_{rs} \Gamma_{ij} < 0 \qquad (3.3.18)$$

for all positive semi-definite Γ (since $\Sigma = \hat{\Sigma} + \Gamma$ must be positive definite). Thus the ijth element of Γ can be written as

$$\Gamma_{ij} = \sum_{k=1}^{n} L_{ik}\, L_{kj}^{\mathrm{T}} \qquad \text{for some matrix} \quad L \qquad (3.3.19)$$

Substituting (3.3.17) and (3.3.19) into (3.3.18) yields

$$\delta^2 J = -(N/2) \sum_{tk} \left(\sum_{rt} L_{tr}^{\mathrm{T}} [\hat{\Sigma}^{-1}]_{ri}\, L_{ik} \right) \left(\sum_{js} L_{kj}^{\mathrm{T}} [\hat{\Sigma}^{-1}]_{js}\, L_{st} \right)$$

$$= -(N/2) \sum_{tk} M_{tk}^2 \qquad (3.3.20)$$

where

$$M = L^{\mathrm{T}} [\hat{\Sigma}^{-1}] L \qquad (3.3.21)$$

Equation (3.3.20) establishes that $\delta^2 J < 0$ as required. ∇

We note that Eqs. (3.3.5) and (3.3.6) do not have a simple closed form solution. However, they can be solved by a relaxation algorithm as follows:

(i)   Pick any value of $\hat{\Sigma}$ (say I).
(ii)   Solve (3.3.5) for $\hat{\theta}$ using $\hat{\Sigma}$.
(iii)   Solve (3.3.6) for $\hat{\Sigma}$ using $\hat{\theta}$.
(iv)   Stop if converged, otherwise go to (ii).

Unfortunately the authors are not aware of the existence of a proof of global convergence for the above relaxation algorithm. However, computational studies indicate that the algorithm works well in practice. Furthermore, it is readily shown that $\hat{\theta}$ is an unbiased estimator of θ independent of the value of $\hat{\Sigma}$ (see Problem 3.9).

3.4. GENERAL PROPERTIES

A fundamental property of the maximum likelihood estimator is the result that the maximum likelihood estimate of a function of the parameters can be computed by taking the function of the maximum likelihood estimates of the parameters. This property is known as the principle of invariance and can be very useful in practice (see Problem 3.4).

Theorem 3.4.1   (*Principle of invariance*)   If $\hat{\theta}$ is a MLE of $\theta \in \Theta \subset R^p$, then $g(\hat{\theta})$ is a MLE of $g(\theta)$ where g is a function; $g : \Theta \rightarrow Z \subset R^n$ where $n \leq p$.

Proof   See Zehna [42]. ∇

Another interesting property is contained in the following theorem.

Theorem 3.4.2   Whenever there exists an unbiased estimator which achieves the Cramer–Rao lower bound (i.e., an efficient unbiased estimator), then it is also the maximum likelihood estimator.

Proof   We assume that there exists an unbiased efficient estimator $g(y)$ of θ. Then from Theorem 1.3.2, we can write

$$\partial \log p(y|\theta)^{\mathrm{T}}/\partial \theta = M_\theta[g(y) - \theta]$$

Equating to zero yields the required result. ∇

Unfortunately it is the exception rather than the rule that an unbiased efficient estimator can be found for the problems that we shall be considering. Of course, bias is not always undesirable. For example a biased estimator can have a variance lower than the Cramer–Rao lower bound (Theorem 1.3.1).

3.5. ASYMPTOTIC PROPERTIES

In this section we turn attention to large sample properties of the MLE.

Theorem 3.5.1 (*Consistency*) Let $\hat{\theta}_N$ designate a MLE of θ based on N i.i.d. random variables $y = (x_1, \ldots, x_N)^T$, then $\hat{\theta}_N$ converges to θ almost surely, i.e,

$$\hat{\theta}_N \xrightarrow{\text{a.s.}} \theta \qquad (3.5.1)$$

Proof See Zacks [27].

A heuristic justification of the result is as follows: We begin by considering the result

$$\int p(x|\theta)\, dx = 1 \qquad (3.5.2)$$

Assuming sufficient regularity to differentiate inside the integral yields

$$\int \frac{\partial p(x|\theta)}{\partial \theta}\, dx = 0$$

i.e.,

$$\int \frac{\partial \log p(x|\theta)}{\partial \theta} p(x|\theta)\, dx = 0 \qquad (3.5.3)$$

or

$$E_{X|\theta}\left[\frac{\partial \log p(x|\theta)}{\partial \theta}\right] = 0 \qquad (3.5.4)$$

Now we expand $\partial \log p(y|\theta')/\partial \theta'$ about θ (the true parameter value).

$$\frac{\partial \log p(y|\theta')}{\partial \theta'} = \frac{\partial \log p(y|\theta)}{\partial \theta} + (\theta' - \theta)^T \frac{\partial^2 \log p(y|\theta^*)}{\partial \theta^{*2}} \qquad (3.5.5)$$

where θ^* lies on the line joining θ and θ'.
Let θ^* be the maximum likelihood estimate satisfying

$$\partial \log p(y|\hat{\theta})/\partial \hat{\theta} = 0 \qquad (3.5.6)$$

Hence from (3.5.5),

$$(\hat{\theta} - \theta)^T \frac{\partial^2 \log p(y|\theta^*)}{\partial \theta^{*2}} = \frac{-\partial \log p(y|\theta)}{\partial \theta} \qquad (3.5.7)$$

Now since $x_1, \ldots, x_N$ are i.i.d., we have

$$\frac{\partial \log p(y|\theta)}{\partial \theta} = \frac{\partial}{\partial \theta} \log \prod_{i=1}^{N} p(x_i|\theta) = \frac{\partial}{\partial \theta} \sum_{i=1}^{N} \log p(x_i|\theta)$$

$$= \sum_{i=1}^{N} \frac{\partial \log p(x_i|\theta)}{\partial \theta} \tag{3.5.8}$$

Similarly,

$$\frac{\partial^2 \log p(y|\theta)}{\partial \theta^2} = \sum_{i=1}^{N} \frac{\partial^2 \log p(x_i|\theta)}{\partial \theta^2} \tag{3.5.9}$$

It now follows from the strong law of large numbers (Theorem B.3.2) that

$$\frac{1}{N} \sum_{i=1}^{N} \frac{\partial \log p(x_i|\theta)}{\partial \theta} \xrightarrow{\text{a.s.}} E\left[\frac{\partial \log p(x|\theta)}{\partial \theta}\right] = 0 \tag{3.5.10}$$

$$\frac{1}{N} \sum_{i=1}^{N} \frac{\partial^2 \log p(x_i|\theta^*)}{\partial \theta^{*2}} \xrightarrow{\text{a.s.}} E\left[\frac{\partial^2 \log p(x|\theta^*)}{\partial \theta^{*2}}\right] \tag{3.5.11}$$

We now make the assumption that the likelihood function is a concave function of θ for all x (this implies that the likelihood function has only one local maximum which is the global maximum). Clearly, this assumption is over-restrictive but the general proof is beyond the scope of the present discussion. Our assumption implies that $E[\partial^2 \log p(x|\theta^*)/\partial \theta^{*2}]$ is positive definite. Hence substituting (3.5.10) and (3.5.11) into (3.5.7) gives the result

$$\hat{\theta} \xrightarrow{\text{a.s.}} \theta \qquad \triangledown$$

Heuristically, the above theorem shows that $\hat{\theta}$ tends to θ for $N \to \infty$. The next theorem shows that the values of $\hat{\theta}$ from different trials (experiments) are clustered around θ with a normal distribution for N large.

Theorem 3.5.2 (*Asymptotic normality*) Let $\hat{\theta}_N$ designate a MLE of θ based on N i.i.d. variables $y = (x_1, \ldots, x_N)$, then $\hat{\theta}_N$ converges in law to a normal random variable, i.e.,

$$\sqrt{N}(\hat{\theta}_N - \theta) \xrightarrow{\text{law}} \beta$$

where

$$\beta \sim N(0, \overline{M}_\theta^{-1})$$

and $\overline{M}_\theta$ is the average value of Fisher's information matrix per sample.

Proof See Cramer [43].

A heuristic justification of the result is as follows. We expand $\partial \log p(y|\hat{\theta})/\partial\hat{\theta}$ about the true parameter value θ

$$\frac{\partial \log p(y|\hat{\theta})}{\partial\hat{\theta}} = \frac{\partial \log p(y|\theta)}{\partial\theta} + (\hat{\theta} - \theta)^{\mathrm{T}} \frac{\partial^2 \log p(y|\theta)}{\partial\theta^2}$$

$$+ \text{ remainder} \tag{3.5.12}$$

Since $\hat{\theta} \xrightarrow{\text{a.s.}} \theta$, we assume sufficient regularity to neglect the remainder. Since $\hat{\theta}$ is the maximum likelihood estimate, the l.h.s. of (3.5.12) is zero and hence

$$\frac{1}{\sqrt{N}} \frac{\partial \log p(y|\theta)}{\partial\theta} = \sqrt{N}(\hat{\theta} - \theta)^{\mathrm{T}} \left| -\frac{1}{N} \frac{\partial^2 \log p(y|\theta)}{\partial\theta^2} \right| \tag{3.5.13}$$

Now from the strong law of large numbers

$$-\frac{1}{N} \frac{\partial^2 \log p(y|\theta)}{\partial\theta^2} \xrightarrow{\text{a.s.}} E\left[\frac{\partial^2 \log p(x|\theta)}{\partial\theta^2} \right] \tag{3.5.14}$$

It can be readily shown that

$$E\left[\frac{\partial^2 \log p(x|\theta)}{\partial\theta^2} \right] = E\left[\left(\frac{\partial \log p(x|\theta)}{\partial\theta} \right)^{\mathrm{T}} \left(\frac{\partial \log p(x|\theta)}{\partial\theta} \right) \right]$$

$$= \overline{M}_\theta (= (1/N)M_\theta) \tag{3.5.15}$$

where M is Fisher's information matrix (see Eqn. (1.3.5)) and $\overline{M}$ is the average information matrix per sample. Now, taking the l.h.s. of (3.5.13), we have

$$\frac{1}{\sqrt{N}} \frac{\partial \log p(y|\theta)}{\partial\theta} = \frac{1}{\sqrt{N}} \sum_{i=1}^{N} \frac{\partial \log p(x_i|\theta)}{\partial\theta}$$

Also

$$E[\partial \log p(x|\theta)/\partial\theta] = 0$$

and

$$E[(\partial \log p(x|\theta)/\partial\theta)^{\mathrm{T}}(\partial \log p(x|\theta)/\partial\theta)] = \overline{M}_\theta$$

Therefore by the central limit Theorem B.3.3, we have

$$\frac{1}{\sqrt{N}} \frac{\partial \log p(y|\theta)}{\partial\theta} \xrightarrow{\text{law}} \gamma$$

where $\gamma \sim N(0, \overline{M}_\theta)$. Thus from (3.5.15), (3.5.14), and (3.5.13),

$$\sqrt{N}(\hat{\theta} - \theta)^{\mathrm{T}} \overline{M}_\theta \xrightarrow{\text{law}} \gamma$$

Hence from Result A.2.9,

$$\sqrt{N}(\hat{\theta} - \theta) \xrightarrow{\text{law}} \beta$$

where $\beta \sim N(0, \overline{M}_\theta^{-1})$. ∇

Theorem 3.5.3 (*Efficiency*) Within the class of consistent uniformly asymptotically normal estimators, $\hat{\theta}$ is efficient in the sense that asymptotically it attains the Cramer–Rao lower bound.

Proof Follows from Theorems 3.5.2 and 1.3.1. ∇

Theorem 3.5.3 is of practical significance since it shows that the maximum likelihood estimator makes efficient use of the available data for large data lengths.

3.6. THE LIKELIHOOD RATIO TEST

In this section we investigate the testing of hypotheses via the likelihood ratio. The basic idea is to determine the maximum likelihood estimates under both a null and alternative hypothesis. We then define the likelihood ratio $\lambda(y)$ as the ratio of the maximum values of the likelihood function in the two cases, i.e.,

$$\lambda(y) = p(y|\hat{\theta}_A)/p(y|\hat{\theta}_0) \tag{3.6.1}$$

where $\hat{\theta}_A$ is the MLE under the alternative hypothesis and $\hat{\theta}_0$ is the MLE under the null hypothesis.

Clearly, if the ratio is large (>1), then the data are seen to be more plausible under the alternative hypothesis than under the null hypothesis. We would thus be inclined to make a decision against the null hypothesis. Additional motivation for the test has been given in Section 1.5.

We restrict attention to the case where the data y consist of N i.i.d. observations $y_t, t = 1, \ldots, N$. Furthermore, we assume that the null hypothesis is that there are s locally independent restriction of the form $l_i(\theta) = 0$, $i = 1, \ldots, s$ between the parameters.

We denote by $\hat{\theta}$ the unrestricted MLE of θ and by $\tilde{\theta}$ the MLE of θ satisfying $l_i(\tilde{\theta}) = 0$, $i = 1, \ldots, s$. Thus the likelihood ratio is

$$\lambda(y) = \frac{p(y|\hat{\theta})}{p(y|\tilde{\theta})} = \frac{\prod_{i=1}^{N} p(y_i|\hat{\theta})}{\prod_{i=1}^{N} p(y_i|\tilde{\theta})} \tag{3.6.2}$$

An important asymptotic result regarding the distribution of $\lambda(y)$ is described in the following theorem.

Theorem 3.6.1 For $\lambda(y)$ defined by (3.6.2), then under the null hypothesis H_0 that $l_i(\theta) = 0$, $i = 1, \ldots, s$, the random variable $2 \log \lambda(y)$ converges in law to a χ^2 random variable on s degrees of freedom, i.e.,

$$2 \log \lambda(y) \xrightarrow{\text{law}} \beta \qquad (3.6.3)$$

where

$$\beta \sim \chi^2(s) \qquad (3.6.4)$$

Proof We present an outline proof for the case when $l_i(\theta) = 0$, $i = 1, \ldots, s$ gives particular values to the first s components of the parameter vector. Thus the equation $l_i(\theta) = 0$ takes the form

$$\alpha - \alpha^* = 0 \qquad (3.6.5)$$

where α^* is a fixed vector of length s and where

$$\theta = \begin{bmatrix} \theta_1 \\ \vdots \\ \theta_s \\ \hline \theta_{s+1} \\ \vdots \\ \theta_n \end{bmatrix} = \begin{bmatrix} \alpha \\ \hline \beta \end{bmatrix} \qquad (3.6.6)$$

Now $\log p(y|\theta)$ is given by

$$\log p(y|\theta) = \sum_{i=1}^{N} \log p(y_i|\theta) \qquad (3.6.7)$$

Hence

$$\frac{1}{N} \log p(y|\theta) = \frac{1}{N} \sum_{i=1}^{N} \log p(y_i|\theta)$$

$$= \frac{1}{N} \sum_{i=1}^{N} \log p(y_i|\theta) \Big|_{\theta=\hat{\theta}} + \frac{1}{N} \sum_{i=1}^{N} \frac{\partial \log p(y_i|\theta)}{\partial \theta} \Big|_{\theta=\hat{\theta}} (\theta - \hat{\theta})$$

$$+ \frac{1}{2N} \sum_{i=1}^{N} (\theta - \hat{\theta})^{\mathrm{T}} \frac{\partial^2 \log p(y_i|\theta)}{\partial \theta^2} \Big|_{\theta=\hat{\theta}} (\theta - \hat{\theta})$$

$$+ \text{remainder} \qquad (3.6.8)$$

where $\hat{\theta}$ is assumed to be the unrestricted maximum likelihood estimate of θ.

If θ is near $\hat{\theta}$, then the remainder can be ignored in (3.6.8). Furthermore, using similar arguments to those used in the proof of Theorem 3.5.2, we have

$$
\frac{1}{N} \sum_{i=1}^{N} \left. \frac{\partial^2 \log p(y_i|\theta)}{\partial \theta^2} \right|_{\theta=\theta} \rightarrow E_{y_i|\theta_0} \left[\frac{\partial^2 \log p(y_i|\theta)}{\partial \theta^2} \right] \tag{3.6.9}
$$

where θ_0 is the true parameter value.

Also

$$
E_{y_i|\theta_0} \left[\frac{\partial^2 \log p(y_i|\theta)}{\partial \theta^2} \right] = - E_{y_i|\theta_0} \left[\left(\frac{\partial \log p(y_i|\theta)}{\partial \theta} \right)^{\mathrm{T}} \left(\frac{\partial \log p(y_i|\theta)}{\cdot \ \partial \theta} \right) \right]
$$

$$
= - \overline{M}_{\theta_0} \tag{3.6.10}
$$

where $\overline{M}_{\theta_0}$ is the average information matrix per sample.

The second term in Eq. (3.6.8) is zero by virtue of the fact that $\hat{\theta}$ is the unrestricted maximum likelihood estimator and thus Eq. (3.6.8) can be expressed as

$$
\frac{1}{N} \log p(y|\theta) \simeq \frac{1}{N} \log p(y|\hat{\theta}) - \frac{1}{2} (\theta - \hat{\theta})^{\mathrm{T}} \overline{M}_{\theta_0} (\theta - \hat{\theta})
$$

$$
= \frac{1}{N} \log p(y|\hat{\theta})
$$

$$
- \frac{1}{2} \begin{bmatrix} \alpha - \hat{\alpha} \\ \beta - \hat{\beta} \end{bmatrix}^{\mathrm{T}} \begin{bmatrix} \overline{M}_{\alpha\alpha} & \overline{M}_{\alpha\beta} \\ \overline{M}_{\alpha\beta}^{\mathrm{T}} & \overline{M}_{\beta\beta} \end{bmatrix} \begin{bmatrix} \alpha - \hat{\alpha} \\ \beta - \hat{\beta} \end{bmatrix} \tag{3.6.11}
$$

We can now optimize the above expression with respect to β with α held at α^*. This gives the restricted maximum likelihood estimate $\tilde{\beta}$ for β. Putting $\alpha = \alpha^*$ and completing the square in (3.6.11) gives

$$
\frac{1}{N} \log p(y|\theta) \simeq \frac{1}{N} \log p(y|\hat{\theta}) - \frac{1}{2} (\alpha^* - \hat{\alpha})^{\mathrm{T}} \overline{M}_{\alpha\alpha} (\alpha^* - \hat{\alpha})
$$

$$
- \frac{1}{2} [\beta - \hat{\beta} + \overline{M}_{\beta\beta}^{-1} \overline{M}_{\alpha\beta}^{\mathrm{T}} (\alpha^* - \hat{\alpha})]^{\mathrm{T}} \overline{M}_{\beta\beta}
$$

$$
\times [\beta - \hat{\beta} + \overline{M}_{\beta\beta}^{-1} \overline{M}_{\alpha\beta}^{\mathrm{T}} (\alpha^* - \hat{\alpha})]
$$

$$
+ \frac{1}{2} (\alpha^* - \hat{\alpha})^{\mathrm{T}} \overline{M}_{\alpha\beta} \overline{M}_{\beta\beta}^{-1} \overline{M}_{\alpha\beta}^{\mathrm{T}} (\alpha^* - \hat{\alpha}) \tag{3.6.12}
$$

It immediately follows from (3.6.12) that the optimum value of β is

$$
\tilde{\beta} \simeq \hat{\beta} - \overline{M}_{\beta\beta}^{-1} \overline{M}_{\alpha\beta}^{\mathrm{T}} (\alpha^* - \hat{\alpha}) \tag{3.6.13}
$$

and substituting (3.6.13) in (3.6.12) gives

$$(1/N) \log p(y|\hat{\theta}) \simeq (1/N) \log p(y|\theta)$$
$$- \tfrac{1}{2}(\alpha^* - \hat{\alpha})^{\mathrm{T}}[\overline{M}_{\alpha\alpha} - \overline{M}_{\alpha\beta}\overline{M}_{\beta\beta}^{-1}\overline{M}_{\alpha\beta}^{\mathrm{T}}](\alpha^* - \hat{\alpha}) \quad (3.6.14)$$

Hence

$$2 \log \lambda(y) = 2 \log p(y|\hat{\theta}) - 2 \log p(y|\tilde{\theta})$$
$$\simeq N(\alpha^* - \hat{\alpha})^{\mathrm{T}}[\overline{M}_{\alpha\alpha} - \overline{M}_{\alpha\beta}\overline{M}_{\beta\beta}^{-1}\overline{M}_{\alpha\beta}^{\mathrm{T}}][\alpha^* - \hat{\alpha}] \quad (3.6.15)$$

From Theorem 3.5.2, the asymptotic distribution of $\sqrt{N}(\hat{\theta} - \theta_0)$ is normal with zero mean and covariance $\overline{M}^{-1}$. Now from Corollary A.2.4, the marginal distribution for $\sqrt{N}(2 - \alpha_0)$ is also normal with zero mean and covariance equal to the upper-left partition of $\overline{M}^{-1}$. Thus using Result E.1.1 from Appendix E,

$$\hat{\alpha} \sim N\left[\alpha_0, \frac{1}{N}(\overline{M}_{\alpha\alpha} - \overline{M}_{\alpha\beta}M_{\beta\beta}^{-1}\overline{M}_{\alpha\beta}^{\mathrm{T}})^{-1}\right] \quad (3.6.16)$$

Now introduce x related to $\hat{\alpha}$ via

$$x = \sqrt{N}\,A(\hat{\alpha} - \alpha^*) \quad (3.6.17)$$

where A is nonsingular and where

$$A^{\mathrm{T}}A = \overline{M}_{\alpha\alpha} - \overline{M}_{\alpha\beta}\overline{M}_{\beta\beta}^{-1}\overline{M}_{\alpha\beta}^{\mathrm{T}} \quad (3.6.18)$$

Then using Result A.2.4,

$$x \sim N[\sqrt{N}A(\alpha_0 - \alpha^*), I] \quad (3.6.19)$$

and Eq. (3.6.15) becomes

$$2 \log \lambda(y) \simeq x^{\mathrm{T}}x \quad (3.6.20)$$

Under the null hypothesis, the true value of α_0 is α^* and hence Eqs. (3.6.19) and (3.6.20) show that (asymptotically) $2 \log \lambda(y)$ is a sum of squares of s independent normal variables with zero mean and unit variance. The theorem then follows from Results A.3.1 and A.3.2. $\quad\triangledown$

The significance of Theorem 3.6.1 is that we can compare the observed value of $2 \log \lambda(y)$ with the value k_α obtained from the cumulative χ^2 distribution on s degrees of freedom where k_α is such that $100(1 - \alpha)\%$ of the distribution lies to the left of k_α. If $2 \log \lambda(y)$ is greater than k_α (for $\alpha = 0.05$, say), we would say that this was strong evidence against the null hypothesis.

So far we have considered only the significance level of the likelihood ratio test (see Section 1.5). However, in Chapter 6 we shall be interested in the properties of experiments that are "good" in some sense for hypothesis testing. In this case, we shall be interested in the power of the likelihood ratio test. As mentioned in Section 1.5, the power of the test is related to the probability of accepting the null hypothesis when it is actually false. We therefore consider a specific alternative hypothesis H_A, and assume that actually the true value for α_0 is α_A not α^*. We then have the following theorem.

Theorem 3.6.2   The same conditions prevail as used in the proof of Theorem 3.6.1. Then with a specific alternative hypothesis, H_A, namely, that the true value of α_0 is α_A not α^*, then random variable $2 \log \lambda(y)$ converges in law to a noncentral χ^2 distribution on s degrees of freedom and with noncentrality parameter

$$h = N(\alpha_A - \alpha^*)^{\mathrm{T}}[\overline{M}_{\alpha\alpha} - \overline{M}_{\alpha\beta}\,\overline{M}_{\beta\beta}^{-1}\overline{M}_{\alpha\beta}^{\mathrm{T}}](\alpha_A - \alpha^*) \qquad (3.6.21)$$

Proof   The result follows immediately by applying Result A.7.1 to Eqs. (3.6.19) and (3.6.20). ∇

The two distributional results described in Theorems 3.6.1 and 3.6.2 are depicted diagrammatically in Fig. 3.6.1.

It is known (see ref. [15], Section 2.4.5) that $\chi'^2(s, h)$ can be approximated for h large by a normal distribution with mean h and variance $4h$. Since h in Eq. (3.6.21) is proportional to the number of data points N, it is clear from Fig. 3.6.1 that the power of the likelihood ratio test can be increased by increasing the number of data points. We shall see later in Chapter 6 that the experimental conditions also influence h via $\overline{M}$ in Eq. (3.6.21) and thus for a fixed number of data points, the experimental conditions can be chosen to give a "high powered" test.

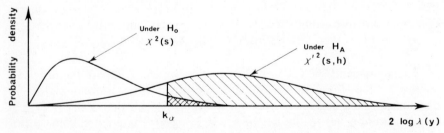

Figure 3.6.1. Distributions for likelihood ratio statistic. //// area = significant level of test; \\\\ area = power of test.

3.7. CONCLUSIONS

The maximum likelihood estimator is used widely in practice largely because of its conceptual simplicity and optimal properties for large data lengths. In subsequent chapters we shall see that the optimal properties in many cases carry over to the dynamic case. In the next chapter we look at suitable dynamic models for identification.

PROBLEMS

3.1 Obtain an expression for the bias of the estimator in Eq. (3.2.3) of Example 3.2.1.

3.2. Consider the following model with "colored disturbances":

$$y_t = \theta u_t + e_t + c e_{t-1}$$

where $\{e_t, t \in (1, \ldots, N)\}$ is a sequence of i.i.d. normal random variables with zero mean and unit variance. Obtain an expression for the MLE of θ assuming c is known.

3.3. Prove Result 3.3.2 (Eqs. (3.3.3) and (3.3.4)).

3.4. Let $Y_1, \ldots, Y_N$ be i.i.d. random variables, each having a log normal distribution: $\log Y_t \sim N(\mu, \sigma^2)$ for $t \in (1, \ldots, N)$.

(a) Show that for any Y_t

$$\xi = E[Y_t] = \exp\left\{\mu + \frac{\sigma^2}{2}\right\} \quad \text{and} \quad v = E[Y_t - \xi]^2 = \xi^2(\exp\{\sigma^2\} - 1)$$

(b) Show that the maximum likelihood estimator of (ξ, v) is

$$\hat{\xi} = \exp\left\{\frac{1}{N}\sum_{t=1}^{N} X_t + \frac{1}{2}S\right\}, \qquad \hat{v} = \hat{\xi}^2(\exp\{S\} - 1)$$

where

$$S = \frac{1}{N}\sum_{t=1}^{N}\left(X_t - \frac{1}{N}\sum_{k=1}^{N}X_k\right)^2$$

where

$$X_k = \log Y_k, \qquad t = 1, \ldots, N$$

3.5. Consider the problem in Result 3.3.3. Show that if Eq. (3.3.6) is used to eliminate Σ from the log likelihood function, then the result is

$$\log p(y|\hat{\theta}) = -(N/2)m(\log(2\pi) + 1) - (N/2)\log \det R(\hat{\theta})$$

where

$$R(\hat{\theta}) = \frac{1}{N} \sum_{t=1}^{N} (y_t - X_t \hat{\theta})(y_t - X_t \hat{\theta})^{\mathsf{T}}$$

3.6. Motivated by Problem 3.5, consider the following cost function:

$$J = \log \det R(\theta)$$

where

$$R(\theta) = \frac{1}{N} \sum_{t=1}^{N} (y_t - X_t \theta)(y_t - X_t \theta)^{\mathsf{T}}$$

Show that

$$\frac{\partial J}{\partial \theta} = -2 \sum_{k=1}^{N} (y_k - X_k \theta)^{\mathsf{T}} R(\theta)^{-1} X_k$$

Hint: $\partial \log \det R / \partial R_{ij} = ji$ element of R^{-1} (see Appendix E).

3.7. Let the probability of a success in a simple Bernoulli trial be μ. (e.g., tossing a coin). In N independent trials a result of the form H, H, T, T, T, ... would be obtained where H denotes head (success) and T denotes tails (failure).

(a) Derive an expression for the likelihood function for a given set of N observations.

(b) Obtain an expression for the maximum likelihood estimate of μ.

(c) Is the maximum likelihood estimator biased in this case?

(d) Derive an expression for the Cramer–Rao lower bound on the covariance of any unbiased estimator of μ.

3.8. Show that $\hat{\theta}$ given in (3.3.5) is an unbiased estimator for θ whether or not $\hat{\Sigma}$ satisfies (3.3.6).

CHAPTER

4

Models for Dynamic Systems

4.1. INTRODUCTION

Our objective in the next chapter is to extend the parameter estimation procedures described in the first three chapters to dynamic systems. We therefore devote this chapter to various models suitable for parameter estimation in the dynamic case.

We begin with a quick revision of models for deterministic systems. In particular, we mention difference equation and state space models. It is true, in general, that many different models can give rise to the same input/output characteristics, i.e., the same external behavior. We are thus led to define T-equivalence and canonical forms under T-equivalence. For linear dynamic systems we describe appropriate canonical forms in both matrix fraction description (MFD) and state space forms (SSF).

In many situations a deterministic model is inadequate to describe a system and we are naturally led to consider the system output as being a realization of a stochastic process. We describe prediction error models (PEM) that are appropriate in the stochastic case. For covariance stationary stochastic processes, we show by use of a spectral factorization theorem that it is natural to use innovations models that are a special case of the more general class of PEM models.

Finally, we show that commonly used models arising from physical or other reasoning can be converted to the PEM form.

4.2. DETERMINISTIC MODELS

We consider causal systems in which the output may be expressed in terms of the past inputs and outputs of the system. For discrete time systems we consider models of the form

$$y_t = f(y_{t-1}, y_{t-2}, \ldots, u_t, u_{t-1}, u_{t-2}, \ldots, t) \tag{4.2.1}$$

where $\{y_t\}$ and $\{u_t\}$ denote the output and input sequences, respectively. A model in the form of (4.2.1) will be said to be a general difference equation description. Alternatively we can model the system in state space form as follows:

$$x_{t+1} = a(x_t, u_t, t) \tag{4.2.2}$$

$$y_t = c(x_t, u_t, t) \tag{4.2.3}$$

where x_t is a state vector.

In the case of linear systems, Eq. (4.2.1) can be expressed as

$$y_t = -\sum_{k=1}^{m} F_{t,k} y_{t-k} + \sum_{k=0}^{m} G_{t,k} u_{t-k} \tag{4.2.4}$$

In the time invariant case, the matrices $F_{t,1}, \ldots, F_{t,m}$, and $G_{t,0}, \ldots, G_{t,m}$ do not depend upon t. In this case we can write

$$y_t + \sum_{k=1}^{m} F_k y_{t-k} = \sum_{k=0}^{m} G_k u_{t-k} \tag{4.2.5}$$

In the sequel we shall frequently wish to refer to equations of the form of (4.2.5). We therefore simplify the notation by introducing the unit delay operator q^{-1} defined by

$$q^{-1} x_t = x_{t-1} \tag{4.2.6}$$

Equation (4.2.5) can then be expressed in operator form as

$$F(q) y_t = G(q) u_t \tag{4.2.7}$$

where $F(q)$ and $G(q)$ are matrix polynomials in q^{-1} defined by

$$F(q) = \sum_{k=0}^{m} F_k q^{-k} \tag{4.2.8}$$

$$G(q) = \sum_{k=0}^{m} G_k q^{-k} \tag{4.2.9}$$

If the system described in Eq. (4.2.7) is original at rest (i.e., $u_t = 0$, $y_t = 0$ for $t < 0$), then it is readily seen [250] that the z-transform of $\{y_t\}$ is related to the z-transform of $\{u_t\}$ by

$$F^*(z)Y(z) = U^*(z)U(z) \tag{4.2.10}$$

where $F^*(z)$ and $G^*(z)$ are matrix polynomials in z given by

$$F^*(z) = \sum_{k=0}^{m} F_k z^{m-k} \tag{4.2.11}$$

$$G^*(z) = \sum_{k=0}^{m} G_k z^{m-k} \tag{4.2.12}$$

To ensure that $\{y_t\}$ is uniquely determined by $\{u_t\}$, we require that $\det F^*(z) \neq 0$. Hence Eq. (2.10) can be expressed as

$$Y(z) = H(z)U(z) \tag{4.2.13}$$

where

$$H(z) = [F^*(z)]^{-1}[G^*(z)] \tag{4.2.14}$$

The rational matrix $H(z)$ will be called a matrix transfer function. We shall call $[F^*(z)]^{-1}[G^*(z)]$ a left *matrix fraction description* (MFD) for $H(z)$. It follows from (4.2.8), (4.2.9), (4.2.11), and (4.2.12) that

$$F^*(z) = z^m F(z) \tag{4.2.15}$$

$$G^*(z) = z^m G(z) \tag{4.2.16}$$

Hence we have

$$H(z) = [F^*(z)]^{-1}[G^*(z)] = [F(z)]^{-1}[G(z)] \tag{4.2.17}$$

Thus Eq. (4.2.13) can be expressed as

$$F(z)Y(z) = G(z)U(z) \tag{4.2.18}$$

Comparing Eq. (4.2.18) with Eq. (4.2.7), we see that the relationship between $Y(z)$ and $U(z)$ is the same as the operator relationship between $\{y_t\}$ and $\{u_t\}$ with the unit delay operator q^{-1} replaced by z^{-1}. In the light of the above reasoning we shall use the one symbol z in place of q and z. Whether z^{-1} should be interpreted as a unit delay operator or the inverse of the z-transform variable will be clear from the context.

Another form of model that we shall use to describe systems is the *state space model* where $\{y_t\}$ is related to $\{u_t\}$ via

$$x_{t+1} = Ax_t + Bu_t \tag{4.2.19}$$

$$y_t = Cx_t + Du_t \tag{4.2.20}$$

If the initial state is zero (i.e., $x_0 = 0$), then the z-transform of $\{y_t\}$ is related to the z-transform of $\{u_y\}$ by

$$Y(z) = H(z)U(z) \tag{4.2.21}$$

where

$$H(z) = C(zI - A)^{-1}B + D \tag{4.2.22}$$

4.3.   CANONICAL MODELS

It is often the case that more than one model gives rise to the same transfer function. We define such systems to be T equivalent, written **T**. This is a special type of equivalence relation. Before proceeding we present some general properties of equivalence relations [3].

Definition 4.3.1   An equivalence relation E on a set S is any relation satisfying

   (i) $s \, E \, s$ for all $s \in S$;
  (ii) $s_1 \, E \, s_2$ implies $s_2 \, E \, s_1$ for all $s_1, s_2 \in S$;
 (iii) $s_1 \, E \, s_2$ and $s_2 \, E \, s_3$ implies $s_1 \, E \, s_3$ for all $s_1, s_2, \in S$. ∇

Definition 4.3.2   An equivalence class is a subset $\mathscr{S} \subset S$ such that $s_1 \in \mathscr{S}$ and $s_1 \, E \, s_2$ implies $s_2 \in \mathscr{S}$ and conversely. ∇

Definition 4.3.3   The equivalence relation E on S generates a set of equivalence classes and the set of all such equivalence classes is called the quotient set S/E. ∇

Definition 4.3.4   A set of canonical forms for equivalence E on S is a subset C of S such that for each $s \in S$ there exists one and only one $c \in C$ for which $s \, E \, c$. ∇

Clearly there is one and only one element of C in each element of the quotient set and we therefore have an isomorphism between C and S/E.

For linear time invariant state-space models, we have that two representations (A, B, C, D) and (A', B', C', D') are T-equivalent if and only if

$$C(zI - A)^{-1}B + D = C'(zI - A')^{-1}B' + D' \tag{4.3.1}$$

It is well known from linear system theory [5, 39, 77] that many quadruples (A, B, C, D) can be found satisfying a relation of the form (4.3.1) for a particular (A', B', C', D'). Obviously there are many ways of defining a unique member from each equivalence class, i.e., a set of canonical forms. We discuss examples of canonical forms below. See also references [7–9] and [80]. We stress that our examples are for illustrative purposes only and that there are many other possibilities.

Canonical Form SSF1 Given any quadruple (A, B, C, D), the rule for finding the SSF1 canonical form in the same equivalence class as (A, B, C, D) is

(i) Form $C(zI - A)^{-1}B$.
(ii) Reduce each element so that the numerator and denominator are coprime.
(iii) Find the least common denominator of all the elements.
(iv) Express

$$C(zI - A)^{-1}B \qquad \text{as} \qquad \frac{1}{f^*(z)} G^*(z)$$

where

$$f^*(z) = z^n + a_1 z^{n-1} + \cdots + a_n$$
$$G^*(z) = G_1 z^{n-1} + \cdots + G_n$$

(v) Put

$$A^{(1)} = \begin{bmatrix} 0 & I & 0 & & & & 0 \\ 0 & 0 & I & & & & \cdot \\ \cdot & & & & & & \cdot \\ \cdot & & & & & & 0 \\ 0 & 0 & & & & 0 & I \\ -a_n I & -a_{n-1}I & & & & & -a_1 I \end{bmatrix} \qquad (4.3.2)$$

$$B^{(1)} = [0 \quad 0 \quad \cdot \quad \cdot \quad \cdot \quad I]^{\mathsf{T}} \qquad (4.3.3)$$

$$C^{(1)} = [G_n \quad G_{n-1} \quad \cdot \quad \cdot \quad G_1] \qquad (4.3.4)$$

$$D^{(1)} = D \qquad (4.3.5)$$

The quadruple $(A^{(1)}, B^{(1)}, C^{(1)}, D^{(1)})$ is the SSF1 canonical form. Of course it is not true that all systems of the structure (4.3.2) to (4.3.5) are canonical forms. Given a system of this structure one would still need to apply steps (i)–(v). This will either leave the system unchanged or result in a smaller value for n. If n is known to be the degree of the least common denominator, then there will be no change. $\triangledown$

Canonical form SSF1 does not have the property of minimality in that there will exist, in general, systems with smaller state dimension. The properties of minimal state space representations are contained in the following result:

Result 4.3.1 Any minimal realization (A, B, C, D) has the following properties:

(i) (A, B) is completely controllable.
(ii) (C, A) is completely observable.
(iii)
$$\text{rank}\begin{bmatrix} CB & CAB & \cdots & CA^{n-1}B \\ \vdots & & & \vdots \\ CA^{n-1}B & & \cdots & CB \end{bmatrix} = n \quad \text{where } n = \dim A$$
$$(4.3.6)$$

(iv) If (A', B', C', D') is any other minimal realization of the same transfer function, then (A, B, C, D) and (A', B', C', D') are related by a similarity transformation, i.e., there exists a nonsingular matrix T such that

$$A' = TAT^{-1} \tag{4.3.7}$$

$$B' = TB \tag{4.3.8}$$

$$C' = CT^{-1} \tag{4.3.9}$$

$$D' = D \tag{4.3.10}$$

Proof See Desoer [77]. ▽

We describe minimal canonical state space forms below.

Canonical Form SSF2[8, 5, 80, 93, 94] Given any quadruple $(\overline{A}, \overline{B}, \overline{C}, \overline{D})$, the rule for finding the SSF2 canonical form in the same equivalence class as $(\overline{A}, \overline{B}, \overline{C}, \overline{D})$ is

(i)   Convert to SSF1 canonical form (A, B, C, D).
(ii) Construct the observability matrix for the new system

$$P_0 = [C^T, A^T C^T, \ldots, (A^{n-1})^T C^T] \tag{4.3.11}$$

(iii) We search the columns of P_0 for linearly independent columns in the order

$$c_1^T c_2^T, \ldots, c_r^T, A^T c_1^T, A^T c_2^T, \ldots, (A^{n-1})^T c_r^T \tag{4.3.12}$$

Having completed the search, we rearrange the resulting $n' \leq n$ linearly independent columns in the order

$$Q^{(2)} = [c_1^T, A^T c_1^T, \ldots, (A^{v_i-1})^T c_1^T, c_2^T, \ldots, (A^{v_i-1})^T c_i^T, \ldots, (A^{v_r-1})^T c_r^T, R] \tag{4.3.13}$$

where $v_1 + v_2 + \cdots + v_r = n'$, and where $R(n \times (n - n'))$ is arbitrary as long as $Q^{(2)}$ is nonsingular.

(iv) Form

$$\overline{A}^{(2)} = Q^{\mathsf{T}} A Q^{-\mathsf{T}} \tag{4.3.14}$$

$$\overline{B}^{(2)} = Q^{\mathsf{T}} B \tag{4.3.15}$$

$$\overline{C}^{(2)} = C Q^{-\mathsf{T}} \tag{4.3.16}$$

$$D^{(2)} = D = \overline{D} \tag{4.3.17}$$

Take $A^{(2)}$ as upper left $n' \times n'$ submatrix from $\overline{A}^{(2)}$, $C^{(2)}$ as first n' columns of $\overline{C}^{(2)}$, $B^{(2)}$ as first n' rows of $\overline{B}^{(2)}$. $(A^{(2)}, B^{(2)}, C^{(2)}, D^{(2)})$ will then be the canonical form SSF2. The form of $Q^{(2)}$ clearly imposes a special structure on the matrices $A^{(2)}$, $C^{(2)}$ as illustrated below for a simple case (see Problem 4.8). (Note the crosses denote nonzero elements.)

$$(4.3.18)$$

$$(4.3.19)$$

The indices $v_1, \ldots, v_r$ are called the *Kronecker invariants* for the system. As for SSF1 it is not true that all systems having the structure of

(4.3.19) and (4.3.20) are canonical forms. However, if $v_1, \ldots, v_r$ are known, then anything in this form will be canonical. $\triangledown$

Canonical Form SSF3 [93, 95, 8, 5]   The procedure is as for SSF2 save that the columns of the observability matrix are searched in the following order

$$c_1^{\mathrm{T}}, A^{\mathrm{T}}c_1^{\mathrm{T}}, \ldots, (A^{n-1})^{\mathrm{T}}c_1^{\mathrm{T}}, c_2^{\mathrm{T}}, A^{\mathrm{T}}c_2^{\mathrm{T}}, \ldots, (A^{n-1})^{\mathrm{T}}c_r^{\mathrm{T}} \qquad (4.3.20)$$

The transformation $Q^{(3)}$ is then formed from the resulting $n' \leq n$ linearly independent columns in the order

$$Q^{(3)} = [c_1^{\mathrm{T}}, \ldots, (A^{\lambda_1-1})^{\mathrm{T}}c_1^{\mathrm{T}}, c_2^{\mathrm{T}}, \ldots, (A^{\lambda_r-1})^{\mathrm{T}}c_r^{\mathrm{T}}, R] \qquad (4.3.21)$$

where

$$\lambda_1 + \cdots + \lambda_r = n' \qquad (4.3.22)$$

and where $R(n \times (n - n'))$ is arbitrary so long as $Q^{(3)}$ is nonsingular.

The form of $Q^{(3)}$ clearly imposes a special structure on the matrices $A^{(3)}$, $C^{(3)}$ as illustrated below for a special case (see SSF2 for method of determining $A^{(3)}$, $B^{(3)}$, $C^{(3)}$, $D^{(3)}$). (Note the crosses denote nonzero elements.)

$$C^{(3)} = \begin{bmatrix} 1 & 0 & \cdots & & & \cdots & & & \cdots & & 0 \\ 0 & & \cdots & 0 & 1 & 0 & \cdots & & & \cdots & 0 \\ 0 & & \cdots & & & & \cdots & 0 & 1 & 0 & \cdots & 0 \end{bmatrix} \qquad (4.3.24)$$ $\triangledown$

We are able to define canonical matrix fraction descriptions (MFDs) in an analogous fashion to canonical SSFs.

Canonical Form MFD1 Analogously to SSF1 we can express the MFD in the canonical form

$$H(z) = [F^{(1)*}(z)]^{-1}G^{(1)*}(z) \qquad (4.3.25)$$

where

$$F^{(1)*}(z) = \text{diag}[f^{(1)*}(z) \cdots f^{(1)*}(z)] \qquad (4.3.26)$$

$$f^{(1)*}(z) = [z^n + a_1 z^{n-1} + \cdots + a_n] \qquad (4.3.27)$$

$$G^{(1)*}(z) = [G_0 z^n + G_1 z^{n-1} + \cdots + G_n] \qquad (4.3.28)$$

and n is necessarily the degree of the least common denominator. ∇

Canonical form MFD1 does not have the property of irreducibility. We say that $F^*(z)$ and $G^*(z)$ are irreducible if they have no common left factors except unimodular matrices (polynomial matrices with constant determinant, i.e., possessing polynomial inverses). Properties of irreducible realizations are described in the following result.

Result 4.3.2 Any irreducible realization $(F^*(z), G^*(z))$ has the following properties

(i) The matrix $[F^*(z), G^*(z)]$ has full rank.

(ii) If $(\tilde{F}^*(z), \tilde{G}^*(z))$ is any other irreducible realization of the same transfer function, then these realizations are related by a unimodular transformation $U^*(z)$,

$$[F^*(z), G^*(z)] = U^*(z)[\tilde{F}^*(z), \tilde{G}^*(z)] \qquad (4.3.29)$$

(iii) If $(\tilde{F}^*(z), \tilde{G}^*(z))$ is any other realization of the same transfer function, then

$$\text{degree det}[F^*(z)] \leq \text{degree det}[\tilde{\tilde{F}}^*(z)]$$

with equality iff $(\tilde{F}^*(z), \tilde{G}^*(z))$ is also irreducible.

(iv) degree det$[F^*(z)]$ = dimension of minimal state space representation of the transfer function

Proof See [87–91]. ∇

Irreducible MFD canonical forms are described below.

Canonical Form MFD2 [89, 91]   Analogously to SSF2 we can express an MFD in the following canonical form (often called the *Echelon canonical form*)

$$H(z) = [F^{(2)*}(z)]^{-1} G^{(2)*}(z) \qquad (4.3.30)$$

There is a set of pivot indices $\{s_i, i\}$ defined so that if $F^{(2)*}(z)$ has ijth element $f^{(2)*}_{ij}(z)$, then

   (i) $v_1 \geq v_2 \cdots \geq v_r$;
   (ii) degree $f^{(2)*}_{ij}(z) = v_i$;
   (iii) $f^{(2)*}_{is_i}(z)$ is a monic polynomial;
   (iv) degree $f^{(2)*}_{is_j}(z) < v_j$ for $i \neq j$;
   (v)   degree $f^{(2)*}_{ij}(z) < v_i$ for $j > s_i$;
   (vi) if $v_i = v_j$ and $i > j$, then $s_i > s_j$.

The integers $v_1, \ldots, v_r$ are the Kronecker invariants as in the description of canonical form SSF1. $\triangledown$

Canonical Form MFD3 [87, 88, 92, 72]   Analogously to SSF3 we can express an MFD in the following canonical form (often called the *Hermite canonical form*)

$$H(z) = [F^{(3)*}(z)]^{-1} [G^{(3)*}(z)]$$

where $F^{(3)*}(z)$ is lower triangular with monic polynomials as diagonal elements and with each diagonal element having greatest degree in its column. $\triangledown$

We should point out, at this stage, that our interest in canonical forms is motivated by their possible application in the modeling of dynamic systems for parameter estimation. From input/output measurements we can only hope to make statements about which element of the quotient set is consistent with the data and we therefore cannot hope to distinguish between elements of an equivalence class. It is useful, though clearly not necessary, to characterize each equivalence class by a unique member, i.e., a canonical form. Conceptually it does not matter which canonical form we choose although this may be dictated by practical considerations such as numerical problems.

A further important point is that the set of canonical forms C must be a subset of the allowable set S. Thus a set C_1 which is canonical for one set S_1 is not necessarily canonical for another set S_2. For example, if physical reasoning leads to constraints on the allowable model structures, then the elements of the set of canonical forms must satisfy these constraints. Thus the specific "canonical" forms described in this chapter would not, in general, be canonical, although they could be used to assign a unique label to each equivalence class. However, the constraints must be borne in mind to prevent the occurrence of models that lie in an equivalence class containing no element satisfying the given constraints. See [3], [96], and [97] for further discussion.

Another difficulty is that, in real situations, modeling always involves some element of approximation since all real systems are, to some extent, nonlinear, time varying, and distributed. Thus it is highly improbable that any set of models will contain the "true" system structure. All that can be hoped for is a model which provides an acceptable level of approximation, as measured by the use to which the model will be put. For example, a transistor model for low frequency applications need not be as complex as a model for high frequency applications.

In summary the following points are relevant to the specification of model structures:

(1)   The model should reflect all physical knowledge regarding the operation of the system.

(2)   The model should have sufficient flexibility to exhibit a wide range of possible external behavior within the constraints imposed by (1).

(3)   The model should have a unique parameterization in so far that, excepting for singular situations (e.g., having the order too high) different parameter values should give different external model characteristics.

(4)   The model should provide an acceptable level of approximation as measured by the ultimate model use or operating range.

(5)   The model should, if possible, lead to a simple and robust estimation algorithm (see Chapter 5 for further discussion).

4.4.   STOCHASTIC MODELS (The Covariance Stationary Case)

The class of models that we shall consider in this section, and in subsequent chapters, differs from the class considered in Section 4.2 in that we introduce another component that is random in nature.

Initially we shall consider systems whose outputs may be adequately modeled by covariance stationary stochastic processes with time varying means which are the outputs of deterministic models driven by the system inputs. We make use of the following spectral factorization result [4].

Theorem 4.4.1 Let $\Phi(z)$ be an $n \times n$ discrete rational spectral density matrix having full normal rank. Then there exists a unique $n \times n$ rational matrix $H(z)$ and a unique positive definite real symmetric matrix Σ satisfying

(i) $\Phi(z) = H(z)\Sigma H^\mathrm{T}(z^{-1})$;
(ii) $H(z)$ is analytic outside and on the unit circle, i.e., for $|z| \geq 1$;
(iii) $H^{-1}(z)$ is analytic outside the unit circle, i.e., for $|z| > 1$;
(iv) $\lim_{z \to \infty} H(z) = I$.

Proof See [72], [73], [4], and [1]. ∇

Interpreting Theorem 4.4.1 in the light of Result C.1.2 of Appendix C leads to the conclusion that a zero mean stochastic process with spectral density $\Phi(z)$ may be modeled as the output of a linear system driven by white noise.

The case where $\Phi(z)$ does not have full normal rank is discussed by Hannan [72]. We shall henceforth restrict attention to the more usual case where $\Phi(z)$ has full normal rank and it thus follows that a general model for the class of systems we are currently considering is

$$y_t = \bar{y}_t + \eta_t \tag{4.4.1}$$

where $\bar{y}_t$ is the output of a deterministic model of the form considered in Section 4.2 and η_t is the output of a linear filter with transfer function $H(z)$ driven by white noise with covariance Σ.

If required, canonical forms for modeling $\{\eta_t\}$ may be obtained by using the canonical forms discussed in Section 4.3 together with the additional constraints (arising from the conditions given in Theorem 4.4.1):

(i)   (MFD form)

$$H(z) = [F^*(z)]^{-1}[G^*(z)]$$

where

(a) $F^*(z) = F_0 z^n + F_1 z^{n-1} + \cdots + F_n$
$G^*(z) = G_0 z^n + G_1 z^{n-1} + \cdots + G_n$

with $\lim_{z \to \infty} H(z) = I$,

(b) $\det[G^*(z)]$ is nonzero for all $|z| > 1$

(ii) (SSF description)

$$H(z) = C(zI - A)^{-1}B + D$$

where

(a) $D = I$

(b) $\det(zI - A + BC)$ is nonzero for all $|z| > 1$

(This ensures stability of $H^{-1}(z)$.)

For linear time-invariant systems, we thus arrive at any of the following models (among others):

(i) (MFD)

$$y_t = F_1^{-1}(z)G_1(z)u_t + F_2^{-1}(z)G_2(z)\varepsilon_t \qquad (4.4.2)$$

or

$$A(z)y_t = B(z)u_t + C(z)\varepsilon_t \qquad (4.4.3)$$

where A, B, C, F_1, F_2, G_1, and G_2 are polynomials in z^{-1} and $\{\varepsilon_t\}$ is a white noise sequence with covariance Σ.

(ii) (SSF)

$$y_t = C_1 x_t^{1} + D_1 u_t + \eta_t \qquad (4.4.4)$$

$$\eta_t = C_2 x_t^{2} + \varepsilon_t \qquad (4.4.5)$$

$$x_{t+1}^{1} = A_1 x_t^{1} + B_1 u_t \qquad (4.4.6)$$

$$x_{t+1}^{2} = A_2 x_t^{2} + B_2 \varepsilon_t \qquad (4.4.7)$$

or

$$y_t = C x_t + D u_t + \varepsilon_t \qquad (4.4.8)$$

$$x_{t+1} = A x_t + B u_t + K \varepsilon_t \qquad (4.4.9)$$

All four of the above models may be written in the general form

$$y_t = H_1(z)u_t + H_2(z)\varepsilon_t \qquad (4.4.10)$$

where

$$\lim_{z \to \infty} H_2(z) = I \qquad (4.4.11)$$

and where $\{\varepsilon_t\}$ is a white sequence with covariance Σ. All that differs is the way in which H_1 and H_2 are parameterized. We shall subsequently use a model of the form (4.4.10) with (4.4.11) satisfied as a general description of time invariant linear systems with covariance stationary disturbances.

In the next section we discuss a more general class of models which includes the models described in this section as special cases.

4.5. STOCHASTIC MODELS (Prediction Error Formulation)

In this section we consider a general class of models known as prediction error models (PEM) [98]. These models have the following general form:

$$y_t = f[Y_{t-1}, U_t, t] + \varepsilon_t \qquad (4.5.1)$$

where Y_{t-1} denotes the set $\{y_{t-1}, y_{t-2}, \ldots\}$, and U_t denotes the set $\{u_t, u_{t-1}, \ldots\}$, and $\{\varepsilon_t\}$ is an *innovations* sequence (Definition D.2.3) having the property that its conditional mean given Y_{t-1}, U_t is zero, i.e.,

$$E_{\varepsilon_t}\big|_{Y_{t-1}, U_t}[\varepsilon_t] = 0 \qquad (4.5.2)$$

We shall find in the next chapter that PEM models are particularly suitable for finding the likelihood function for the data and hence this class of models has particular relevance for parameter estimation.

We now show that many commonly occurring models, including those described in the last section, can be readily transformed to the PEM structure.

4.5.1 *Linear models with covariance stationary disturbances*   In Section 4.4 we saw that linear systems having covariance stationary disturbances can be modeled in the form

$$y_t = H_1(z)u_t + H_2(z)\varepsilon_t \qquad (4.5.3)$$

where $H_1(z)$ and $H_2(z)$ are stable rational transfer functions; $H_2(z)$ having the additional properties

(i) $H_2(z)^{-1}$ is stable;
(ii) $\lim_{z \to x} H_2(z) = I$.

$\{\varepsilon_t\}$ is a white sequence with covariance Σ.

Properties (i) and (ii) of $H_2(z)$ imply that (4.5.3) can be written in the form

$$y_t = L_1(z)y_{t-1} + L_2(z)u_t + \varepsilon_t \qquad (4.5.4)$$

where $L_1(z)$ and $L_2(z)$ are stable transfer functions given by

$$L_1(z) = z[I - H_2^{-1}(z)] \qquad (4.5.5)$$

$$L_2(z) = H_2^{-1}(z)H_1(z) \qquad (4.5.6)$$

Comparing (4.5.4) with (4.5.1) shows that (4.5.4) is in the PEM form.

4.5.2   *Linear time invariant state space models*   A model frequently used to describe linear time invariant stochastic systems is

$$x_{t+1} = Ax_t + Bu_t + \omega_t \tag{4.5.7}$$
$$y_t = Cx_t + Du_t + v_t \tag{4.5.8}$$

with A stable and (A, C) observable and where $\{\omega_t\}$ and $\{v_t\}$ are two uncorrelated white sequences with covariances Q and R, respectively.

By superposition the model can be expressed in terms of two components as follows

$$y_t = \bar{y}_t + \eta_t$$

where $\{\bar{y}_t\}$ is modeled by

$$\bar{x}_{t+1} = A\bar{x}_t + Bu_t \tag{4.5.9}$$
$$\bar{y}_t = C\bar{x}_t + Du_t \tag{4.5.10}$$

and where η_t is a zero mean stochastic process having spectral density

$$\Phi_\eta(z) = C(zI - A)^{-1}Q(z^{-1}I - A^{\mathrm{T}})^{-1}C^{\mathrm{T}} + R \tag{4.5.11}$$

(See Eq. (C.1.16).)

We now proceed, as in Theorem 4.4.1, to determine a spectral factorization for $\Phi_\eta(z)$. This can be achieved as follows:

$$\Phi_\eta(z) = H(z)\Sigma H^{\mathrm{T}}(z^{-1}) \tag{4.5.12}$$

where

$$H(z) = C(zI - A)^{-1}K + I \tag{4.5.13}$$

K and Σ can be determined as follows:

$$K = APC^{\mathrm{T}}(CPC^{\mathrm{T}} + R)^{-1} \tag{4.5.14}$$
$$\Sigma = CPC^{\mathrm{T}} + R \tag{4.5.15}$$

where P is the unique positive definite symmetric solution of the following algebraic Riccati equation

$$P = APA^{\mathrm{T}} - APC^{\mathrm{T}}(CPC^{\mathrm{T}} + R)^{-1}CPA^{\mathrm{T}} + Q \tag{4.5.16}$$

The above spectral factorization can be verified by substitution (see Problem 4.9). $H(z)$, as found above, can be shown to have the properties given in the statement of Theorem 4.4.1 [76, 78, 99, 100, 101].

It can be seen from (4.5.9), (4.5.10), (4.5.12), and (4.5.13) that $\{y_t\}$ has a representation of the form

$$\bar{x}_{t+1} = A\bar{x}_t + Bu_t \tag{4.5.17}$$
$$\tilde{x}_{t+1} = A\tilde{x}_t + K\varepsilon_t \tag{4.5.18}$$
$$y_t = C\bar{x}_t + C\tilde{x}_t + Du_t + \varepsilon_t \tag{4.5.19}$$

where $\{\varepsilon_t\}$ is a white sequence with covariance Σ.

Equations (4.5.17)–(4.5.19) can be written more concisely by putting $\hat{x}_t = \bar{x}_t + \tilde{x}_t$. Then

$$\hat{x}_{t+1} = A\hat{x}_t + Bu_t + K\varepsilon_t \tag{4.5.20}$$

$$y_t = C\hat{x}_t + Du_t + \varepsilon_t \tag{4.5.21}$$

Furthermore, Eqs. (4.5.20) and (4.5.21) can be expressed as

$$y_t = H_1(z)u_t + H_2(z)\varepsilon_t \tag{4.5.22}$$

where

$$H_1(z) = C(zI - A)^{-1}B + D \tag{4.5.23}$$

$$H_2(z) = C(zI - A)^{-1}K + I \tag{4.5.24}$$

which is readily expressible in the PEM form as described in Eqs. (4.5.4)–(4.5.6).

4.5.3 *Continuous system with nonuniform sampling* We consider the problem of sampling a linear time invariant continuous time system at times $t_1, t_2, \ldots, t_N$ where the sampling intervals

$$\Delta_k = t_{k+1} - t_k \tag{4.5.25}$$

$$k = 0, 1, \ldots, N - 1 \tag{4.5.26}$$

are not necessarily equal (t_0 corresponds to the beginning of the experiment).

In general, the data from the system will be passed through a linear filter prior to sampling. The purpose of the filter is to remove aspects of the data which are detrimental to subsequent analysis (see Chapter 6).

A general model for a continuous linear time invariant system is (see Appendix C)

$$dx'(t) = A'x'(t)\, dt + B'u(t)\, dt + d\eta(t) \tag{4.5.27}$$

$$dy'(t) = C'x'(t)\, dt + d\omega(t) \tag{4.5.28}$$

where $y'(t)$: $T \to R^r$ is the output prior to filtering and sampling, $u(t)$: $T \to R^m$ is the input vector, $x'(t)$: $T \to R^{n'}$ is a state vector and $\eta(t)$, $\omega(t)$ are wide sense stationary processes with uncorrelated increments. The incremental covariance is assumed to be

$$E\left\{ \begin{vmatrix} d\eta(t) \\ d\omega(t) \end{vmatrix} [d\eta^T(t)\, d\omega^T(t)] \right\} = \begin{vmatrix} R_{11} & R_{12} \\ R_{12}^T & R_{22} \end{vmatrix} dt \tag{4.5.29}$$

The above model is based on the usual assumptions of rational input/output transfer function and rational output noise power density spectrum.

We assume that the system output y' is passed through a linear filter of the following form:

$$(d/dt)x''(t) = A''x''(t) + B''u(t) + K''y'(t) \qquad (4.5.30)$$

$$y(t) = C''x''(t) + D''y'(t) \qquad (4.5.31)$$

where $y(t)$ is the filter output.

Equations (4.5.27) and (4.5.28) can be combined with Eq. (4.5.30) and (4.5.31) to produce a composite model for the system and filter of the following form:

$$dx(t) = Ax(t)\, dt + Bu(t)\, dt + K d\varepsilon(t) \qquad (4.5.32)$$

$$y(t) = Cx(t) \qquad (4.5.33)$$

where $x(t)$ is the state of the composite model and $\varepsilon(t)$ is a process with wide sense stationary uncorrelated increments having incremental covariance $\Sigma\, dt$.

For simplicity of exposition, the class of allowable inputs is now restricted to piecewise constant functions, that is

$$u(t) = u_k, \qquad t_k \le t < t_{k+1} \qquad (4.5.34)$$

This class of inputs has the added advantage of ease of implementation with the specified sampling scheme. However, the analysis can be readily extended to other classes of inputs, e.g., absolutely continuous functions.

A discrete time model for the sampled observations is now obtained. Integration of (4.5.32) and (4.5.33) gives

$$x_{k+1} = \phi(\Delta_k, 0)x_k + \psi(\Delta_k, 0)u_k + \lambda_k \qquad (4.5.35)$$

$$y_k = Cx_k \qquad (4.5.36)$$

where

$$\phi(\tau_2, \tau_1) = \exp[A(\tau_2 - \tau_1)] \qquad (4.5.37)$$

$$\psi(\tau_2, \tau_1) = \int_{\tau_1}^{\tau_2} \phi(\tau_2, \tau)B\, d\tau \qquad (4.5.38)$$

$$\lambda_k = \int_0^{\Delta_k} \phi(\Delta_k, \tau)K\, d\varepsilon(\tau) \qquad (4.5.39)$$

and

$$\Delta_k = t_{k+1} - t_k \qquad (4.5.40)$$

The sequence $\{\lambda_k\}$ is a sequence of zero mean uncorrelated random variables having zero mean and covariance at time t_k given by

$$E\{\lambda_k \lambda_k^{\mathrm{T}}\} = \int_0^{\Delta_k} \phi(\Delta_k, \tau)K\Sigma K^{\mathrm{T}}\phi^{\mathrm{T}}(\Delta_k, \tau)\, d\tau = Q_k \qquad \text{(by definition)} \qquad (4.5.41)$$

It now follows from Kalman filtering theory that the model of Eq. (4.5.35), (4.5.36), and (4.5.41) admits an alternative innovations representation [6, 14] of the form

$$\hat{x}_{k+1} = \phi_k \hat{x}_k + \psi_k u_k + \Gamma_k v_k \qquad (4.5.42)$$

$$y_k = C\hat{x}_k + v_k \qquad (4.5.43)$$

where the notation $\phi_k = \phi(\Delta_k, 0)$; $\psi_k = (\Delta_k, 0)$ has been used and where $v_k = [y_k - C\hat{x}_k]$ is the innovations sequence having the properties

$$E_{v_k|Y_{k-1}, u_k}[v_k] = 0 \qquad (4.5.44)$$

$$E_{v_k|Y_{k-1}, u_k}[v_k v_k^T] = S_k \qquad (4.5.45)$$

where Γ_k in Eq. (4.5.42) and S_k in Eq. (4.5.45) are given by

$$\Gamma_k = \phi_k P_k C^T (C P_k C^T)^{-1} \qquad (4.5.46)$$

$$S_k = C P_k C^T \qquad (4.5.47)$$

and P_k (the conditional covariance of x_k) is the solution of a matrix Riccati equation

$$P_{k+1} = \phi_k P_k \phi_k^T + Q_k - \Gamma_k (C P_k C^T) \Gamma_k^T \qquad (4.5.48)$$

Finally, substituting (4.5.43) into (4.5.42) yields the following model for $\{y_k\}$:

$$y_k = \bar{y}_k + v_k \qquad (4.5.49)$$

where

$$\bar{y}_k = C\hat{x}_k \qquad (4.5.50)$$

$$\hat{x}_{k+1} = (\phi_k - \Gamma_k C)\hat{x}_k + \psi_k u_k + \Gamma_k y_k \qquad (4.5.51)$$

Clearly, $\{\bar{y}_k\}$ depends upon past values of $\{y_k\}$ and past and present values of $\{u_k\}$. Thus Eq. (4.5.44) guarantees that (4.5.49) is a PEM model of the general form (4.5.1).

4.6. CONCLUSIONS

In this chapter we have discussed models that can be used to describe the input/output behavior of dynamical systems. A number of specific situations have been studied in detail to illustrate the methodology of constructing models in standard form. However, the discussion has been by no means exhaustive and the reader is encouraged to keep an open mind on the question of model structure, since this is greatly influenced by the practicalities of the system under study and the purpose for which the model is required.

In the next chapter we turn to the problem of estimating the parameters within a model of a dynamic system using observations made on the inputs and outputs over some specified time period.

PROBLEMS

4.1. Consider the following model of a stochastic process $\{y_k\}$

$$y_k = -\alpha y_{k-1} + v_k + w_k + \beta w_{k-1}$$

where $\{v_k\}$ and $\{w_k\}$ are unmeasured stochastic processes having zero mean and joint covariance given by

$$E\left\{\left|\begin{matrix} v_k \\ w_k \end{matrix}\right| [v_j \, w_j]\right\} = \left|\begin{matrix} Q & C \\ C & R \end{matrix}\right| \delta_{kj}$$

where δ_{kj} is 1 for $k = j$ and zero for $k \neq j$.

(a) Derive an expression for the spectral density of the process $\{y_k\}$, assuming $|\alpha| < 1$.

(b) Show that the spectral density found in part (a) can be achieved with a model of the following form:

$$y_k = a y_{k-1} + \eta_k + \gamma \eta_{k-1}$$

where $\{\eta_k\}$ is a white noise sequence having zero mean and covariance Σ.

(c) Express the quantities a, γ, Σ in terms of the quantities α, β, Q, C, R for equality of the spectral densities.

(d) Briefly comment on the merits or otherwise of the two models for the process $\{y_k\}$ for identification purposes.

4.2. Find the spectral density matrix for the continuous stationary stochastic process whose covariance matrix is given by

$$D(\tau) = e^{-|\tau|}$$

4.3. What is the spectral density matrix of the stationary stochastic process modeled by the following stochastic differential equation:

$$dx = -x \, dt + d\omega, \qquad y = x$$

where ω is a Wiener process having $E\{d\omega \, d\omega^T\} = dt$.

4.4. Consider the following linear system

$$y_k = \varepsilon_k + c\varepsilon_{k-1}$$

where ε_k is a white noise sequence having zero mean and unit variance.

(a)   What is the covariance function of the process $\{y_k\}$?

(b)   What is the spectral density of the process $\{y_k\}$?

4.5.   Find a stable, minimum phase system which, when driven by white noise, produces the following spectral density at the output

$$\Phi_Y(e^{j\omega}) = (10 + 6 \cos \omega)/(5 + 4 \cos \omega)$$

4.6.   Consider a normal stationary process $\{y_k\}$ which is given by the model $y_k = x_k^{(1)} + x_k^{(2)}$ where

$$x_{k+1}^{(1)} = 0.2x_k^{(1)} + \varepsilon_k^{(1)}$$
$$x_{k+1}^{(2)} = 0.5x_k^{(2)} + \varepsilon_k^{(2)}$$

The sequences $\{\varepsilon_k^{(1)}\}$ and $\{\varepsilon_k^{(2)}\}$ are assumed to be uncorrelated, zero mean and to have variances $v^{(1)}$ and $v^{(2)}$, respectively. Show that the stochastic process described below has the same spectral properties:

$$y_k = \{(z + c)/(z - 0.2)(z - 0.5)\}\varepsilon_k$$

where z^{-1} is the unit delay operator and $\{\varepsilon_k\}$ is a white sequence having zero mean and variance λ. What are the values of the constants c and λ?

4.7.   Let $P = [p_1 \cdots p_n]$ be a nonsingular square matrix. If we define

$$A' = P^{-1}AP \qquad A'' = P^T A(P^T)^{-1}$$
$$B' = P^{-1}B \qquad C'' = C(P^T)^{-1}$$

(i)   Show that the ith column of A' contains the coordinates of Ap_i with respect to the basis $p_1 \cdots p_n$.

(ii)   Show that the ith column of B' contains the coordinates of the ith column of B with respect to the basis $p_1 \cdots p_n$.

(iii)   Show that the ith row of A'' contains the coordinates of $p_i^T A$ with respect to the basis $p_1^T \cdots p_n^T$.

(iv)   Show that the ith row of C'' contains the coordinates of the ith row of C with respect to the basis $p_1^T \cdots p_n^T$.

4.8.   Using the results of Problem 4.7, or otherwise, show that the procedures leading to canonical forms SSF2 and SSF3 are minimal realization algorithms and that the procedures result in A and C matrices having the structure indicated in Eq. (4.3.18), (4.3.19), (4.3.23), and (4.3.24).

4.9.   Following the development of Section 4.5.2 in the text, consider the following spectral density matrix:

$$\Phi_n(z) = C(zI - A)^{-1}Q(z^{-1}I - A^T)^{-1}C^T + R$$

where the pair $[C, A]$ is assumed to be completely observable and R and Q are positive definite and positive semidefinite, respectively. Show that $\Phi_n(z)$ can be spectrally factored as follows:

$$\Phi_n(z) = H(z)\Sigma H^T(z^{-1})$$

where $H(z) = C(zI - A)^{-1}K + I$. K and Σ are given by

$$K = APC^\mathsf{T}(CPC^\mathsf{T} + R)^{-1}, \qquad \Sigma = CPC^\mathsf{T} + R$$

where P is the unique positive definite symmetric solution of the following matrix Riccati equation (see [222] for existence results).

$$P = APA^\mathsf{T} - APC^\mathsf{T}(CPC^\mathsf{T} + R)^{-1}CPA^\mathsf{T} + Q$$

Hint: Establish the following sequence of results:

$$\begin{aligned}
H(z)\Sigma H^\mathsf{T}(z^{-1}) = {}& C(zI - A)^{-1}APC^\mathsf{T}(CPC^\mathsf{T} + R)^{-1}CPA^\mathsf{T}(z^{-1}I - A^\mathsf{T})^{-1}C^\mathsf{T} \\
& + C(zI - A)^{-1}APC^\mathsf{T} + CPA^\mathsf{T}(zI - A^\mathsf{T})^{-1}C^\mathsf{T} \\
& + CPC^\mathsf{T} + R
\end{aligned}$$

Now use the matrix Riccati equation to show

$$\begin{aligned}
H(z)\Sigma H^\mathsf{T}(z^{-1}) = {}& C(zI - A)^{-1}\{APA^\mathsf{T} - P + Q\}(z^{-1}I - A^\mathsf{T})^{-1}C^\mathsf{T} \\
& + C(zI - A)^{-1}APC^\mathsf{T} + CPA^\mathsf{T}(zI - A^\mathsf{T})^{-1}C^\mathsf{T} \\
& + CPC^\mathsf{T} + R
\end{aligned}$$

Collect together the first three terms to give

$$\begin{aligned}
H(z)\Sigma^{-1}H^\mathsf{T}(z^{-1}) = {}& C(zI - A)^{-1}\{-P + Q + z^{-1}AP + zPA^\mathsf{T} - APA^\mathsf{T}\} \\
& \times (z^{-1}I - A^\mathsf{T})^{-1}C^\mathsf{T} + CPC^\mathsf{T} + R
\end{aligned}$$

Factorize the first term to give

$$\begin{aligned}
H(z)\Sigma^{-1}H^\mathsf{T}(z^{-1}) = {}& C(zI - A)^{-1}\{Q - (zI - A)P(z^{-1}I - A^\mathsf{T})\} \\
& \times (z^{-1}I - A^\mathsf{T})^{-1}C^\mathsf{T} + CPC^\mathsf{T} + R \\
= {}& \Phi_n(z)
\end{aligned}$$

CHAPTER

5

Estimation for Dynamic Systems

5.1. INTRODUCTION

In this chapter we turn to the problem of estimating the parameters within the models described in Chapter 4. For these models we describe algorithms that are direct analogs of the algorithms developed in Chapters 2 and 3. We also point to differences between the static and dynamic estimation problems. These differences lead us to develop extensions of the basic algorithms to enhance their performance in the dynamic case. In Section 5.5 we discuss asymptotic properties of the resulting estimators, viz., consistency, normality, and efficiency. Finally, in Section 5.6 we discuss the problem of estimation in closed loop. In particular we obtain conditions under which the open loop transfer functions can be found. We also discuss direct and indirect identification methods. We defer treatment of recursive or on-line methods to Chapter 7.

5.2. LEAST SQUARES FOR LINEAR DYNAMIC SYSTEMS

In Chapter 4, it was shown that a general model for a large class of single-input, single-output systems is (cf. Eq. (4.4.10)):

$$y_t = H_1(z)u_t + H_2(z)\xi_t, \qquad t = 1, 2, \ldots, N \tag{5.2.1}$$

where $\{y_t\}$ and $\{u_t\}$ are the output and input sequences, respectively, and $\{\xi_t\}$ is a sequence of uncorrelated zero mean random variables with common variance σ^2. H_1 and H_2 are transfer functions in z^{-1} the unit delay.

For the case where H_1 is parameterized as $h_1 z^{-1} + h_2 z^{-2} + \cdots + h_n z^{-n}$ and $H_2 \equiv 1$, the least squares procedure is directly applicable (see Example 2.2.1). The estimator is known to be BLUE in this case. (See Theorem 2.3.1.)

We next consider a slightly more general situation where H_1 and H_2 are parameterized as $B(z)/A(z)$ and $1/A(z)$, respectively, where B and A are polynomials in z^{-1}:

$$A(z) = 1 + a_1 z^{-1} + \cdots + a_n z^{-n} \tag{5.2.2}$$

$$B(z) = b_1 z^{-1} + b_2 z^{-1} + \cdots + b_n z^{-n} \tag{5.2.3}$$

The above model can be expressed in the form of (4.4.10) as

$$A(z)y_t = B(z)u_t + \xi_t \tag{5.2.4}$$

or

$$y_t = -a_1 y_{t-1} \cdots - a_n y_{t-n} + b_1 u_{t-1} + \cdots + b_n u_{t-n} + \xi_t \tag{5.2.5}$$

$$= x_t^{\mathrm{T}}\theta + \xi_t, \qquad t = 1, 2, \ldots, N \tag{5.2.6}$$

where

$$\theta^{\mathrm{T}} = (a_1, a_2, \ldots, a_n, b_1, b_2, \ldots, b_n) \tag{5.2.7}$$

$$x_t^{\mathrm{T}} = (-y_{t-1}, -y_{t-2}, \ldots, -y_{t-n}, u_{t-1}, u_{t-2}, \ldots, u_{t-n}) \tag{5.2.8}$$

Equation (5.2.6) can be expressed in vector form as

$$Y = X\theta + \Xi \tag{5.2.9}$$

where

$$Y^{\mathrm{T}} = [y_1, \ldots, y_n] \tag{5.2.10}$$

$$X^{\mathrm{T}} = [x_1, \ldots, x_N] \tag{5.2.11}$$

$$\Xi^{\mathrm{T}} = [\xi_1, \ldots, \xi_N] \tag{5.2.12}$$

Equation (5.2.9) suggests that the least squares procedure can again be applied. The resulting estimator for θ is

$$\hat{\theta} = (X^{\mathrm{T}}X)^{-1}X^{\mathrm{T}}Y \tag{5.2.13}$$

The properties of the above estimator differ from those discussed in Chapter 2 since X depends upon the data whereas this possibility was previously excluded. The properties of the estimator of Eq. (5.2.13) are discussed below.

Theorem 5.2.1   Under suitable conditions, the least squares estimator of the parameter vector θ in the model of Eq. (5.2.4) is strongly consistent, i.e., $\hat{\theta} \xrightarrow{\text{a.s.}} \theta$ as $N \to \infty$.

Proof   The proof of strong consistency is technically involved, relying upon martingale convergence theory. An outline proof is given in Section 5.5. $\triangledown$

Remark 5.2.2   Convergence of $\hat{\theta}$ to θ implies that $\hat{\theta}$ is asymptotically unbiased. It is not true, however, that $\hat{\theta}$ is unbiased for finite data lengths. (See Problem 5.1 for a counterexample.) It is also not true, in general, that $\hat{\theta}$ is a consistent estimator for more general models compared with (5.2.4). In the next section we investigate modifications to the basic least squares procedure which leads to consistent estimation for general linear systems. $\triangledown$

5.3.   CONSISTENT ESTIMATORS FOR LINEAR DYNAMIC SYSTEMS

The standard proofs of consistency for the least squares estimator rely upon the uncorrelatedness of the sequence $\{\xi_t\}$. We now show by means of a counterexample that the estimates obtained by use of the least squares estimator will not in general be consistent when the sequence $\{\xi_t\}$ is correlated.

Example 1.1.1   Consider the following model

$$y_t = a y_{t-1} + b u_t + \xi_t \tag{5.3.1}$$

where $|a| < 1$ and $\{\xi_t\}$ and $\{u_t\}$ are zero mean wide sense stationary process (possessing exponentially bounded fourth moments). $\{\xi_t\}$ is the noise that is not necessarily uncorrelated and $\{u_t\}$ is the system input which is assumed statistically independent of $\{\xi_t\}$.

The least squares estimator for $\theta = (a, b)^{\mathrm{T}}$ is

$$\hat{\theta} = \begin{bmatrix} \hat{a} \\ \hat{b} \end{bmatrix} = \begin{bmatrix} \dfrac{1}{N} \sum_{t=1}^{N} y_{t-1}^2 & \dfrac{1}{N} \sum_{t=1}^{N} u_t y_{t-1} \\ \dfrac{1}{N} \sum_{t=1}^{N} y_{t-1} u_t & \dfrac{1}{N} \sum_{t=1}^{N} u_t^2 \end{bmatrix}^{-1} \begin{bmatrix} \dfrac{1}{N} \sum_{t=1}^{N} y_t y_{t-1} \\ \dfrac{1}{N} \sum_{t=1}^{N} y_t u_t \end{bmatrix} \tag{5.3.2}$$

Stability of the system (together with exponentially bounded fourth moments for $\{\xi_t\}$ and $\{u_t\}$) imply that the summations in Eq. (5.3.2) converge in quadratic mean and hence in probability to their expected values [26], i.e.,

$$\begin{bmatrix} \dfrac{1}{N}\sum\limits_{t=1}^{N} y_{t-1}^2 & \dfrac{1}{N}\sum\limits_{t=1}^{N} u_t\, y_{t-1} \\ \dfrac{1}{N}\sum\limits_{t=1}^{N} y_{t-1}u_t & \dfrac{1}{N}\sum\limits_{t=1}^{N} u_t^{\,2} \end{bmatrix} \xrightarrow{\text{prob}} \begin{bmatrix} R_{yy}(0) & R_{uy}(1) \\ R_{uy}(1) & R_{uu}(0) \end{bmatrix}$$

$$\begin{bmatrix} \dfrac{1}{N}\sum\limits_{t=1}^{N} y_t\, y_{t-1} \\ \dfrac{1}{N}\sum\limits_{t=1}^{N} y_t u_t \end{bmatrix} \xrightarrow{\text{prob}} \begin{bmatrix} R_{yy}(1) \\ R_{uy}(0) \end{bmatrix}$$

where

$$R_{\alpha\beta}(\tau) = E[\alpha_t\, \beta_{t-\tau}]$$

Hence from Frechèt's theorem (Theorem B.3.4)

$$\begin{bmatrix} \hat{a} \\ \hat{b} \end{bmatrix} \xrightarrow{\text{prob}} \begin{bmatrix} R_{yy}(0) & R_{uy}(1) \\ R_{uy}(1) & R_{uu}(0) \end{bmatrix}^{-1} \begin{bmatrix} R_{yy}(1) \\ R_{uy}(0) \end{bmatrix}$$

i.e.,

$$\begin{bmatrix} \hat{a} \\ \hat{b} \end{bmatrix} \xrightarrow{\text{prob}} \frac{1}{\Delta} \begin{bmatrix} R_{uu}(0)R_{yy}(1) - R_{uy}(1)R_{uy}(0) \\ -R_{uy}(1)R_{yy}(1) + R_{yy}(0)R_{uy}(0) \end{bmatrix} \tag{5.3.3}$$

where

$$\Delta = R_{yy}(0)R_{uu}(0) - R_{uy}^2(1) \tag{5.3.4}$$

For simplicity, we assume that $R_{\xi\xi}(\tau)$ is zero for $|\tau| \geq 2$. Then, from Eq. (5.3.1)

$$R_{yy}(1) = aR_{yy}(0) + bR_{uy}(1) + R_{\xi\xi}(1) \tag{5.3.5}$$

$$R_{uy}(0) = aR_{uy}(1) + bR_{uu}(0) \tag{5.3.6}$$

Substituting (5.3.5) and (5.3.6) into (5.3.3) yields

$$\begin{bmatrix} \hat{a} \\ \hat{b} \end{bmatrix} \xrightarrow{\text{prob}} \begin{bmatrix} a + R_{\xi\xi}(1)R_{uu}(0)/\Delta \\ b - R_{\xi\xi}(1)R_{uy}(1)/\Delta \end{bmatrix} \tag{5.3.7}$$

Thus, $\hat{a}$ and $\hat{b}$ do not converge in probability (and thus not a.s. or q.m.) to the true parameters a and b, unless the noise is uncorrelated, i.e., $R_{\xi\xi}(1)$ is zero.

Modifications of the basic least squares procedure that are aimed at overcoming the inconsistency problem are described below.

5.3.1 *Instrumental variables* Consider the least squares estimator

$$\hat{\theta} = [X^T X]^{-1} X^T Y \tag{5.3.8}$$

If Y is given by (5.2.9), then it can be readily seen that

$$\hat{\theta} = \theta + \left[\frac{1}{N} X^T X\right]^{-1} \left[\frac{1}{N} X^T \Xi\right] \tag{5.3.9}$$

We note that $(1/N)X^T X$ and $(1/N)X^T \Xi$ are given by

$$\frac{1}{N} X^T X = \frac{1}{N} \sum_{t=1}^{N} x_t x_t^T \tag{5.3.10}$$

$$\frac{1}{N} X^T \Xi = \frac{1}{N} \sum_{t=1}^{N} x_t \xi_t \tag{5.3.11}$$

Under suitable regularity conditions, it can be shown that the expressions indicated in Eqs. (5.3.10) and (5.3.11) converge to their expected values, i.e.,

$$X^T X/N \xrightarrow{\text{prob}} \psi_{xx} \triangleq E[x_t x_t^T] \tag{5.3.12}$$

$$X^T \Xi/N \xrightarrow{\text{prob}} \psi_{x\xi} \triangleq E[x_t \xi_t] \tag{5.3.13}$$

Hence, applying Frechèt's theorem (Theorem B.3.4) to Eq. (5.3.9) yields

$$\hat{\theta} \xrightarrow{\text{prob}} \theta + [\psi_{xx}]^{-1}[\psi_{x\xi}] \tag{5.3.14}$$

Thus, it is clear that if $\psi_{x\xi}$ is zero, then $\hat{\theta}$ is a weakly consistent estimator for θ. For the case $\{\xi_t\}$ uncorrelated, we have that x_t and ξ_t are uncorrelated since x_t depends upon past values of ξ_t. Thus $\psi_{x\xi}$ is zero and hence $\hat{\theta} \xrightarrow{\text{prob}} \theta$ (cf. Theorem 5.2.1). However, if $\{\xi_t\}$ is correlated, then $\psi_{x\xi}$ will not, in general, be zero and hence $\hat{\theta}$ will not converge to θ.

One way around the above problem is to replace the least squares estimator by the following estimator

$$\bar{\theta} = [Z^T X]^{-1} Z^T Y \tag{5.3.15}$$

where Z is a matrix, closely related to X in the sense that

$$Z^T X/N \xrightarrow{\text{prob}} \psi_{zx} \tag{5.3.16}$$

where

$$\det \psi_{zx} \neq 0 \tag{5.3.17}$$

and with the additional property that

$$Z^T \Xi/N \xrightarrow{\text{prob}} \psi_{z\xi} \equiv 0 \tag{5.3.18}$$

Any matrix Z having properties (5.3.16)–(5.3.18) is called an *instrumental variable matrix* [113]. The estimator

$$\bar{\theta}_N = [Z^T X]^{-1} Z^T Y \tag{5.3.19}$$

is called an *instrumental variable estimator*.

It is readily verified that this estimator is weakly consistent, see Problem 5.4. It is also possible to establish strong consistency results in certain circumstances, see [107] and [81] for further discussion.

In practice, the matrix Z may be constructed by using an auxiliary model which is an approximation to the system. The kth row of Z is then given by

$$z_k^T = (-\hat{y}_{k-1}, -\hat{y}_{k-2}, \ldots, -\hat{y}_{k-n}, u_k, u_{k-1}, \ldots, u_{k-n}) \tag{5.3.20}$$

where

$$\hat{A}(z)\hat{y}_k = \hat{B}(z)u_k \tag{5.3.21}$$

and $\hat{A}$, $\hat{B}$ are polynomials in z^{-1}. $\hat{A}$ and $\hat{B}$ may be obtained from an initial least squares fit. However, conditions (5.3.16)–(5.3.18) are difficult to check and thus consistent estimation cannot be guaranteed in general.

The technique has been used successfully in practical applications, especially in its recursive form [108]. Relationships with similar algorithms are discussed in [109]. An alternative method for overcoming the inconsistency problem is described in the next subsection.

5.3.2   *Generalized least squares*   If the sequence $\{\xi_t\}$ is correlated but the autocorrelation function is known, then it is possible to model ξ_t as follows (see Theorem 4.4.1):

$$\xi_t = H(z)\varepsilon_t \tag{5.3.22}$$

where $\{\varepsilon_t\}$ is an uncorrelated sequence and where $H(z)$ is invertible. Hence Eq. (5.2.4) may be written as

$$A(z)y_t = B(z)u_t + H(z)\varepsilon_t \tag{5.3.23}$$

or equivalently,

$$A(z)y_t^* = B(z)u_t^* + \varepsilon_t \tag{5.3.24}$$

where

$$y_t^* = H^{-1}(z)y_t \tag{5.3.25}$$
$$u_t^* = H^{-1}(z)u_t \tag{5.3.26}$$

If $\{y_t^*\}$ and $\{u_t^*\}$ can be calculated, then least squares applied to Eq. (5.3.24) will yield a consistent estimate of the parameters in $A(z)$ and $B(z)$.

The problem is, however, that we usually do not know $H(z)$. We thus might attempt to estimate $H(z)$ along with $A(z)$ and $B(z)$. A simple algorithm is obtained if $H(z)$ is parameterized in the form $1/R(z)$ where $R(z)$ is a polynomial in z^{-1}. A plausible algorithm is

   (1)   Set $\hat{R}(z) = 1$
   (2)   Form $y_t^* = \hat{R}(z)y_t$ and $u_t^* = \hat{R}(z)u_t$
   (3)   Obtain the least squares estimate for θ (A and B parameters) using y_t^* and u_t^*. Denote the estimated A and B polynomials by $\hat{A}$ and $\hat{B}$.
   (4)   Calculate

$$e_t = \hat{A}(z)y_t - \hat{B}(z)u_t$$

   (5)   For the model $R(z)e_t = \varepsilon_t$ (ε_t uncorrelated), obtain the least squares estimate of R (denoted by $\hat{R}$). (Note that $R(z)e_t = \varepsilon_t$ is in the usual least squares format with zero input.)
   (6)   If converged—stop, otherwise go to (2). ∇

The above algorithm is commonly called *generalized least squares* [52]. The technique has been shown to work well in practice [51, 110]. Furthermore, for normal disturbances, the algorithm can be interpreted as a relaxation solution to the maximum likelihood problem. We defer discussion on this point to Section 5.4.5.

5.4. PREDICTION ERROR FORMULATION AND MAXIMUM LIKELIHOOD

5.4.1 *Methods based on prediction error covariance*   In Section 4.5 the following general prediction error model was introduced:

$$y_t = f[Y_{t-1}, U_t, t; \theta] + \varepsilon_t \tag{5.4.1}$$

where y_t denotes the output vector at time t, Y_{t-1} the set $\{y_{t-1}, y_{t-2}, \ldots\}$, U_t the set $\{u_t, u_{t-1}, \ldots\}$, $\{\varepsilon_t\}$ is an innovations sequence, and θ is a vector of parameters. For a particular value of θ, say $\theta = \hat{\theta}$, the *prediction error* w_t is given by

$$w_t(\hat{\theta}) = y_t - f[Y_{t-1}, U_t, t; \hat{\theta}] \tag{5.4.2}$$

The sample covariance of $\{w_t(\hat{\theta})\}$ is given by

$$D(\hat{\theta}) = \frac{1}{N} \sum_{t=1}^{N} w_t(\hat{\theta}) w_t(\hat{\theta})^{\mathsf{T}} \tag{5.4.3}$$

It seems reasonable to propose that a good model should have small prediction errors. This suggests the use of a positive scalar function of $D(\hat{\theta})$

as an estimation criterion. Commonly used choices for the scalar function are

$$J_1(\hat{\theta}) = \text{trace}[WD(\hat{\theta})] \tag{5.4.4}$$

where W is positive definite

$$J_2(\hat{\theta}) = \log \det(D(\hat{\theta})) \tag{5.4.5}$$

An alternative motivation for the use of the above criteria will be given in Section 5.4.3. Minimization of a scalar function such as J_1 and J_2 leads to an estimate of the parameters θ. A discussion of some of the general properties of the estimator is given in Section 5.5.

So far, we have not appealed to a knowledge of the joint probability density function of the data in motivating the estimation criteria, J_1 and J_2. In the next section we assume knowledge of the conditional distribution function of the innovations sequence to motivate the maximum likelihood criterion.

5.4.2   Maximum likelihood   Consider again the prediction error model

$$y_t = f[Y_{t-1}, U_t, t; \theta] + \varepsilon_t \tag{5.4.6}$$

where the innovations form an independent, though not necessarily identically distributed, sequence.

We assume for the moment, that the input sequence $\{u_t\}$ is exogenous, i.e., generated independently of $\{y_t\}$. We treat the more general case, e.g., when feedback is present, in Section 5.6.

The likelihood function for the data is thus given by Bayes' rule as

$$p(Y_N \mid U_N, \theta) = \prod_{t=1}^{N} p(y_t \mid Y_{t-1}, U_t; \theta) \tag{5.4.7}$$

Now using Eq. (5.4.6) and the rule for transformation of random variables [10], we can relate the conditional distribution of y_t to that of ε_t as follows:

$$p(y_t \mid Y_{t-1}, U_t; \theta) = p_{\varepsilon_t}(w_t(\theta) \mid \theta) \mid \det[\partial \varepsilon_t / \partial y_t] \mid \tag{5.4.8}$$

where the Jacobian $\mid \det[\partial \varepsilon_t / \partial y_t] \mid = 1$, $w_t(\theta)$ is given by

$$w_t(\theta) = y_t - f[Y_{t-1}, U_t, t; \theta] \tag{5.4.9}$$

and $p_{\varepsilon_t}(\cdot \mid \theta)$ is the conditional density function for ε_t given Y_{t-1}, U_t and may depend upon θ.

Finally using Eq. (5.4.7), the likelihood function is given by

$$p(Y_N \mid U_N, \theta) = \prod_{t=1}^{N} p_{\varepsilon_t}(w_t(\theta) \mid \theta) \tag{5.4.10}$$

We present a simple example showing how an expression for the likelihood function may be obtained in a typical situation.

Example 5.4.1 Consider the problem of a nonuniformly sampled linear time invariant continuous time system. This problem was discussed in detail in Section 4.5.3 where it was shown that the appropriate PEM model (see Eqs. (4.5.49)–(4.5.51)) was

$$y_t = \hat{y}_t + v_t \tag{5.4.11}$$

where $\hat{y}_t$ is a function of past data and $\{v_t\}$ is an innovations sequence. $\hat{y}_t$ is given by

$$\hat{y}_t = C\hat{x}_t \tag{5.4.12}$$

$$\hat{x}_{t+1} = (\phi_t - \Gamma_t C)\hat{x}_t + \psi_t u_t + \Gamma_t y_t \tag{5.4.13}$$

We now assume that the continuous time system is influenced by noise sources which are Wiener processes, i.e., normal (see Appendix C). Then the sequence $\{v_t\}$ is an independent normal sequence with zero mean and time varying covariance S_t given by Eq. (4.5.47), i.e.,

$$p_{v_t}(v_t \mid \theta) = [(2\pi)^m \det S_t]^{-1/2} \exp\{-\tfrac{1}{2} v_t^{\mathsf{T}} S_t^{-1} v_t\} \tag{5.4.14}$$

Hence the likelihood function for the data is

$$p(Y_N \mid U_N, \theta) = \prod_{t=1}^{N} ([(2\pi)^m \det S_t]^{-1/2} \exp\{-\tfrac{1}{2} w_t^{\mathsf{T}} S_t^{-1} w_t\}) \tag{5.4.15}$$

where substitution of (5.4.11) into (5.4.9) yields

$$w_t = y_t - C\bar{x}_t \tag{5.4.16}$$

$$\bar{x}_{t+1} = (\phi_t - \Gamma_t C)\bar{x}_t + \psi_t u_t + \Gamma_t y_t \tag{5.4.17}$$

where $\{\phi_t\}$, $\{\Gamma_t\}$, $\{\psi_t\}$, $\{S_t\}$, and C in Eqs. (5.4.15)–(5.4.17) depend upon θ via the relationships derived in Section 4.5. ▽

The maximum likelihood estimator is obtained as the value of θ which maximizes the likelihood function given in Eq. (5.4.10). The explicit form of the estimator depends upon the distribution of the innovations. We now consider the Gaussian case in more detail.

5.4.3 *The multivariable Gaussian case* When the innovations are independent and normally distributed with common covariance Σ, the likelihood function (Eq. (5.4.10)) becomes

$$p(Y_N \mid U_N, \theta) = \prod_{t=1}^{N} ([(2\pi)^m \det \Sigma]^{-1/2} \exp\{-\tfrac{1}{2}w_t^{\mathrm{T}}\Sigma^{-1}w_t\})$$

$$= [(2\pi)^m \det \Sigma]^{-N/2} \exp\left\{-\frac{1}{2}\sum_{t=1}^{N} w_t^{\mathrm{T}}\Sigma^{-1}w_t\right\} \quad (5.4.18)$$

where

$$w_t(\theta) = y_t - f(Y_{t-1}, U_t, t; \theta) \quad (5.4.19)$$

For the case where Σ is known, maximization of the likelihood function is equivalent to minimization of $J_1(\theta)$ where

$$J_1(\theta) = \frac{1}{N}\sum_{t=1}^{N} w_t^{\mathrm{T}}\Sigma^{-1}w_t = \text{trace } \Sigma^{-1}D(\theta) \quad (5.4.20)$$

where $D(\theta)$ is the sample covariance of the prediction errors, i.e.,

$$D(\theta) = \frac{1}{N}\sum_{t=1}^{N} w_t w_t^{\mathrm{T}} \quad (5.4.21)$$

The cost function $J_1(\theta)$ given in Eq. (5.4.20) will be seen to be the same as the scalar cost function previously introduced in Eq. (5.4.4) with $W = \Sigma^{-1}$, i.e., maximum likelihood with Σ known is a particular prediction error method where the J_1 cost function is used.

For the case where Σ is included in the parameter vector, then maximization of (5.4.18) is equivalent to minimization of

$$J(\theta, \Sigma) = \frac{1}{2}mN \log 2\pi + \frac{1}{2}N \log \det \Sigma + \frac{1}{2}\sum_{t=1}^{N} w_t^{\mathrm{T}}\Sigma^{-1}w_t \quad (5.4.22)$$

Differentiating w.r.t. Σ (see Appendix E for matrix differentiation results) gives

$$\frac{\partial J}{\partial \Sigma} = \frac{N}{2}\Sigma^{-1} - \frac{1}{2}\Sigma^{-1}\left(\sum_{t=1}^{N} w_t w_t^{\mathrm{T}}\right)\Sigma^{-1} \quad (5.4.23)$$

Setting $\partial J/\partial \Sigma$ equal to zero gives

$$\Sigma = \frac{1}{N}\sum_{t=1}^{N} w_t w_t^{\mathrm{T}} = D(\theta) \quad (5.4.24)$$

The problem thus splits into two parts. We see that for any fixed value of θ, Eq. (5.4.24) maximizes the likelihood function with respect to Σ (see also Result 3.3.4). The dimension of the optimization problem can therefore be reduced by simply imposing the constraint that $\Sigma = D(\theta)$ for all θ. Substituting (5.4.24) into (5.4.22) yields

$$J_c(\theta) = \frac{1}{2} mN \log 2\pi + \frac{1}{2} N \log \det D(\theta) + \frac{1}{2} \sum_{t=1}^{N} w_t^T D(\theta)^{-1} w_t \quad (5.4.25)$$

Simplification of (5.4.25) then yields (see Problem 5.8):

$$J_c(\theta) = \tfrac{1}{2} mN(\log 2\pi + 1) + \tfrac{1}{2} N \log \det D(\theta) \quad (5.4.26)$$

Minimization of $J_c(\theta)$ with respect to θ is equivalent to minimization of

$$J_2(\theta) = \log \det D(\theta) \quad (5.4.27)$$

Equation (5.4.27) will be seen to be identical to the estimation criterion previously given in Eq. (5.4.5) indicating that maximum likelihood is again a prediction error method with criterion J_2.

5.4.4 *Optimization algorithms* Estimation criteria $J_1(\theta)$ or $J_2(\theta)$ may be maximized by standard function optimization techniques, see for example [48] and [49]. Many of the algorithms use the gradient of the cost function which can be computed as follows (see Problem 5.9):

$$\frac{\partial J_1}{\partial \theta_i} = \frac{2}{N} \sum_{t=1}^{N} w_t^T \Sigma^{-1} \frac{\partial w_t}{\partial \theta_i}, \qquad i = 1, \ldots, p \quad (5.4.28)$$

$$\frac{\partial J_2}{\partial \theta_i} = \frac{2}{N} \sum_{t=1}^{N} w_t^T D(\theta)^{-1} \frac{\partial w_t}{\partial \theta_i}, \qquad i = 1, \ldots, p \quad (5.4.29)$$

where $\{\partial w_t/\partial \theta_i\}$ is calculated from the following *sensitivity equation*

$$(\partial w_t/\partial \theta_i) = -(\partial f/\partial \theta_i)(Y_{t-1}, U_t, t; \theta) \quad (5.4.30)$$

The second derivatives of the cost function can be computed as follows:

$$\frac{\partial^2 J_1}{\partial \theta_i \, \partial \theta_j} = \frac{2}{N} \sum_{t=1}^{N} \frac{\partial w_t^T}{\partial \theta_i} \Sigma^{-1} \frac{\partial w_t}{\partial \theta_j} + \frac{2}{N} \sum_{t=1}^{N} w_t^T \Sigma^{-1} \frac{\partial^2 w_t}{\partial \theta_i \, \partial \theta_j} \quad (5.4.31)$$

$$\frac{\partial^2 J_2}{\partial \theta_i \, \partial \theta_j} = \frac{2}{N} \sum_{t=1}^{N} \frac{\partial w_t^T}{\partial \theta_i} D(\theta)^{-1} \frac{\partial w_t}{\partial \theta_j} + \frac{2}{N} \sum_{t=1}^{N} w_t^T D(\theta)^{-1} \frac{\partial^2 w_t}{\partial \theta_i \, \partial \theta_j}$$

$$- \frac{2}{N^2} \sum_{t=1}^{N} \sum_{k=1}^{N} w_t^T D^{-1}(\theta) \left| w_k \frac{\partial w_k^T}{\partial \theta_i} + \frac{\partial w_k}{\partial \theta_i} w_k^T \right| D^{-1}(\theta) \frac{\partial w_t}{\partial \theta_j} \quad (5.4.32)$$

The second derivative sequence $\{\partial^2 w_t/\partial \theta_i \, \partial \theta_j\}$ can be computed from

$$\partial^2 w_t/\partial \theta_i \, \partial \theta_j = -(\partial^2 f/\partial \theta_i \, \partial \theta_j)(Y_{t-1}, U_t, t; \theta) \quad (5.4.33)$$

It is not advisable, in general, to use Eqs. (5.4.31) or (5.4.32) to calculate the second derivatives. One reason for this is that it increases the computational burden. A second reason is that the resulting second derivative matrices $[\partial^2 J_1/\partial\theta_i\,\partial\theta_j]$ or $[\partial^2 J_2/\partial\theta_i\,\partial\theta_j]$ may not be positive definite and special precautions must then be included into the optimization procedure (see [50] for discussion). A third reason is that for θ near the true parameter value, the terms containing $\partial^2 w_t/\partial\theta_i\,\partial\theta_j$ in (5.4.31) and (5.4.32) approach their expected value of zero. It can also be shown that the third term in Eq. (5.4.32) approaches zero as $N \to \infty$ (see Problem 5.8). The above line of reasoning suggests that the following approximate second derivative matrices should be used:

$$\frac{\partial^2 J_1^*}{\partial\theta_i\,\partial\theta_j} = \frac{2}{N}\sum_{t=1}^{N}\frac{\partial w_t}{\partial\theta_i}\Sigma^{-1}\frac{\partial w_t}{\partial\theta_j} \tag{5.4.34}$$

$$\frac{\partial^2 J_2^*}{\partial\theta_i\,\partial\theta_j} = \frac{2}{N}\sum_{t=1}^{N}\frac{\partial w_t}{\partial\theta_i}D(\theta)^{-1}\frac{\partial w_t}{\partial\theta_j} \tag{5.4.35}$$

If the estimation criteria J_1 or J_2 are approximately quadratic in the parameters, then the maximum likelihood estimates for θ can be approached in one step as follows:

$$\hat{\theta} = \theta_0 - (\partial^2 J^*/\partial\theta^2)^{-1}(\partial J/\partial\theta) \tag{5.4.36}$$

where θ_0 is an initial guess and $(\partial J/\partial\theta)$ and $(\partial^2 J^*/\partial\theta^2)$ are evaluated at θ_0 (J stands for either J_1 or J_2).

If the initial guess for θ is poor and the criterion nonquadratic, then an iterative procedure is required. For example, one might use the following algorithm:

(i)   Guess on initial value for θ, say θ_0, set $k = 0$.
(ii)   Evaluate $[\partial J/\partial\theta]$ and $[\partial^2 J^*/\partial\theta^2]$ at θ_k.
(iii)   Perform a linear search using scalar search constant s, so that $J(\theta_k - s[\partial^2 J^*/\partial\theta^2]^{-1}[\partial J/\partial\theta])$ is minimized for $s = s_k^*$.
(iv)   Set $\theta_{k+1} = \theta_k - s_k^*[\partial^2 J^*/\partial\theta^2]^{-1}[\partial J/\partial\theta]$.
(v)   If $J(\theta_{k+1}) - J(\theta_k)$ is suitably small, stop, otherwise set $k = k + 1$ and go to (ii).

The above simple algorithm has been found to work satisfactorily in practice [50]. However, many other related algorithms are available [48, 49], often with special features for special classes of problems. We illustrate this for the linear multivariable Gaussian case below.

5.4.5   *Relaxation algorithm*   We consider a multiple-input, multiple-output linear model in the following form:

$$A(z)y_t = B(z)u_t + \xi_t \tag{5.4.37}$$

where $\{u_t\}$ $\{y_t\}$, and $\{\xi_t\}$ are the input, output, and disturbance sequences, respectively. For the sake of being specific, we assume that (A, B) is in canonical form MFD1 (see Chapter 4):

$$A(z) = \text{diag}[a(z), \ldots, a(z)] \tag{5.4.38}$$

where

$$a(z) = 1 + a_1 z^{-1} + \cdots + a_n z^{-n} \tag{5.4.39}$$

$$B(z) = B_1 z^{-1} + B_2 z^{-2} + \cdots + B_n z^{-n} \tag{5.4.40}$$

We also assume that $\{\xi_t\}$ is modeled by an autoregressive linear model, i.e.,

$$F(z)\xi_t = \varepsilon_t \tag{5.4.41}$$

where $\{\varepsilon_t\}$ is an independent Gaussian sequence with zero mean and common covariance Σ and $F(z)$ is given by

$$F(z) = I + F_1 z^{-1} + \cdots + F_q z^{-q} \tag{5.4.42}$$

For the above model the log likelihood function is

$$J = -\frac{1}{2} mN \log 2\pi - \frac{1}{2} N \log \det \Sigma - \frac{1}{2} \sum_{t=1}^{N} w_t^{\mathrm{T}} \Sigma^{-1} w_t \tag{5.4.43}$$

where $\{w_t\}$ is given by (see Eq. (5.4.9))

$$w_t = F(z)\eta_t \tag{5.4.44}$$

and $\{\eta_t\}$ satisfies

$$\eta_t = A(z)y_t - B(z)u_t \tag{5.4.45}$$

Thus it is seen that J is a nonquadratic function of the parameters in A, B, F, and Σ and so maximization of J requires some iterative procedure. We note, however, that if F and Σ are fixed, then J is quadratic in the parameters α, in $A(z)$ and $B(z)$. In fact, from Eqs. (5.4.44) and (5.4.45), we can express w_t as a linear function of α:

$$w_t = \bar{y}_t - X_t \alpha \tag{5.4.46}$$

where

$$\alpha^{\mathrm{T}} = (a^{\mathrm{T}}, b_{10}^{\mathrm{T}}, \ldots, b_{rn}^{\mathrm{T}}] \tag{5.4.47}$$

$$a^{\mathrm{T}} = (a_1, a_2, \ldots, a_n) \tag{5.4.48}$$

$$b_{ij} = i\text{th column of } B_j \tag{5.4.49}$$

and

$$X_t = [\bar{y}_{t-1}, \ldots, \bar{y}_{t-n}; F \cdot u_{t-1}^1, F \cdot u_{t-1}^2, \ldots, F \cdot u_{t-n}^r] \tag{5.4.50}$$

where

$$\bar{y}_t = F(z)y_t \tag{5.4.51}$$

and $F \cdot u_{t-i}^j$ denotes a matrix given by $\sum_{k=0}^{q} u_{t-i-k}^j F_k$ and u_t^j denotes the scalar which is the jth component of the vector u_t.

Substitution of (5.4.46) into (5.4.43) and maximizing J with respect to α leads to

$$\hat{\alpha} = \left| \frac{1}{N} \sum_{t=1}^{N} X_t^T \Sigma^{-1} X_t \right|^{-1} \left| \frac{1}{N} \sum_{t=1}^{N} X_t^T \Sigma^{-1} \bar{y}_t \right| \tag{5.4.52}$$

Similarly we note that if $A(z)$ and $B(z)$ are fixed, then J has a unique maximum with respect to the remaining parameters in $F(z)$ and Σ. To show this, we note that for fixed A and B, w_t can be expressed as a linear function of $F(z)$ as

$$w_t = \eta_t - \mathcal{F}\gamma_t \tag{5.4.53}$$

where

$$\eta_t = A(z)y_t - B(z)u_t \tag{5.4.54}$$

$$\mathcal{F} = [F_1, F_2, \ldots, F_q] \tag{5.4.55}$$

$$\gamma_t^T = [\eta_{t-1}^T, \eta_{t-2}^T, \ldots, \eta_{t-q}^T] \tag{5.4.56}$$

Substituting (5.4.53) into (5.4.43) yields

$$J = -\frac{mN}{2} \log 2\pi - \frac{N}{2} \log \det \Sigma - \frac{1}{2} \sum_{t=1}^{N} (\eta_t - \mathcal{F}\gamma_t)\Sigma^{-1}(\eta_t - \mathcal{F}\gamma_t) \tag{5.4.57}$$

Differentiating with respect to Σ and $\mathcal{F}$ and equating to zero leads to the following maximizing values for $\mathcal{F}$ and Σ:

$$\hat{\mathcal{F}} = \left| \frac{1}{N} \sum_{t=1}^{N} \eta_t \gamma_t^T \right| \left| \frac{1}{N} \sum_{t=1}^{N} \gamma_t \gamma_t^T \right|^{-1} \tag{5.4.58}$$

$$\hat{\Sigma} = \frac{1}{N} \sum_{t=1}^{N} (\eta_t - \hat{\mathcal{F}}\gamma_t)(\eta_t - \hat{\mathcal{F}}\gamma_t)^T \tag{5.4.59}$$

The above discussion motivates the following relaxation algorithm:

Initialization   Set $\Sigma = I$, $F(z) = I$.
Stage 1   Compute the coefficients in $A(z)$ and $B(z)$ using Eq. (5.4.52).
Stage 2   Compute the coefficients in $F(z)$ using Eq. (5.4.58).
Stage 3   Compute Σ using Eq. (5.4.59).
Convergence Check   Stop if converged; otherwise go to Stage 1.   ▽

The above algorithm is seen to be an extension to the multivariable case of the generalized least squares procedure described in Section 5.3.2. For the case $F(z) = I$, the algorithm is identical to the one described in Section 3.3 and Problem 3.6. The main advantage of the algorithm is simplicity. It has been found to work well in practice [114, 115]. To illustrate the performance of the algorithm, we present a simple two-input, two-output example.

Example 5.4.2   Consider the following two-input, two-output linear system:

$$\begin{pmatrix} y_1 \\ y_2 \end{pmatrix}_k = a_1 \begin{pmatrix} y_1 \\ y_2 \end{pmatrix}_{k-1} + a_2 \begin{pmatrix} y_1 \\ y_2 \end{pmatrix}_{k-2} + \begin{bmatrix} b_1^{11} & b_1^{12} \\ b_1^{21} & b_1^{22} \end{bmatrix} \begin{pmatrix} u_1 \\ u_2 \end{pmatrix}_{k-1}$$

$$+ \begin{bmatrix} b_2^{11} & b_2^{12} \\ b_2^{21} & b_2^{22} \end{bmatrix} \begin{pmatrix} u_1 \\ u_2 \end{pmatrix}_{k-2} + \begin{pmatrix} \eta_1 \\ \eta_2 \end{pmatrix}_k \qquad (5.4.60)$$

with

$$\begin{pmatrix} \eta_1 \\ \eta_2 \end{pmatrix}_k + \begin{bmatrix} f_1^{11} & f_1^{12} \\ f_1^{21} & f_1^{22} \end{bmatrix} \begin{pmatrix} \eta_1 \\ \eta_2 \end{pmatrix}_{k-1} = \begin{pmatrix} \varepsilon_1 \\ \varepsilon_2 \end{pmatrix}_k \qquad (5.4.61)$$

where $\{(\varepsilon_1, \varepsilon_2)_k\}$ is a sequence of i.i.d. Gaussian random variables having zero mean and covariance Σ, i.e.,

$$\begin{pmatrix} \varepsilon_1 \\ \varepsilon_2 \end{pmatrix}_k \sim N\left[\begin{pmatrix} 0 \\ 0 \end{pmatrix}, \begin{pmatrix} \Sigma_0^{11} & \Sigma_0^{12} \\ \Sigma_0^{21} & \Sigma_0^{22} \end{pmatrix} \right] \qquad (5.4.62)$$

Estimation results are shown in Table 5.4.1. Typical convergence for one of the parameters is shown in Fig. 5.4.1.

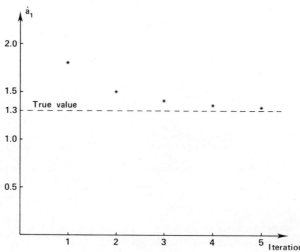

Figure 5.4.1. Typical convergence for one of the parameters in Example 5.4.2.

TABLE 5.4.1[a]

Estimation Results for a Two-Input, Two-Output System

Parameter	Value used in simulation	Least squares estimate (1st iteration)	Relaxation algorithm estimate (9th iteration)
a_1	1.3	1.663 ± 0.003	1.301 ± 0.012
a_2	-0.6	-0.721 ± 0.013	-0.619 ± 0.020
b_1^{11}	0.8	0.788 ± 0.045	0.779 ± 0.034
b_1^{21}	0.2	0.220 ± 0.045	0.201 ± 0.040
b_1^{12}	0.2	0.328 ± 0.045	0.309 ± 0.034
b_1^{22}	0.4	0.451 ± 0.045	0.430 ± 0.038
b_2^{11}	0.3	0.019 ± 0.047	0.331 ± 0.036
b_2^{21}	0.3	0.250 ± 0.045	0.319 ± 0.043
b_2^{12}	0.3	0.291 ± 0.045	0.355 ± 0.032
b_2^{22}	0.3	0.451 ± 0.045	0.350 ± 0.042
f_1^{11}	0.6	0.385	0.583
f_1^{21}	0.4	0.265	0.483
f_1^{12}	0.3	0.187	0.320
f_1^{22}	0.6	0.329	0.536
Σ_0^{11}	1.0	1.309	0.902
Σ_0^{21}	0.5	0.652	0.521
Σ_0^{12}	0.5	0.652	0.521
Σ_0^{22}	1.0	1.438	1.093

[a] Input signal: 500 bit independent PRBS into each input; number of data points: 500; SNR in each output: 1.18, 0.43.

5.5. ASYMPTOTIC PROPERTIES

Consistency and other asymptotic properties of maximum likelihood estimators were presented in Chapter 3 for the case of independent and identically distributed data. It is perhaps not surprising that the results can be extended to the dynamic case. It also turns out that many of the asymptotic properties apply to a large class of prediction error estimators. The interested reader is referred to [45], [15], [82], [83], [85], [223], [118], [120], and [224] for full details. Here we shall briefly describe two key results; namely, the consistency and asymptotic normality of prediction error identification methods.

The general prediction error model was described in Section 4.5. The model can be expressed as

$$y_t = f[Y_{t-1}, U_t, t; \theta] + \varepsilon_t \tag{5.5.1}$$

where y_t denotes the vector output at time t. (In the sequel we shall assume for simplicity that y_t is stationarity and that f is time invariant.) Y_{t-1} denotes the set $\{y_{t-1}, y_{t-2}, \ldots\}$ and U_t denotes the set $\{u_t, u_{t-1}, \ldots\}$. $\{\varepsilon_t\}$ is a sequence of zero mean and independent random variables having covariance Σ. ($\{\varepsilon_t\}$ is the innovations sequence.)

For a particular value of θ, say $\theta = \hat{\theta}$, the prediction error is given by

$$w_t(\hat{\theta}) = y_t - f[Y_{t-1}, U_t, \theta] \tag{5.5.2}$$

Let θ^0 denote the true parameter value, then

$$w_t(\theta_0) = y_t - f[Y_{t-1}, U_t, \theta_0] = \varepsilon_t \tag{5.5.3}$$

As in Section 5.4, we introduce the following two estimation criteria:

$$J_1(\theta, N) = \text{trace}[WD_N(\theta)], \qquad W > 0 \tag{5.5.4}$$

$$J_2(\theta, N) = \log \det[D_N(\theta)] \tag{5.5.5}$$

where $D_N(\theta)$ is the sample value of the prediction error covariance, i.e.,

$$D_N(\theta) = \frac{1}{N} \sum_{t=1}^{N} w_t(\theta) w_t(\theta)^\mathsf{T} \tag{5.5.6}$$

We assume that θ_0 is unique in the sense that there exists no other parameter θ_1 with $\theta_1 \neq \theta_0$ such that $f[Y_{t-1}, U_t, \theta_0] = f[Y_{t-1}, U_t, \theta_1]$ with probability 1. This implies that the model parameterization is unique (see Section 4.3) and that there are certain weak restrictions on the input (e.g., $\{u_t\}$ cannot be generated by certain types of noise-free linear feedback—see Section 5.6 for further discussion on this point).

We then have the following result regarding the strong consistency of the prediction error estimator.

Theorem 5.5.1 Subject to mild regularity conditions, the prediction error estimator $\hat{\theta}_N$ obtained by minimizing either $J_1(\theta, N)$ or $J_2(\theta, N)$ is strongly consistent, i.e.,

$$\hat{\theta}_N \xrightarrow{\text{a.s.}} \theta_0 \tag{5.5.7}$$

Proof See [120] or [224].

We shall present a heuristic proof. We first note that

$$D_N(\theta) = \frac{1}{N} \sum_{t=1}^{N} w_t(\theta) w_t(\theta)^\mathsf{T}$$

$$= \frac{1}{N} \sum_{t=1}^{N} (y_t - f[Y_{t-1}, U_t, \theta])(y_t - f[Y_{t-1}, U_y, \theta])^\mathsf{T} \tag{5.5.8}$$

$$= \frac{1}{N} \sum_{t=1}^{N} (\varepsilon_t + \Delta f_t[\theta, \theta_0])(\varepsilon_t + \Delta f_t[\theta, \theta_0])^\mathsf{T} \tag{5.5.9}$$

where

$$\Delta f_t[\theta, \theta_0] = f[Y_{t-1}, U_t, \theta] - f[Y_{t-1}, U_t, \theta_0] \qquad (5.5.10)$$

Substituting (5.5.10) into (5.5.9) yields

$$D_N(\theta) = \frac{1}{N} \sum_{t=1}^{N} \varepsilon_t \varepsilon_t^T + \frac{1}{N} \sum_{t=1}^{N} \varepsilon_t \, \Delta f_t[\theta, \theta_0]^T$$

$$+ \frac{1}{N} \sum_{t=1}^{N} \Delta f_t[\theta, \theta_0] \varepsilon_t^T + \frac{1}{N} \sum_{t=1}^{N} \Delta f_t[\theta, \theta_0] \, \Delta f_t[\theta, \theta_0]^T \quad (5.5.11)$$

The stationarity assumption implies convergence of each of the above terms to its expected value, i.e.,

$$D_N(\theta) \to \Sigma + E\{\Delta f_t[\theta, \theta_0] \, \Delta f_t[\theta, \theta_0]^T\} \qquad (5.5.12)$$

Substituting (5.5.12) into (5.5.4) and (5.5.5) gives

$$J_1(\theta, N) \to \text{trace}[W(\Sigma + E\{\Delta f_t[\theta, \theta_0] \, \Delta f_t[\theta, \theta_0]^T\})] \qquad (5.5.13)$$

$$J_2(\theta, N) \to \log \det[\Sigma + E\{\Delta f_t[\theta, \theta_0] \, \Delta f_t[\theta, \theta_0]^T\}] \qquad (5.5.14)$$

We now note that $\Delta f_t[\theta, \theta_0] \neq 0$ with probability 1 unless $\theta = \theta_0$. It therefore follows from (5.5.13) and (5.5.14) that if $\hat{\theta}_N$ minimizes either $J_1(\theta, N)$ or $J_2(\theta, N)$, then $\hat{\theta}_N \to \theta_0$, $D_N(\hat{\theta}_N) \to \Sigma$, $J_1(\hat{\theta}_N, N) \to \text{trace } W\Sigma$ and $J_2(\hat{\theta}_N, N) \to \log \det \Sigma$. $\quad \triangledown$

We also immediately have the following corollary.

Corollary 5.5.1 Subject to mild regularity conditions, the maximum likelihood estimator for the Gaussian noise case is strongly consistent.

Proof The result follows from the relationship between maximum likelihood and prediction error methods established in Section 5.4.3. $\quad \triangledown$

The following asymptotic normality result also holds for prediction error estimators.

Theorem 5.5.2 Subject to mild stationarity and regularity conditions, the prediction error estimator $\hat{\theta}_N$ obtained by minimizing either $J_1(\theta, N)$ or $J_2(\theta, N)$ is asymptotically normally distributed in the sense that

$$\sqrt{N}\,(\hat{\theta}_N - \theta_0) \xrightarrow{\text{law}} \gamma \qquad (5.5.15)$$

where

$$\gamma \sim N(0, P) \qquad (5.5.16)$$

For the criterion $J_1(\theta, N)$, P takes the value

$$P_1 = (EZ^TWZ)^{-1}(EZ^TW\Sigma WZ)(EZ^TWZ)^{-1} \qquad (5.5.17)$$

For the criterion $J_2(\theta, N)$, P takes the value

$$P_2 = (EZ^T\Sigma^{-1}Z)^{-1} \qquad (5.5.18)$$

where E denotes expected value, Z is the matrix $(\partial/\partial\theta)f(Y_{t-1}, U_t; \theta)$ evaluated at θ_0, and Σ is the covariance of the innovations.

Proof Full details may be found in [223].
We present a simplified version.
Let $J(\theta)$ denote $J_1(\theta, N)$ or $J_2(\theta, N)$. Then expanding $(\partial J/\partial\theta)(\hat{\theta}_N)$ about θ_0 gives

$$(\partial J/\partial\theta)(\hat{\theta}_N) = (\partial J/\partial\theta)(\theta_0) + (\hat{\theta}_N - \theta_0)^T(\partial^2 J/\partial\theta^2)(\theta_0) + \text{remainder} \qquad (5.5.19)$$

The remainder can be ignored since we know $\hat{\theta}_N \to \theta_0$. Also $(\partial J/\partial\theta)(\hat{\theta}_N)$ is zero since $\hat{\theta}_N$ minimizes $J(\theta)$. Hence from (5.5.19)

$$(\hat{\theta}_N - \theta_0) = [(\partial^2 J/\partial\theta^2)(\theta_0)]^{-1}[(\partial J/\partial\theta)(\theta_0)]^T \qquad (5.5.20)$$

Now, using the chain rule for differentiation:

$$\partial J/\partial\theta_k = (\partial J/\partial(D_N)_{ij})(\partial(D_N)_{ij}/\partial\theta_k) \qquad (5.5.21)$$

where the Einstein dummy suffix notation has been used (i.e., a repeated suffix automatically implies summation with respect to that suffix).

$$\frac{\partial(D_N)_{ij}}{\partial\theta_k} = \frac{1}{N}\sum_{t=1}^{N}\left[(w_t)_i\frac{\partial(w_t)_j}{\partial\theta_k} + (w_t)_j\frac{\partial(w_t)_i}{\partial\theta_k}\right] \qquad (5.5.22)$$

Hence

$$\frac{\partial J}{\partial\theta_k} = \frac{\partial J}{\partial(D_N)_{ij}}\frac{2}{N}\sum_{t=1}^{N}(w_t)_i\frac{\partial(w_t)_j}{\partial\theta_k} \qquad (5.5.23)$$

Also

$$\frac{\partial^2 J}{\partial\theta_k\,\partial\theta_l} = \frac{\partial J}{\partial(D_N)_{ij}}\frac{2}{N}\sum_{t=1}^{N}\left[(w_t)_i\frac{\partial^2(w_t)_j}{\partial\theta_k\,\partial\theta_l} + \frac{\partial(w_t)_i}{\partial\theta_l}\frac{\partial(w_t)_j}{\partial\theta_k}\right]$$

$$+ \frac{\partial^2 J}{\partial(D_N)_{ij}\partial(D_N)_{mn}}\frac{2}{N}\sum_{t=1}^{N}(w_t)_i\frac{\partial(w_t)_j}{\partial\theta_k}\frac{2}{N}\sum_{t=1}^{N}(w_t)_m\frac{\partial(w_t)_n}{\partial\theta_l} \qquad (5.5.24)$$

The above expressions should be evaluated for $\theta = \theta_0$, but at θ_0, $w_t = \varepsilon_t$ which is independent of Y_{t-1} and U_t and hence of $\partial w_t/\partial\theta = -(\partial f/\partial\theta)[Y_{t-1},$

$U_t; \theta]$. This fact together with the stationarity assumption gives

$$(\partial J/\partial\theta_k) \to 0 \tag{5.5.25}$$

$$(\partial^2 J/\partial\theta_k \, \partial\theta_l) \to (\partial J/\partial(D_N)_{ij})2E\{Z_{il} Z_{jk}\} \tag{5.5.26}$$

where Z is the matrix with jkth element $\partial(w_t)_j/\partial\theta_k|_{\theta_0}$. (Note, in deriving (5.5.25) and (5.5.26) we have used the independence of ε_t and Y_{t-1}, U_t.)

Also using the matrix differentiation results given in Appendix E,

$$\partial J_1/\partial(D_N)_{ij} = W_{ij} \tag{5.5.27}$$

$$\partial J_2/\partial(D_N)_{ij} = [D_N]_{ij}^{-1} \to [\Sigma]_{ij}^{-1} \tag{5.5.28}$$

Now for $\theta = \theta_0$, it is readily seen that the quantity $\sum_{t=1}^{N} (w_t)_i(\partial(w_t)_j/\partial\theta_k)$ is a Martingale (see Appendix D). Thus from the central limit theorem for Martingales [231], we have that the quantity $(2/\sqrt{N}) \sum_{t=1}^{N} (w_t)_i(\partial(w_t)_j/\partial\theta_k)$ tends in law to a normal distribution with zero mean and variance $4E\{(w_t)_i(\partial(w_t)_j/\partial\theta_k)\}^2 = 4\Sigma_{ii} E\{Z_{jk} Z_{jk}\}$.

Hence it follows from (5.5.23) that $\sqrt{N} \, \partial J^T/\partial\theta$ tends in law to a normal distribution with zero mean and covariance $4E\{Z^T(\partial J/\partial D)\Sigma(\partial J/\partial D)Z\}$. Finally, from (5.5.20) and Result A.2.4 we have

$$\sqrt{N} \, (\hat{\theta}_N - \theta_0) \xrightarrow{\text{law}} \gamma \tag{5.5.29}$$

$$\gamma \sim N(0, P) \tag{5.5.30}$$

where

$$P = 4 \left[\frac{\partial^2 J}{\partial\theta^2}\right]^{-1} \left[EZ^T \frac{\partial J}{\partial D} \Sigma \frac{\partial J}{\partial D} Z\right] \left[\frac{\partial^2 J}{\partial\theta^2}\right]^{-1} \tag{5.5.31}$$

It follows from (5.5.27) that for the criterion $J \equiv J_1(\theta, N)$ that

$$\partial J_1/\partial D = W \tag{5.5.32}$$

Also from (5.5.26) we have

$$\partial^2 J_1/\partial\theta^2 \to 2EZ^T W Z \tag{5.5.33}$$

Substituting (5.5.32) and (5.5.33) into (5.5.31) gives

$$P_1 = (EZ^T W Z)^{-1}(EZ^T W\Sigma W Z)(EZ^T W Z)^{-1} \tag{5.5.34}$$

Similarly, from (5.5.28) for the criterion $J \equiv J_2(\theta, N)$

$$\partial J_2/\partial D \to \Sigma^{-1} \tag{5.5.35}$$

Also from (5.5.26) we have

$$\partial^2 J_2/\partial\theta^2 \to 2Z^T\Sigma^{-1}Z \tag{5.5.36}$$

Substituting (5.5.35) and (5.5.36) into (5.5.31) gives

$$P_2 = (EZ^T\Sigma^{-1}Z)^{-1}(EZ^T\Sigma^{-1}\Sigma\Sigma^{-1}Z)(EZ^T\Sigma^{-1}Z) = (EZ^T\Sigma^{-1}Z)^{-1} \quad \triangledown$$

$$\tag{5.5.37}$$

The above theorem gives a means of determining the accuracy of prediction error estimates. This is very important since the estimates by themselves are of little value without some measure of their accuracy. In practice, the expected value operations indicated in (5.5.17) and (5.5.18) can be replaced by sample averages computed at $\hat{\theta}_N$.

There is also an interesting accuracy result [223] which plays a role somewhat akin to the concept of efficiency (cf. Definition 1.3.5). We shall be interested in the following question: What is the best value for W in the criterion $J_1(\theta, N)$ and what is the relationship between estimates obtained using $J_1(\theta, N)$ and those obtained using $J_2(\theta, N)$? This question is studied in the following result.

Theorem 5.5.3 When $J_1(\theta, N)$ is used as the estimation criterion, then the smallest possible asymptotic parameter covariance is obtained by using $W = \Sigma^{-1}$, where Σ is the covariance of the innovations. Moreover, the same value for the asymptotic parameter covariance is obtained when $J_2(\theta, N)$ is used without prior knowledge of Σ.

Proof Our proof follows [223].
Consider the following matrix inequality

$$\begin{vmatrix} EZ^T\Sigma^{-1}Z & EZ^TWZ \\ EZ^TWZ & EZ^TW\Sigma WZ \end{vmatrix} \geq 0 \tag{5.5.38}$$

(This follows, for instance, from $E(Za + \Sigma WZb)^T\Sigma^{-1}(\Sigma WZb + a) \geq 0$ for real vectors a and b.)
Hence

$$EZ^T\Sigma^{-1}Z \geq (EZ^TWZ)(EZ^TW\Sigma WZ)^{-1}(EZ^TWZ) \tag{5.5.39}$$

with equality iff $W = \Sigma^{-1}$. Thus $W = \Sigma^{-1}$ yields the largest value of P_1 and hence the smallest covariance.

Note that the second part of the theorem is immediate since $P_2 = P_1$ with $W = \Sigma^{-1}$. $\triangledown$

Remark 5.5.1 We shall see in the next chapter that the matrix P_2^{-1} is actually the average information matrix for the case when the innovations are Gaussian. (Compare Eq. (5.5.18) with Eq. (6.3.23).) Moreover, it was shown in Section 5.4.3 that the maximum likelihood estimator in the Gaussian case can be achieved by using $J_2(\theta, N)$ as an estimation criterion. Hence Theorem 5.5.2 establishes the asymptotic efficiency of the maximum likelihood estimator in the case when the innovations are Gaussian. The theorem also shows that the same asymptotic covariance is achieved whenever the $J_2(\theta, N)$ criterion is used whether or not the innovations are Gaussian. Of course, in the non-Gaussian case it may be possible to achieve greater estimation accuracy by the use of other estimation criteria. However, Theorem 5.5.2 is of considerable importance because it shows that the maximum likelihood method devised for Gaussian innovations can be applied to general distributions without any of the essential properties being lost. ∇

We conclude this section by presenting a proof of strong consistency of the least squares identification method. This result could be passed on a first reading. However, we are motivated to include this proof by four factors. First, the proof is particularly simple in the least squares case since many of the difficulties associated with the general prediction error case are avoided. Second, the least squares method is very widely used in practice and therefore there is likely to be interest in its properties. Third, the proof illustrates the general philosophy of proofs of this type. Finally, the proof allows for general (nonlinear, time varying, adaptive) feedback and is therefore applicable in a wide variety of circumstances. We shall pursue the question of estimation under closed loop conditions more thoroughly in the next section.

Consider a model in the least squares structure:

$$A(z)y_t = B(z)u_t + \varepsilon_t \tag{5.5.40}$$

where $\{y_t\}$ is the n-dimensional output sequence, $\{u_t\}$ the r-dimensional input sequence, $\{\varepsilon_t\}$ an innovations sequence, and

$$A(z) = I + A_1 z^{-1} + \cdots + A_p z^{-p} \tag{5.5.41}$$

$$B(z) = B_1 z^{-1} + \cdots + B_q z^{-q} \tag{5.5.42}$$

Equation (5.5.40) can be written in the form

$$y_t = \theta^T X_t + \varepsilon_t \tag{5.5.43}$$

where

$$\theta^T = (A_1, \ldots, A_p, B_1, \ldots, B_q) \tag{5.5.44}$$

$$X_t^T = (-y_{t-1}^T, \ldots, -y_{t-p}^T, u_{t-1}^T, \ldots, u_{t-q}^T) \tag{5.5.45}$$

The least squares estimate of θ up to time N is determined by minimization of

$$J_1(\theta, N) = \text{trace } D_N(\theta) \tag{5.5.46}$$

where

$$D_N(\theta) = \frac{1}{N} \sum_{t=1}^{N} w_t(\theta) w_t(\theta)^{\mathrm{T}} \tag{5.5.47}$$

$$w_t(\theta) = y_t - \theta^{\mathrm{T}} X_t \tag{5.5.48}$$

The following conditions are now imposed:

(i) $\{\varepsilon_t\}$ is an innovations sequences satisfying

$$E\{\varepsilon_t \mid \varepsilon_{t-1}, \varepsilon_{t-2}, \ldots\} = 0 \tag{5.5.49}$$

$$E\{\varepsilon_t^{\mathrm{T}} \varepsilon_t \mid \varepsilon_{t-1}, \varepsilon_{t-2}, \ldots\} < C < \infty \tag{5.5.50}$$

(ii)   The input u_t is independent of $\varepsilon_{t+1}, \varepsilon_{t+2}, \ldots$.
(iii) $\lim_{N \to \infty} \sup 1/N \sum_{t=1}^{N} y_t^{\mathrm{T}} y_t + u_t^{\mathrm{T}} u_t < \infty$ with probability 1.

Note that condition (ii) allows for feedback in the generation of u_t. The only restriction is that the closed loop system has the stability property given in (iii) (actually the stability condition can also be removed—see [221] and [232] for details).
We now have the following result.

Theorem 5.5.4   Consider the system (5.5.43) satisfying assumptions (i), (ii), and (iii).

(a)   If the minimization of (5.5.46) is restricted to a finite set of parameter vectors $\theta \in D_F = \{\theta_1, \theta_2, \ldots, \theta_M\}$ with $\theta_0 \in D_F$, then

$$\hat{\theta}_N \xrightarrow{\text{a.s.}} D_I^F = \left\{ \theta \mid \theta \in D_F; \theta = \theta_0 + \tilde{\theta}; \sum_{t=1}^{\infty} |\tilde{\theta}^{\mathrm{T}} X_t|^2 < \infty \right\} \tag{5.5.51}$$

(b)   If the minimization of (5.5.46) is carried out over an infinite set of parameter vectors, then

$$\hat{\theta}_N = \theta_0 + \tilde{\theta}_N \tag{5.5.52}$$

where

$$\tilde{\theta}_N / (1 + \| \tilde{\theta}_N \|) \xrightarrow{\text{a.s.}} \tilde{D}_I \tag{5.5.53}$$

where $\| \cdot \|$ denotes matrix norm (see Remark D.1.1) and

$$\tilde{D}_I = \left\{ \tilde{\theta} \mid |\tilde{\theta}| \leq 1; \lim_{N \to \infty} \inf \frac{1}{N} \sum_{t=1}^{N} |\tilde{\theta}^{\mathrm{T}} X_t|^2 = 0 \right\} \tag{5.5.54}$$

(c)   If $(1/N) \sum_{t=1}^{N} X_t X_t^{\mathrm{T}}$ tends to a positive definite matrix with probability 1, then

$$\hat{\theta}_N \xrightarrow{\text{a.s.}} \theta_0 \tag{5.5.55}$$

Proof   We present an outline version of the proof given by Ljung [232].

(a)   Introduce $\tilde{\theta} = \theta - \theta_0$. Using (5.5.46) to (5.5.48), we obtain

$$NJ_1(\theta, N) = \sum_{t=1}^{N} |\varepsilon_t - \tilde{\theta}^{\mathrm{T}} X_t|^2$$

$$= \sum_{t=1}^{N} |\varepsilon_t|^2 - 2\sum_{t=1}^{N} \varepsilon_t^{\mathrm{T}} \tilde{\theta}^{\mathrm{T}} X_t + \sum_{t=1}^{N} X_t^{\mathrm{T}} \tilde{\theta} \tilde{\theta}^{\mathrm{T}} X_t$$

$$= \sum_{t=1}^{N} |\varepsilon_t|^2$$

$$+ \mathrm{trace}\left\{ \tilde{\theta}^{\mathrm{T}} \left(\sum_{t=1}^{N} X_t X_t^{\mathrm{T}} \right) \tilde{\theta} - \tilde{\theta}^{\mathrm{T}} \left(\sum_{t=1}^{N} X_t \varepsilon_t^{\mathrm{T}} \right) - \left(\sum_{t=1}^{N} \varepsilon_t X_t^{\mathrm{T}} \right) \tilde{\theta} \right\}$$

$$(5.5.56)$$

Let A_t be the sequence of σ-algebras generated by $\{\varepsilon_0, u_0, \ldots, \varepsilon_{t-1}, u_{t-1}\}$. Let

$$s(t, \theta) = 1 + \sum_{k=1}^{t} (X_k^{\mathrm{T}} \tilde{\theta} \tilde{\theta}^{\mathrm{T}} X_k) \qquad (5.5.57)$$

Consider

$$z(t; \theta) = \sum_{k=1}^{t} \varepsilon_k^{\mathrm{T}} \tilde{\theta}^{\mathrm{T}} X_k / s(k, \theta), \qquad z(0; \theta) = 0 \qquad (5.5.58)$$

It follows from assumptions (i) and (ii) that $z(t; \theta)$ is a Martingale with respect to $\{A_t\}$ (see Definition D.2.2). Also, we can readily see that $z(t; \theta)$ has bounded variance since

$$Ez^2(N; \theta) = E\sum_{k=1}^{N} E\{z^2(k; \theta) - z^2(k-1; \theta) \,|\, A_k\}$$

$$= E\sum_{k=1}^{N} E\{(\varepsilon_k^{\mathrm{T}} \tilde{\theta}^{\mathrm{T}} X_k / s(k, \theta))^2 \,|\, A_k\}$$

$$= E\sum_{k=1}^{N} E\{(\varepsilon_k^{\mathrm{T}} \tilde{\theta}^{\mathrm{T}} X_k / s(k, \theta))^2 \,|\, A_k\}$$

$$< CE\sum_{k=1}^{N} X_k^{\mathrm{T}} \tilde{\theta} \tilde{\theta}^{\mathrm{T}} X_k / s^2(k, \theta)$$

$$= CE\sum_{k=1}^{N} \frac{s(k, \theta) - s(k-1, \theta)}{s^2(k, \theta)}$$

$$\leq CE\sum_{k=1}^{N} \frac{s(k, \theta) - s(k-1, \theta)}{s(k, \theta)s(k-1, \theta)}$$

$$\leq C \qquad \text{since} \quad 1 \geq 1/s_{k-1} \geq 1/s_k \geq 0 \qquad (5.5.59)$$

Hence applying Theorem D.2.3, $z(N; \theta)$ converges almost surely. Applying the Kronecker lemma (Lemma D.1.2) to (5.5.7) gives

$$\frac{1}{s(N, \theta)} \sum_{t=1}^{N} \varepsilon_t^T \tilde{\theta}^T X_t \xrightarrow{\text{a.s.}} 0 \qquad (5.5.60)$$

for any θ such that $s(N, \theta) \to \infty$, i.e., in particular for $\theta \notin D_I^F$, and, since the second line of (5.5.56) can be written as

$$NJ_1(\theta, N) = \sum_{t=1}^{N} |\varepsilon_t|^2 + s(N, \theta) \left[\frac{2}{s(N, \theta)} \sum_{t=1}^{N} \varepsilon_t^T \tilde{\theta}^T X_t + 1 - \frac{1}{s(N, \theta)} \right] \qquad (5.5.61)$$

we see that almost surely

$$NJ_1(\theta, N) - NJ_1(\theta_0, N) \to \infty \qquad \text{for} \quad \theta \notin D_I^F \qquad (5.5.62)$$

This establishes part (a).

(b)   Here we introduce $\tilde{s}(N, \theta) = \max(s(N, \theta), N)$. Obviously the quantity $\tilde{z}(t, \theta)$ defined analogously to (5.5.58) is still a Martingale with bounded variance. Therefore, as before

$$\frac{1}{\tilde{s}(N, \theta)} \sum_{t=1}^{N} \varepsilon_t^T \tilde{\theta}^T X_t \xrightarrow{\text{a.s.}} 0 \qquad (5.5.63)$$

Assumption (iii) implies that $\limsup \tilde{s}(N, \theta)/N < \infty$ with probability 1 for all θ, and therefore (5.5.63) implies that the matrix

$$\frac{1}{N} \sum_{t=1}^{N} X_t \varepsilon_t^T \xrightarrow{\text{a.s.}} 0 \qquad (5.5.64)$$

However, the minimizing element $\hat{\theta}_N = \theta_0 + \tilde{\theta}_N$ satisfies

$$\left(\frac{1}{N} \sum_{t=1}^{N} X_t X_t^T \right) \tilde{\theta}_N = \frac{1}{N} \sum_{t=1}^{N} X_t \varepsilon_t^T \qquad (5.5.65)$$

Multiplying this equation by $\overset{*}{\theta}_N^T = \tilde{\theta}_N^T/(1 + \|\tilde{\theta}\|)$ and using (5.5.64) gives

$$\frac{1}{N} \sum_{t=1}^{N} |\overset{*}{\theta}_N^T X_t|^2 \xrightarrow{\text{a.s.}} 0 \qquad (5.5.66)$$

Equation (5.5.66) and condition (iii) implies that all cluster points of the sequence $\{\overset{*}{\theta}_N\}$ must be in $\tilde{D}_I$. This completes part (b).

(c)   The given condition implies that

$$\tilde{D}_I = \{0\} \qquad \text{or} \qquad D_I^F = \{\theta_0\} \quad \text{a.s.} \qquad (5.5.67)$$

Hence

$$\hat{\theta}_N \xrightarrow{\text{a.s.}} \theta_0 \qquad \triangledown$$

The above result is very general and applies, for example, when the system input is generated by adaptive feedback control, where the behavior of the regulator may be realization dependent. This type of algorithm is discussed in Section 7.8. A very special case of the result is when the input is exogenous, i.e., generated in open loop. For this case, a sufficient condition for (c) to be satisfied is that the input is persistently exciting of some finite order. This is discussed in Theorem 6.4.5.

5.6 ESTIMATION IN CLOSED LOOP

In this section we consider parameter estimation problems for systems operating under closed loop conditions. This is of considerable practical importance since it is frequently impossible, or at least highly undesirable, to operate a system in open loop.

The feedback may occur naturally as in sociological, biological, and economic modeling problems. Alternatively, the feedback may be purposefully introduced to achieve some acceptable level of system performance. For example, the output may be required to meet normal production constraints. An extreme case is when the system is open loop unstable.

We consider the class of feedback systems modeled as in Fig. 5.6.1.

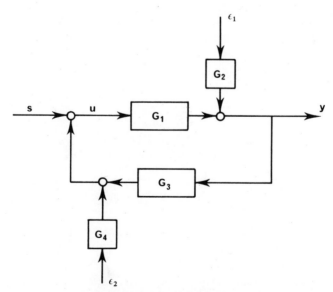

Figure 5.6.1. Closed loop system.

In Fig. 5.6.1 we have $s \in R^r$: externally applied set point perturbation, $u \in R^r$: system input, $y \in R^n$: process output, $\varepsilon_1 \in R^n$: noise in forward path, $\varepsilon_2 \in R^r$: noise in feedback path, and G_1, G_2, G_3, G_4: transfer functions of appropriate dimensions.

For simplicity, we assume that the white noise sequences $\{\varepsilon_1\}$ and $\{\varepsilon_2\}$ have a joint Gaussian distribution with zero mean and joint covariance

$$E\left[\binom{\varepsilon_1}{\varepsilon_2} (\varepsilon_1^{\mathrm{T}} \quad \varepsilon_2^{\mathrm{T}}) \right] = \begin{vmatrix} \Sigma_{11} & \Sigma_{12} \\ \Sigma_{21} & \Sigma_{22} \end{vmatrix} = \Sigma \tag{5.6.1}$$

It is assumed that measurements of $\{s_t\}$, $\{u_t\}$, and $\{y_t\}$ are available to the experimenter.

If prior knowledge dictates that a model structure of the form depicted in Fig. 5.6.1 is appropriate, then the question arises whether or not it is possible to extract meaningful estimates of the transfer functions G_1, G_2, G_3, and G_4 from the measurements. We consider two cases, viz., noisy feedback $(G_4 \not\equiv 0)$ and noise-free feedback $(G_4 \equiv 0)$.

5.6.1 *Noisy feedback* In this section we consider the case where the joint process (y_t, u_t) is a full rank stochastic process. We assume that the process (y_t, u_t) is stationary and we denote the joint spectral density by Φ_{yu}. From Fig. 5.6.1, Φ_{yu} is given by

$$\Phi_{yu}(z) = K(z) \Sigma K^{\mathrm{T}}(z^{-1}) \tag{5.6.2}$$

where Σ is defined in Eq. (5.6.1) and where

$$K = \begin{vmatrix} K_{11} & K_{12} \\ \hline K_{21} & K_{22} \end{vmatrix} = \begin{vmatrix} (I - G_1 G_3)^{-1} G_2 & (I - G_1 G_3)^{-1} G_1 G_4 \\ \hline (I - G_3 G_1)^{-1} G_3 G_2 & (I - G_3 G_1)^{-1} G_4 \end{vmatrix} \tag{5.6.3}$$

We now have the following result regarding the relationship between K and G_1–G_4.

Lemma 5.6.1 There is a one-to-one correspondence between K, as defined in Eq. (5.6.3) and the quadruple (G_1, G_2, G_3, G_4). Furthermore the values of G_1–G_4 are uniquely expressible in terms of K as

$$G_1 = K_{12} K_{22}^{-1} \tag{5.6.4}$$

$$G_2 = K_{11} - K_{12} K_{22}^{-1} K_{21} \tag{5.6.5}$$

$$G_3 = K_{21} K_{11}^{-1} \tag{5.6.6}$$

$$G_4 = K_{22} - K_{21} K_{11}^{-1} K_{12} \tag{5.6.7}$$

Proof It is obvious that K is uniquely determined by G_1–G_4. To establish the converse, we assume that there exists another set $(H_1, H_2,$

H_3, H_4) giving rise to K. Then from Eq. (5.6.3) we have

$$K_{11} = (I - G_1 G_3)^{-1} G_2 = (I - H_1 H_3)^{-1} H_2 \tag{5.6.8}$$

$$K_{12} = (I - G_1 G_3)^{-1} G_1 G_4 = (I - H_1 H_3)^{-1} H_1 H_4 \tag{5.6.9}$$

$$K_{21} = (I - G_3 G_1)^{-1} G_3 G_2 = (I - H_3 H_1)^{-1} H_3 H_2 \tag{5.6.10}$$

$$K_{22} = (I - G_3 G_1)^{-1} G_4 = (I - H_3 H_1)^{-1} H_4 \tag{5.6.11}$$

Now using the fact that $(I - AB)^{-1} A = A(I - BA)^{-1}$ (see Result E.1.4), it follows that Eqs. (3.7)–(3.10) have a unique solution (see Problem 5.11).

$$G_1 = H_1 = K_{12} K_{22}^{-1} \tag{5.6.12}$$

$$G_2 = H_2 = K_{11} - K_{12} K_{22}^{-1} K_{21} \tag{5.6.13}$$

$$G_3 = H_3 = K_{21} K_{11}^{-1} \tag{5.6.14}$$

$$G_4 = H_4 = K_{22} - K_{21} K_{11}^{-1} K_{12} \qquad \triangledown \tag{5.6.15}$$

Lemma 5.6.1 indicates that it is possible to uniquely recover G_1–G_4 from K. However, there is a fundamental and unavoidable nonuniqueness associated with the spectral factorization given in Eq. (5.6.2). Thus for a given Φ_{yu}, there will be many transfer function matrices G_1–G_4 corresponding to the different spectral factorizations of Φ_{yu}. Of course, unique values for G_1–G_4 can be determined by using a particular factorization of Φ_{yu}. For example, one might specify the unique stable minimum phase spectral factor satisfying $\lim_{z \to \infty} K(z) = I$ (see Theorem 4.4.1). However, it is not obvious that this will necessarily lead to physical meaningful values for G_1–G_4. The conditions under which the stable minimum phase spectral factor leads to physical meaningful values of G_1–G_4 are studied in the following theorem.

Theorem 5.6.1 Consider the process depicted in Fig. 5.6.1 where we assume $\lim_{z \to \infty} G_1 = 0$, $\lim_{z \to \infty} G_3 = 0$, $\lim_{z \to \infty} G_2 = I$, and $\lim_{z \to \infty} G_4 = I$. Without loss of generality, we assume that $[G_1 \vdots G_2]$ is represented as a left MFD of the form $A^{*-1}[B^* \vdots C^*]$ and that $[G_3 \vdots G_4]$ is represented by a left MFD of the form $E^{*-1}[F^* \vdots L^*]$. The dynamic equations for the system then become

$$A y_t = B u_t + C \varepsilon_{1t} \tag{5.6.16}$$

$$E u_t = G y_t + L \varepsilon_{2t} \tag{5.6.17}$$

where A, B, C, E, F, and L are polynomial matrices in z^{-1}. (For the relationship between A and A^*, see the discussion in Section 4.2). We further assume that ε_{1t} and ε_{2t} are uncorrelated, i.e.,

$$\Sigma = E\left\{ \begin{pmatrix} \varepsilon_{1t} \\ \varepsilon_{2t} \end{pmatrix} \begin{pmatrix} \varepsilon_{1t}^T & \varepsilon_{2t}^T \end{pmatrix} \right\} = \begin{vmatrix} \Sigma_{11} & 0 \\ 0 & \Sigma_{22} \end{vmatrix} \tag{5.6.18}$$

Then

(i)   a sufficient condition for the joint process (y_t, u_t) to be stationary is that

$$\det \begin{vmatrix} A^* & -B^* \\ -F^* & E^* \end{vmatrix} \neq 0 \qquad \text{for all} \quad |z| \geq 1 \qquad (5.6.19)$$

(ii)   Given that (5.6.19) is satisfied, then the power spectrum of Φ_{yu} is given by

$$\Phi_{yu}(z) = K(z)\Sigma K^{\mathrm{T}}(z^{-1}) \qquad (5.6.20)$$

where

$$K = \begin{vmatrix} A^* & -B^* \\ -F^* & E^* \end{vmatrix}^{-1} \begin{vmatrix} C^* & 0 \\ 0 & L^* \end{vmatrix} \qquad (5.6.21)$$

(iii)   Given that (5.6.19) is satisfied, and if $\det C^* \neq 0$, $\det L^* \neq 0$ in $|z| \geq 1$, then $K(z)$ is stable and minimum phase.

(iv)   All possible sextuplets $(A^*, B^*, C^*, E^*, F^*, L^*)$ such that (5.6.19), (5.6.20), and (5.6.21) are satisfied have the property that

$$K(z) = \tilde{K}(z) \begin{vmatrix} V_{11}(z) & 0 \\ 0 & V_{22}(z) \end{vmatrix} \qquad (5.6.22)$$

where $\tilde{K}(z)$ is the unique stable minimum phase spectral factor of Φ_{yu} satisfying $\lim_{z \to \infty} \tilde{K}(z) = I$ and where $V_{11}(z)$ and $V_{22}(z)$ are scaled paraunitary matrices satisfying $V_{11}(z)\Sigma_{11}V_{11}^{\mathrm{T}}(z^{-1}) = \tilde{\Sigma}_{22}$ and $V_{22}(z)\Sigma_{22}V_{22}^{\mathrm{T}}(z^{-1}) = \tilde{\Sigma}_{22}$ for real positive semidefinite matrices $\tilde{\Sigma}_{11}$, and $\tilde{\Sigma}_{22}$.

(v)   Given any Φ_{yu}, then provided (3.18) is satisfied, $G_1 = A^{-1}B$ and $G_3 = E^{-1}F$ are uniquely computable from the stable minimum phase spectral factor of Φ_{yu} via Eqs. (5.6.4) and (5.6.6). Furthermore, $G_2 = A^{-1}C$ and $G_4 = D^{-1}L$ are uniquely computable to within right multiplication by scaled paraunitary matrices.

Proof (i) Clearly (5.6.19) implies the asymptotic stability of the closed loop system and hence the stationarity of (y_t, u_t).

(ii)   Follows by direct substitution.

(iii)   We have

$$K^{-1} = \begin{vmatrix} C^* & 0 \\ 0 & L^* \end{vmatrix}^{-1} \begin{vmatrix} A^* & -B^* \\ -F^* & E^* \end{vmatrix} \qquad (5.6.23)$$

and hence $\det C^* \neq 0$, $\det L^* \neq 0$ in $|z| \geq 1$ implies K^{-1} is asymptotically stable, i.e., K is minimum phase.

(iv) Let $(A^*, B^*, C^*, E^*, F^*, L^*)$ be any set of polynomial matrices in z satisfying (5.6.19). It is known [1, 4, 222] that there exists rational matrices $V_{11}(z)$ and $V_{22}(z)$ such that

$$[A^*(z)]^{-1}[C^*(z)] = M_1(z)V_{11}(z) \tag{5.6.24}$$

$$[E^*(z)]^{-1}[L^*(z)] = M_2(z)V_{22}(z) \tag{5.6.25}$$

where $M_1(z)$ and $M_2(z)$ are minimum phase $\lim_{z\to\infty} M_1(z) = I$, $\lim_{z\to\infty} M_2(z) = I$, and where $V_{11}(z)$ and $V_{22}(z)$ satisfy

$$V_{11}(z)\Sigma_{11}V_{11}^{\mathrm{T}}(z^{-1}) = \tilde{\Sigma}_{11} \tag{5.6.26}$$

$$V_{22}(z)\Sigma_{22}V_{22}^{\mathrm{T}}(z^{-1}) = \tilde{\Sigma}_{22} \tag{5.6.27}$$

with $\tilde{\Sigma}_{11}$ and $\tilde{\Sigma}_{22}$ real positive definite matrices.

Now consider a new model for the process defined as follows:

$$y_t = (A^*)^{-1}B^*u_t + M_1\varepsilon_{1t} \tag{5.6.28}$$

$$u_t = (E^*)^{-1}F^*y_t + M_2\varepsilon_{2t} \tag{5.6.29}$$

Equations (3.26) and (3.27) can be expressed in the form of (5.6.16) and (5.6.17) yielding

$$\tilde{A}y_t = \tilde{B}u_t + \tilde{C}\varepsilon_{1t} \tag{5.6.30}$$

$$\tilde{E}u_t = \tilde{F}y_t + \tilde{L}\varepsilon_{2t} \tag{5.6.31}$$

where $\tilde{A}^{-1}\tilde{B} = A^{-1}B$, $\tilde{A}^{-1}\tilde{C} = M_1$, $\tilde{E}^{-1}\tilde{F} = E^{-1}F$, and $\tilde{E}^{-1}\tilde{L} = M_2$.

It is readily seen that Φ_{yu} for the set $(\tilde{A}, \tilde{B}, \tilde{C}, \tilde{E}, \tilde{F}, \tilde{L}, \Sigma)$ is the same as Φ_{yu} for the set $(A, B, C, E, F, L, \Sigma)$. Also by construction, $\tilde{A}^{-1}\tilde{C}$ and $\tilde{E}^{-1}\tilde{L}$ are minimum phase, i.e., $|\tilde{C}^*| \neq 0$ and $|\tilde{L}^*| \neq 0$ for $|z| \geq 1$. Hence using part (iii), K corresponding to $(\tilde{A}, \tilde{B}, \tilde{C}, \tilde{E}, \tilde{F}, \tilde{L})$ is stable and minimum phase. Also $\lim_{z\to\infty} \tilde{K} = I$ and hence $\tilde{K}$ is unique (see Theorem 4.4.1). From (5.6.30) and (5.6.31) we have

$$
\tilde{K} = \begin{vmatrix} \tilde{A} & -\tilde{B} \\ -\tilde{F} & \tilde{E} \end{vmatrix}^{-1} \begin{vmatrix} \tilde{C} & 0 \\ 0 & \tilde{L} \end{vmatrix}
$$

$$
= \begin{vmatrix} I & -\tilde{A}^{-1}\tilde{B} \\ -\tilde{E}^{-1}\tilde{F} & I \end{vmatrix}^{-1} \begin{vmatrix} M_1 & 0 \\ 0 & M_2 \end{vmatrix}
$$

$$
= \begin{vmatrix} I & -A^{-1}B \\ -E^{-1}F & I \end{vmatrix}^{-1} \begin{vmatrix} M_1 & 0 \\ 0 & M_2 \end{vmatrix} \tag{5.6.32}
$$

Hence

$$
\tilde{K} \begin{vmatrix} V_{11} & 0 \\ 0 & V_{22} \end{vmatrix} = \begin{vmatrix} I & -A^{-1}B \\ -E^{-1}F & I \end{vmatrix}^{-1} \begin{vmatrix} M_1 V_{11} & 0 \\ 0 & M_2 V_{22} \end{vmatrix} = K
$$

(v) Using Lemma 5.6.1, G_1–G_4 are uniquely determined by K via Eqs. (5.6.4)–(5.6.7). Substituting (5.6.33) into (5.6.4)–(5.6.7) yields

$$G_1 = (\tilde{K}_{12} V_{22})(\tilde{K}_{22} V_{22})^{-1} = \tilde{K}_{12} \tilde{K}_{22}^{-1} \tag{5.6.34}$$

$$G_2 = (\tilde{K}_{11} V_{11}) - (\tilde{K}_{12} V_{22})(\tilde{K}_{22} V_{22})^{-1}(\tilde{K}_{21} V_{11}) = (\tilde{K}_{11} - \tilde{K}_{12} \tilde{K}_{22}^{-1} \tilde{K}_{21})V_{11} \tag{5.6.35}$$

$$G_3 = (\tilde{K}_{21} V_{11})(\tilde{K}_{11} V_{11})^{-1} = \tilde{K}_{21} \tilde{K}_{11}^{-1} \tag{5.6.36}$$

$$G_4 = (\tilde{K}_{22} V_{22}) - (\tilde{K}_{21} V_{11})(\tilde{K}_{11} \tilde{V}_{11})^{-1}(\tilde{K}_{12} V_{22}) = (\tilde{K}_{22} - \tilde{K}_{21} \tilde{K}_{11}^{-1} \tilde{K}_{12})V_{22} \tag{5.6.37}$$

Hence using Lemma 5.6.1 again, G_1–G_4 are given by

$$G_1 = \tilde{G}_1 \tag{5.6.38}$$

$$G_2 = \tilde{G}_2 V_{11} \tag{5.6.39}$$

$$G_3 = \tilde{G}_3 \tag{5.6.40}$$

$$G_4 = \tilde{G}_4 V_{22} \tag{5.6.41}$$

where $(\tilde{G}_1, \tilde{G}_2, \tilde{G}_3, \tilde{G}_4)$ are uniquely determined by the stable minimum phase spectral factor K via Eqs. (5.6.4)–(5.6.7). ∇

The above theorem shows that, provided the noise in the forward and reverse paths are uncorrelated, then the correct values of G_1 and G_3 can be obtained by applying Eqs. (5.6.4) and (5.6.6) to the stable minimum phase spectral factor of Φ_{yu}. Furthermore the values of G_2 and G_4 obtained from (5.6.5) and (5.6.7) will differ from the "true" values by at most, right multiplication by a scaled paraunitary matrix. This latter ambiguity is of a fundamental nature and occurs even in the open loop case. This ambiguity does not influence the input/output characteristics of either the forward or reverse path and is therefore of no practical importance.

Equations (5.6.4)–(5.6.7) form the basis of the, so-called, joint input/output identification method [235]. The method has been successfully used in a number of practical situations [148, 238, 239]. Dablemont and Gevers [239] point out that an advantage of the method is that it requires a minimum of prior knowledge regarding the model structure.

For the case where the noise in the forward and feedback paths is correlated, then the models obtained for G_1–G_4 from Eqs. (5.6.4)–(5.6.7) will depend upon the particular factorization of Φ_{yu} used. Among all possible stable spectral factors of Φ_{yu}, only one will correspond to the true system. In practice there is no way of knowing which spectral factor should be used. We illustrate the above difficulty by a simple example.

Example 5.6.1 Consider a single-input, single-output feedback system of the form illustrated in Fig. 5.6.1 with the following values for the transfer functions G_1–G_4 and for the covariance Σ.

$$G_1(z) = 1/(z - 0.5) \tag{5.6.42}$$

$$G_2(z) = (z - 2)/(z - 0.5) \tag{5.6.43}$$

$$G_3(z) = -0.5/z \tag{5.6.44}$$

$$G_4(z) = 1 \tag{5.6.45}$$

$$\Sigma = \begin{vmatrix} 1 & 1 \\ 1 & 2 \end{vmatrix} \tag{5.6.46}$$

The corresponding value of K is given by

$$K(z) = \frac{1}{z^2 - 0.5z + 0.5} \begin{vmatrix} z^2 - 2z & z \\ -0.5z + 1 & z^2 - 0.5z \end{vmatrix} \tag{5.6.47}$$

It can be seen that $K(z)$ given above is a stable, but nonminimum phase spectral factor of Φ_{yu}. One method of obtaining a minimum phase spectral factor is to obtain the equivalent innovation model using Kalman filtering ideas.

A state space model for the above system is given by

$$\begin{vmatrix} x_1 \\ x_2 \end{vmatrix}_{k+1} = \begin{vmatrix} 0.5 & 1.0 \\ -0.5 & 0 \end{vmatrix} \begin{vmatrix} x_1 \\ x_2 \end{vmatrix}_k + \begin{vmatrix} -1.5 & 1 \\ -0.5 & 0 \end{vmatrix} \begin{vmatrix} \varepsilon_1 \\ \varepsilon_2 \end{vmatrix}_k \tag{5.6.48}$$

$$\begin{vmatrix} y \\ u \end{vmatrix}_k = \begin{vmatrix} 1 & 0 \\ 0 & 1 \end{vmatrix} \begin{vmatrix} x_1 \\ x_2 \end{vmatrix}_k + \begin{vmatrix} 1 & 0 \\ 0 & 1 \end{vmatrix} \begin{vmatrix} \varepsilon_1 \\ \varepsilon_2 \end{vmatrix}_k \tag{5.6.49}$$

The above model has the general form

$$x_{k+1} = Ax_k + \omega_k \tag{5.6.50}$$

$$y_k = Cx_k + v_k \tag{5.6.51}$$

with

$$E\{\omega_k \omega_k^T\} = Q = \begin{vmatrix} 1.25 & 0.25 \\ 0.25 & 0.25 \end{vmatrix} \tag{5.6.52}$$

$$E\{v_k v_k^T\} = R = \begin{vmatrix} 1 & 1 \\ 1 & 2 \end{vmatrix} \tag{5.6.53}$$

$$E\{\omega_k v_k^T\} = S = \begin{vmatrix} -0.5 & 0.5 \\ -0.5 & -0.5 \end{vmatrix} \tag{5.6.54}$$

The equivalent innovations model can be determined using the approach outlined in Section 4.5.2. The stable minimum phase spectral factor is then given by

$$\tilde{K}(z) = C(zI - A)^{-1}L + I \tag{5.6.55}$$

where L is the Kalman gain satisfying

$$L = (APC^T + S)(CPC^T + R)^{-1} \tag{5.6.56}$$

where P is the positive semidefinite solution of the following steady state matrix Riccati equation

$$P = APA^T + Q - (APC^T + S)(CPC^T + R)^{-1}(APC^T + S)^T \tag{5.6.57}$$

For the above problem, it turns out that P and L are given by

$$P = \begin{vmatrix} 1.5 & 0 \\ 0 & 0 \end{vmatrix}, \quad L = \begin{vmatrix} 0 & 0.25 \\ -0.5 & 0 \end{vmatrix} \tag{5.6.58}$$

Hence substituting (5.6.58) into (5.6.55) yields

$$\tilde{K}(z) = \frac{1}{z^2 - 0.5z + 0.5} \begin{vmatrix} z^2 - 0.5z & 0.25z \\ -0.5z + 0.25 & z^2 - 0.5z + 0.375 \end{vmatrix} \tag{5.6.59}$$

with $\Phi_{yu} = \tilde{K}(z)\tilde{\Sigma}\tilde{K}^T(z^{-1})$ and

$$\tilde{\Sigma} = CPC^T + R = \begin{vmatrix} 2.5 & 1.00 \\ 1.00 & 2.00 \end{vmatrix} \tag{5.6.60}$$

The unique values of $\tilde{G}_1 - \tilde{G}_2$ corresponding to $\tilde{K}$ are

$$\tilde{G}_1 = \frac{0.25z}{z^2 - 0.5z + 0.375} \tag{5.6.61}$$

$$\tilde{G}_2 = \frac{z^2 - 0.5z}{z^2 - 0.5z + 0.375} \tag{5.6.62}$$

$$\tilde{G}_3 = \frac{-0.5}{z} \tag{5.6.63}$$

$$\tilde{G}_4 = 1 \quad \nabla \tag{5.6.64}$$

Comparing Eqs. (5.6.61)–(5.6.64) with Eqs. (5.6.42)–(5.6.45) shows that use of the minimum phase spectral factor when Σ is not block diagonal can give incorrect estimates for G_1–G_4. Of course $\tilde{G}_1$–$\tilde{G}_4$ plus $\tilde{\Sigma}$ are in the equivalence class of systems having the structure depicted in Fig. 5.6.1 and giving rise to the same value of Φ_{yu}. However, the model would lead to incorrect conclusions regarding the forward and reverse path models. This could be important in practice. For example, if one were interested in predicting the effect of changes in the feedback law, then it is important to have the correct forward path model.

Remark 5.6.1 In many practical situations, it is reasonable to assume that the disturbances in the forward and feedback paths are uncorrelated. In these cases, Theorem 5.6.1 shows that it is possible to obtain physical meaningful models from the stable minimum phase spectral factor of Φ_{yu}. This spectral factor can be determined from Φ_{yu} [239] or by identifying a model for the joint process (y, u) using the standard prediction error method [148, 238].

Remark 5.6.2 Theorem 5.6.1 does not depend upon A^* or E^* being stable. Thus there is no difficulty identifying open loop unstable systems provided the closed loop system is stable.

Remark 5.6.3 If it is known that G_2 and G_4 are minimum phase, then part (iii) of Theorem 5.6.1 shows that the stable minimum phase spectral factor yields the correct values of G_1–G_4 (see also [225]). However, the assumption that G_2 and G_4 are minimum phase is restrictive when Σ is not block diagonal and furthermore there appears to be no way of checking the minimum phase condition in practice.

Remark 5.6.4 It is not always true, that the maximum likelihood estimates of G_1 and G_2 can be obtained by analyzing data (y, u) as if these were open loop data with input u and output y. To see this, consider the joint distribution $p(Y_N, U_N|\theta)$ for the data $Y_N = \{y_1, y_2, \ldots, y_N\}$ and $U_N = \{u_1, u_2, \ldots, u_N\}$. The true maximum likelihood estimates for θ are obtained by maximizing $p(Y_N, U_N|\theta)$ with respect to θ. Using Bayes' rule we can express the likelihood function as

$$p(Y_N, U_N|\theta) = \prod_{t=1}^{N} p(y_t, u_t | Y_{t-1}, U_{t-1}; \theta)$$

$$= \prod_{t=1}^{N} p(y_t | Y_{t-1}, U_{t-1}; \theta) p(u_t | Y_{t-1}, U_{t-1}; \theta) \quad (5.6.65)$$

We recognize the term $\prod_{t=1}^{N} p(y_t | Y_{t-1}, U_t; \theta)$ as the usual open loop likelihood function with $\{y_t\}$ the output and $\{u_t\}$ the exogenous input. From Eq. (5.6.18) it will be clear that maximization of $p(Y_N, U_N|\theta)$ is not necessarily equivalent to maximizing $\prod_{t=1}^{N} p(y_t | Y_{t-1}, U_t; \theta)$ with respect to θ. However, the latter approach is valid if the data generating mechanism (i.e., $p(u_t | Y_{t-1}, U_{t-1}; \theta)$) is independent of θ. This will often be the case in practice. An example of an independent data-generating mechanism arises when Σ is block diagonal (i.e., $\Sigma_{12} = \Sigma_{21}^T = 0$ in Eq. (5.6.1)) and there are no parameters in common between (G_1, G_2, Σ_{11}) and (G_3, G_4, Σ_{22}). ∇

5.6.2   *Noise-Free feedback*   We make similar assumptions to those in Section 5.6.1 save that $G_4 \equiv 0$ and the set point sequence $\{s_t\}$ is not necessarily zero. We assume throughout this section that the feedback regulator G_3 is known. (This is reasonable since the feedback path is noise free by assumption.)

Remark 5.6.5   If $\{s_t\}$ is a full rank stochastic process independent of $\{\varepsilon_{1t}\}$ then Theorem 5.6.1 applies without modification. Thus we can obtain identifiability of G_1, G_2, G_3, and Σ_{11} even if $\{s_t\}$ is not measured.

If G_3 is parameterized independently of G_1, G_2, and Σ_{11}, then open loop identification can be applied to (u, y) without loss of accuracy (cf. Remark 5.6.4). ∇

Remark 5.6.6   For the case where the set point perturbations are zero, identifiability of G_1 and G_2 can be achieved by use of a sufficiently complex feedback regulator. This result is proved in [122] and [123] for single-input, single-output systems and in [226] for multiple-input, multiple-output systems. A sufficient condition for identifiability is that the minimum observability index for G_3 be greater than or equal to the maximum observability index for $[G_1 : G_2]$. For simplicity, we proof the single-input single-output version of the theorem.

Consider the following single-input, single-output model

$$A(z)y_t = B(z)u_t + C(z)\varepsilon_t \tag{5.6.66}$$

where $\{\varepsilon_t\}$, $\{u_t\}$, and $\{y_t\}$ are a white sequence, the input and output, respectively, and where A has no factors in common with both B and C. Also

$$A(z) = 1 + a_1 z^{-1} + \cdots + a_{n_1} z^{-n_1} \tag{5.6.67}$$

$$B(z) = b_1 z^{-1} + \cdots + b_{n_1} z^{-n_1} \tag{5.6.68}$$

$$C(z) = 1 + c_1 z^{-1} + \cdots + c_{n_1} z^{-n_1} \tag{5.6.69}$$

The input $\{u_t\}$ is assumed to be generated by the following feedback law:

$$E(z)u_t = F(z)y_t \tag{5.6.70}$$

where $E(z)$ and $F(z)$ are relatively prime and

$$E(z) = 1 + e_1 z^{-1} + \cdots + e_{n_2} z^{-n_2} \qquad (5.6.71)$$

$$F(z) = f_0 + f_1 z^{-1} + \cdots + f_{n_2} z^{-n_2} \qquad (5.6.72)$$

It is assumed that $(EA{-}FB)$ and (EC) are relatively prime (i.e., the feedback does not introduce pole zero cancellations.

Theorem 5.6.2   Under the conditions described above, a sufficient condition for identifiability of $A(z)$, $B(z)$, and $C(z)$ is that the order of the feedback law be greater than or equal to the order of the forward path, i.e., $n_2 \geq n_1$.

Proof   Substituting (5.6.70) into (5.6.66) yields the following closed loop model

$$P(z)y_t = Q(z)\varepsilon_t \qquad (5.6.73)$$

where

$$P(z) = E(z)A(z) - F(z)B(z) \qquad (5.6.74)$$

$$Q(z) = E(z)C(z) \qquad (5.6.75)$$

Standard algorithms (e.g., maximum likelihood) may be used to determine $P(z)$ and $Q(z)$.

Equating the first n_1 coefficients in (5.6.75) yields

$$\begin{bmatrix} 1 & 0 & \cdots & & 0 \\ e_1 & 1 & & & \\ \vdots & & \ddots & & \\ e_{n_2} & & & \ddots & \\ 0 & & e_{n_2} & \cdots & 1 \end{bmatrix} \begin{bmatrix} c_1 \\ \vdots \\ \vdots \\ \\ c_{n_1} \end{bmatrix} = \begin{bmatrix} q_1 \\ \vdots \\ \vdots \\ \\ q_{n_1} \end{bmatrix} \qquad (5.6.76)$$

Clearly, the matrix on the left-hand side of (5.6.76) is nonsingular and thus $C(z)$ can be obtained from $Q(z)$.

Similarly, equating coefficients in (5.6.74) yields

$$\begin{bmatrix} 1 & 0 & f_0 & & 0 \\ e_1 & \ddots & f_1 & \ddots & \\ \vdots & & \vdots & & \\ e_{n_2} & \cdot & 1 & f_{n_2} & \cdot f_0 \\ & \cdot & e_1 & \cdot & f_1 \\ & & \vdots & & \vdots \\ 0 & & e_{n_2} & 0 & f_{n_2} \end{bmatrix} \begin{bmatrix} a_1 \\ a_2 \\ \vdots \\ a_{n_1} \\ b_1 \\ \vdots \\ b_{n_1} \end{bmatrix} = \begin{bmatrix} p_1 - e_1 \\ \vdots \\ p_{n_2} - e_{n_2} \\ \vdots \\ p_{n_p} \end{bmatrix} \qquad (5.6.77)$$

where

$$n_p = n_1 + n_2 \qquad (5.6.78)$$

The matrix on the left-hand side of (5.6.77) is the Sylvester matrix [212] for the polynomials $E(z)$ and $F(z)$. Since $E(z)$ and $F(z)$ are coprime, this matrix has full column rank provided $n_1 \leq n_2$. Therefore a necessary and sufficient condition for identifiability of $A(z)$ and $B(z)$ is

$$n_1 \leq n_2 \qquad \nabla \qquad (5.6.79)$$

Remark 5.6.7 If G_3 contains a long delay (compared with the modes of the system), then the feedback signal contains a component which is approximately independent of the current output. Thus, it is not surprising in the light Remark 5.6.5 that identifiability can be achieved in this case. ∇

Remark 5.6.8 We stress that Theorem 5.6.2 is applicable only when the orders of the polynomials A, B, and C are known. Structural information, such as the system order cannot be determined from closed loop measurements alone. For example, the two systems illustrated in Fig. 5.6.2 give rise to the same external characteristics and are thus indistinguishable from input/output data (see also Problem 5.14).

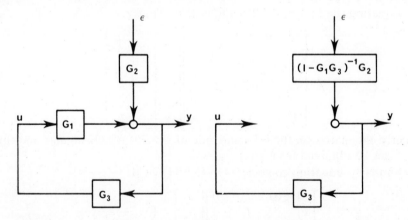

Figure 5.6.2. Two different systems having the same input/output characteristics.

Theorem 5.6.2 depends upon the fact that there is a known upper limit on the model order for the forward path. This requirement can be eliminated by switching between a number of different feedback regulators [124].

Theorem 5.6.3 A sufficient condition for identifiability (for all model orders) is that

$$\text{rank} \begin{vmatrix} I & I & \cdots & I \\ G_3^{(1)} & G_3^{(2)} & & G_3^{(l)} \end{vmatrix} = r + m \quad \text{(a.e. } z) \qquad 5.6.80$$

where r is number of system inputs, m the number of system outputs, and $G^{(1)}, G^{(2)}, \ldots, G^{(l)}$ are l different feedback regulators.

Proof Let $M^{(i)}$ denote the closed loop transfer function from $\{\varepsilon_t\}$ to $\{y_t\}$ when the ith feedback regulator is used, i.e., $G_3 = G_3^{(i)}$. Then $M^{(i)}$ can be determined from closed loop measurements and we have

$$M^{(i)} = [I - G_1 G_3^{(i)}]^{-1} G_2, \qquad i = 1, \ldots, l \qquad (5.6.81)$$

Assuming, as before that $G_2^{-1}(z)$ exists, then Eq. (5.6.81) can be rearranged to give

$$[M^{(i)}]^{-1} = G_2^{-1}[I - G_1 G_3^{(i)}], \qquad i = 1, \ldots, l \qquad (5.6.82)$$

or

$$[M^{(1)} \vdots M^{(2)} \vdots \cdots \vdots M^{(l)}] = [G_2^{-1} \vdots -G_2^{-1} G_1] \begin{vmatrix} I & \vdots & I & \vdots & & I \\ G_3^{(1)} & \vdots & G_3^{(2)} & \vdots & \cdots & G_3^{(l)} \end{vmatrix}$$
$$(5.6.83)$$

It is clear that, if the rank condition given in the statement of the theorem is satisfied, then Eq. (5.6.83) can be solved uniquely for G_2^{-1} and $-G_2^{-1} G_1$ and hence for G_2 and G_1. $\triangledown$

Remark 5.6.9 A necessary condition for (5.6.80) to hold is that the number of columns lr be not less than $r + m$. Hence a necessary condition is that the number of feedback regulators l satisfy

$$l \geq 1 + (m/r) \qquad (5.6.84)$$

For the case $m = r$ (same number of outputs as inputs), then $l = 2$ suffices and condition (5.6.80) becomes simply

$$\det[G_3^{(1)} - G_3^{(2)}] \neq 0 \qquad (5.6.85)$$

The above condition can be readily satisfied, e.g., by two diagonal "proportional" control laws:

$$G_3^{(1)} = \text{diag}[K_1^{(1)}, \ldots, K_r^{(1)}]$$
$$G_3^{(r)} = \text{diag}[K_1^{(2)}, \ldots, K_r^{(2)}]$$

where

$$K_i^{(1)} \neq K_i^{(2)}, \qquad i = 1, \ldots, r \qquad \triangledown \qquad (5.6.86)$$

5.7. CONCLUSIONS

This chapter has been concerned with the problem of estimating the parameters in models of dynamic systems. A large number of algorithms have appeared in the literature. We have attempted to present the key ideas without becoming too involved in the details of individual methods. We have tacitly assumed throughout this chapter that the model structure is known and thus the inference problem is simply one of estimating the values of certain free parameters.

So far we have studied the problem of making inferences from a given set of data. In the next chapter, we study the effect of different types of data on the achievable estimation accuracy.

PROBLEMS

5.1. Consider the model $y_t = ay_{t-1} + \xi_t$ where $\{\xi_t\}$ is an uncorrelated sequence of random variables which can take the value $+1$ or -1 each with probability 0.5.

Show that the least squares estimator for a is biased for the case where $N = 2$ and y_0 is known. Hint: Show that

$$\hat{a} = a + (y_0 \xi_1 + ay_0 \xi_2 + \xi_1 \xi_2)/(y_0^2 + a^2 y_0^2 + 2ay_0 \xi_1 + \xi_1^2)$$

Hence show that the bias is

$$\frac{-2ay_0^2}{(y_0^2 + a^2 y_0^2 + 1)^2 - 4a^2 y_0^2} \neq 0 \quad \text{in general}$$

5.2. Show that $C = A^{-1}xx^T(B^{-1} - A^{-1})$ is positive semidefinite where $A = B + xx^T$ and B is positive definite. Hint: First verify by multiplication that

$$(D - gh^T)^{-1} = \left| I + D^{-1} \frac{gh^T}{1 - h^T D^{-1}g} \right| D^{-1} \qquad (*)$$

Hence show that

$$C = (A^{-1}xx^T A^{-1}xx^T A^{-1})/(1 - x^T A^{-1}x)$$

Then show (using (*) or otherwise) that

$$x^T A^{-1}x = (x^T B^{-1}x)/(1 + x^T B^{-1}x) < 1$$

5.3. Subject to the same conditions as in Section 5.3 and for the case where $E(\xi_t^2) = \sigma^2$, show that $S_N = (1/N)(Y_N - X_N\hat{\theta}_N)^T(Y_N - X_N\hat{\theta}_N)$ is a weakly consistent estimator of σ^2 (i.e., $S_N \xrightarrow{\text{prob}} \sigma^2$). Hint: Use Theorem B.3.4.

5.4. Establish the weak consistence of the instrumental variable estimator given by (5.3.15) (use Theorem B.3.4).

5.5. Consider the system described in Example 5.3.1 of the text. Show that Z_N, defined as follows:

$$Z_N = \begin{bmatrix} u_0 & u_1 \\ \vdots & \vdots \\ u_{t-1} & u_t \\ \vdots & \vdots \\ u_{N-1} & u_N \end{bmatrix}$$

is a suitable instrumental variable matrix. Show that the corresponding instrumental variable estimator is

$$\begin{bmatrix} \bar{a}_N \\ \bar{b}_N \end{bmatrix} = \begin{bmatrix} \dfrac{1}{N}\sum_{t=1}^{N} u_{t-1}y_{t-1} & \dfrac{1}{N}\sum_{t=1}^{N} u_t u_{t-1} \\ \dfrac{1}{N}\sum_{t=1}^{N} u_t y_{t-1} & \dfrac{1}{N}\sum_{t=1}^{N} u_t^2 \end{bmatrix}^{-1} \begin{bmatrix} \dfrac{1}{N}\sum_{t=1}^{N} u_{t-1}y_t \\ \dfrac{1}{N}\sum_{t=1}^{N} u_t y_t \end{bmatrix}$$

Following the argument given in Example 5.3.1, demonstrate that

$$\begin{bmatrix} \bar{a}_N \\ \bar{b}_N \end{bmatrix} \xrightarrow{\text{prob}} \frac{1}{\Delta} \begin{bmatrix} R_{uu}(0)R_{uy}(-1) - R_{uu}(1)R_{uy}(0) \\ -R_{uy}(1)R_{uy}(-1) + R_{uy}^2(0) \end{bmatrix}$$

where $\Delta = R_{uy}(0)R_{uu}(0) - R_{uy}(1)R_{uu}(1)$. Hence, verify directly that

$$\begin{bmatrix} \bar{a}_N \\ \bar{b}_N \end{bmatrix} \xrightarrow{\text{prob}} \begin{bmatrix} a \\ b \end{bmatrix}$$

5.6. For the single-input, single-output system with Gaussian disturbances $A(z)y_t = B(z)u_t + \varepsilon_t$ where $\varepsilon_t \sim N(0, \Sigma)$

$$A(z) = 1 + a_1 z^{-1} + \cdots + a_n z^{-n}$$

$$B(z) = \quad b_1 z^{-1} + \cdots + b_m z^{-m}$$

show that the maximum likelihood estimator for $(\theta^T, \Sigma) = (a_1, \ldots, a_n, b_1, \ldots, b_m, \Sigma)$ is given by

$$\hat{\theta} = (X^TX)^{-1}X^TY \quad \text{and} \quad \hat{\Sigma} = (1/N)(Y - X\hat{\theta})^T(Y - X\hat{\theta})$$

where X has tth row

$$x_t^T = (-y_{t-1}, \ldots, -y_{t-n}, u_{t-1}, \ldots, u_{t-m})$$

and $Y^T = (y_1, \ldots, y_N)$. Hence show that the maximum likelihood estimator for this problem is the least squares estimator described in Section 5.2.

5.7. Show that for the multiple-input, multiple-output system with Gaussian disturbances

$$A(z)y_t = B(z)u_t + \varepsilon_t,$$

where $\varepsilon_t \sim N(0, \Sigma)$ and Σ is known leads to an optimization problem with cost function that is quadratic in the parameters. Hence show that the maximum likelihood estimator of the coefficients of the polynomial matrices A and B may be expressed in the least squares form

$$\hat{\theta} = (X^T X)^{-1} X^T Y$$

for some X and Y.

5.8. Establish Eq. (5.4.26) from (5.4.25). Further show that the expected value of the third term in (5.4.32) tends to zero as $N \to \infty$.

5.9. Verify the gradient expressions given in (5.4.28) and using matrix differentiation results given in Appendix E.

5.10. For the following state space model:

$$x_{k+1} = Ax_k + Bu_k + Ke_k, \qquad y_k = Cx_k + Du_k + e_k$$

where $e_k \sim N(0, \Sigma)$ for $k = 1, \ldots, N$.

(a) Show that the likelihood function is

$$p(Y_N \mid U_N, \theta) = ((2\pi)^m \det \Sigma)^{-N/2} \exp\left\{ -\tfrac{1}{2} \sum_{k=1}^{N} w_k^T \Sigma^{-1} w_k \right\}$$

where

$$w_k = y_k - C\bar{x}_k - Du_k \quad \text{and} \quad \bar{x}_{k+1} = (A - KC)\bar{x}_k + (B - KD)u_k + Ky_k$$

(b) Derive an expression for the gradient of the likelihood function with respect to θ (the entries in A, B, C, D, and K) and Σ.

5.11. Establish Eqs. (5.6.12)–(5.6.15).

5.12. Consider the following first-order system

$$y_k = -ay_{k-1} + bu_{k-1} + \varepsilon_k$$

with feedback control $u_k = gy_k$. Show that a and b are not identifiable.

5.13. Consider a single-input, single-output linear system

$$y_k = -a_1 y_{k-1} - \cdots - a_n y_{k-n} + b_1 u_{k-1} + \cdots + b_n u_{k-n} + \eta_k$$

$$\eta_k = \varepsilon_k + c_1 \varepsilon_{k-1} + \cdots + c_n \varepsilon_{k-n}, \qquad \varepsilon_k \text{ white}$$

Let $\theta = (a_1, \ldots, a_n, b_1, \ldots, b_n, c_1, \ldots, c_n)$ be a parameter vector. Show that θ is not identifiable if $\{u_k\}$ is generated by a minimum variance feedback controller [14].

5.14. Show that the following two feedback systems give rise to the same external characteristics

$$y = G_1 u + G_2 \varepsilon, \qquad u = G_3 y$$

System (a) $G_1 = 0.1Z^{-1}/(1 + 0.5Z^{-1})$, $G_2 = 1$, and $G_3 = 1$.
System (b) $G_1 = 0$, $G_2 = (1 + 0.5Z^{-1})/(1 + 0.4Z^{-1})$, and $G_3 = 1$.

CHAPTER

6

Experiment Design

6.1. INTRODUCTION

The previous chapters of this book have been primarily concerned with the problem of extracting information from a given set of data. However, in most situations there are a number of variables which can be adjusted, subject to certain constraints, so that the information provided by the experiment is maximized [151–157]. We devote this chapter to a study of the design of the experimental conditions so that the experiment is maximally informative.

For dynamic systems, experiment design includes choice of input and measurement ports, test signals, sampling instants, and presampling filters. We shall find that each of these variables has a significant bearing upon the information provided by an experiment. The effects of these variables are, in general, closely interrelated and a joint design is required. However, we shall initially consider the variables separately to give insight into their individual effects upon the information content of the data. Later we study the joint design problem.

Any experimental design must take account of the constraints on the allowable experimental conditions. In fact the real purpose of experiment design is to maximize the information content of the data within the limits imposed by the given constraints.

Typical constraints that might be met in practice are

(1)   amplitude constraints on inputs, outputs or internal variables;
(2)   power constraints on inputs, outputs, or internal variables;
(3)   total time available for the experiment;
(4)   total number of samples that can be taken or analyzed;
(5)   maximum sampling rate;
(6)   availability of transducers and filters; and
(7)   availability of hardware and software for analysis.

Which of the above constraints (or others) are important in a particular experiment will depend upon the situation. In some cases a number of constraints may be relevant and it may not be obvious which constraints should be taken into account during the design of an experiment. A useful approach, leading to simplified designs, is to work with a subset of the constraints and subsequently to check for violation of other constraints.

We begin in Section 6.2 by describing measures of the information provided by an experiment. We then discuss the input design problem with both input and output power constraints. Finally we look at the broader problem of coupled design of input, presampling, filters, and sampling times.

6.2.   DESIGN CRITERIA

To form a basis for the comparison of different experiments, a measure of the "goodness" of an experiment is required. A logical approach is to choose a measure related to the expected accuracy of the parameter estimates to be obtained from the data collected. Clearly the parameter accuracy is a function of both the experimental conditions and the form of the estimator. However, it is sensible to assume that the estimator used is efficient in the sense that the parameter covariance matrix achieves the Cramer–Rao lower bound, i.e.,

$$\text{cov } \hat{\theta} = M^{-1} \tag{6.2.1}$$

where M is Fisher's information matrix given by

$$M = E_{Y|\theta}\left\{\left(\frac{\partial \log p(Y|\theta)}{\partial \theta}\right)^{\mathsf{T}}\left(\frac{\partial \log p(Y|\theta)}{\partial \theta}\right)\right\} \tag{6.2.2}$$

(see Theorem 1.3.1).

A suitable measure of goodness can therefore be defined as

$$J = E_\theta \phi(M) \tag{6.2.3}$$

where ϕ is a suitably chosen scalar function.

The expectation over the prior distribution of θ was introduced in (6.2.3) because of the dependence (in general) of M on the unknown parameter values θ. The dependence on the prior information is natural. Our ability to design a good experiment should depend upon our prior knowledge regarding the nature of the data generating mechanism. In fact, in the light of the unavoidable dependence on prior information, it could be argued that a Bayesian approach is called for. Furthermore, the design criterion J could be a Bayesian risk function that would reflect the ultimate use to which the data from the experiment will be put (see discussion in Chapter 1). For example, if we measure the ultimate model performance by a scalar function $s(\theta, \hat{\theta})$ of the "true state of nature" θ and the estimated parameters $\hat{\theta}$, then prior to the experiment the expected performance is

$$J' = E_{\theta, Y}[s(\theta, \hat{\theta}(Y))] \tag{6.2.4}$$

In principle, experiment design may be carried out by optimizing J' as given in (6.2.4) with respect to the allowable experimental conditions. An example of this approach has been discussed in Problem 1.9. However, for the general class of problems that we shall be concerned with, this approach is more complex than is usually justifiable (see [149]). By making suitable simplifying assumptions we can heuristically relate the above criterion to a criterion of the form of (6.2.3):

$$
\begin{aligned}
J' &= E_{\theta, Y}[s(\theta, \hat{\theta}(Y))] = E_\theta E_{Y|\theta}[s(\theta, \hat{\theta}(Y))] \\
&\simeq E_\theta E_{Y|\theta}[s(\theta, \theta) + (\partial s/\partial \hat{\theta})(\hat{\theta} - \theta) + \tfrac{1}{2}(\hat{\theta} - \theta)^{\mathsf{T}}(\partial^2 s/\partial \hat{\theta}^2)(\hat{\theta} - \theta)] \\
&= E_\theta[s(\theta, \theta) + \tfrac{1}{2}\operatorname{trace}(\partial^2 s/\partial \hat{\theta}^2)M^{-1}]
\end{aligned}
\tag{6.2.5}
$$

where we have assumed that $\hat{\theta}$ is an efficient unbiased estimator.

Optimization of (6.2.5) is thus equivalent to optimizing

$$J'' = E_\theta[\operatorname{trace} WM^{-1}] \tag{6.2.6}$$

where $W = [\partial^2 s/\partial \hat{\theta}^2]$. J'' given in (6.2.6) is of the form of J given in (6.2.3) with $\phi(M)$ equal to the weighted trace of M^{-1}.

In practice, useful designs can be obtained simply by evaluating $\phi(M)$ at a representative parameter value, say the prior mean. The sensitivity of the design to other parameter values may then be checked.

Another criterion which is commonly used for experiment design is

$$J = -\log \det M \tag{6.2.7}$$

This criterion is related to the volume of highest probability density region for the parameters [43]. An interesting property of the determinant criterion is that it is independent of the scaling of the parameters (see Problem 6.1).

Other experiment design criteria are possible, for example, Lindley's measure of the average information provided by an experiment (see Section 1.7). However, the choice of criteria is often not critical since it is usually the case that a good experiment according to one criterion will be deemed good by other criteria. Naturally this will only be true for sensibly chosen criteria. For example, some criteria, such as trace M, are to be viewed with suspicion from a pragmatic viewpoint since they can lead to the design of experiments in which the parameters are unidentifiable (see Problem 6.2).

The above discussion has concentrated on the problem of experiment design for parameter estimation. However, we can also design experiments for maximizing the effectiveness of tests for model structure discrimination. For example, Wald in 1943 [150], proposed the maximization of the determinant of the information matrix as a means of maximizing the local power of the F-ratio test. If the structure test can be posed as a test for zero parameters, then a suitable criterion for experiment design is to maximize the ratio of the determinants of the information matrices for the two competing models. Similar approaches have been followed in [169] and [188]. An advantage of the use of the determinant ratio criterion is that experiment design methods for parameter estimation based on minimization of (6.2.7) can be applied, mutatis mutandis, to the structure discrimination case. This point is taken up in a later section.

In deriving expressions for the information matrix we shall usually make the assumption that the noise has a Gaussian distribution. However, as pointed out in Remark 5.5.1, the inverse of the information matrix based on a Gaussian assumption is equal to the asymptotic covariance of the prediction error estimator (using the $J_2(\theta, N)$ criterion) whether or not the noise is actually Gaussian. This means that the Gaussian assumption is not very restrictive and there are no serious consequences of the noise being non-Gaussian.

In our subsequent analysis we shall use the determinant of the information matrix derived under a Gaussian assumption as the design criterion without further comment. However, the general points made above should be borne in mind.

6.3.   TIME DOMAIN DESIGN OF INPUT SIGNALS

In this section we consider input design for linear systems of the form (cf. discussion in Section 4.4).

$$y_t = G_1(z)u_t + G_2(z)\varepsilon_t \qquad (6.3.1)$$

where $\{u_t\}$ and $\{y_t\}$ are the input and output sequences, respectively, and $\{\varepsilon_t\}$ is a sequence of i.i.d. Gaussian random variables having covariance Σ. G_1 and G_2 are transfer functions.

6.3.1 *Moving average model* [159]   To gain insight into experiment design, we begin by considering a special case of the model of Eq. (6.3.1), viz., a single-input, single-output moving average model with white output noise, i.e.,

$$G_1(z) = b_1 z^{-1} + b_2 z^{-2} + \cdots + b_n z^{-n} \tag{6.3.2}$$

$$G_2(z) = 1 \tag{6.3.3}$$

Models of the form of (6.3.2) and (6.3.3) are frequently used in practical applications of identification, e.g., in equalizers for digital communication channels [201].

The parameter vector is

$$\beta^T = (b_1, b_2, \ldots, b_n, \Sigma) \tag{6.3.4}$$

$$= (\theta^T, \Sigma) \tag{6.3.5}$$

For this model, we have seen previously (see Eq. (2.5.11) and Problem 2.9) that Fisher's information matrix is given by

$$M = \begin{bmatrix} (X^T X)(1/\Sigma) & 0 \\ 0 & N/2\Sigma^2 \end{bmatrix} \tag{6.3.6}$$

where N is the number of data points and

$$X = \begin{bmatrix} u_0 & u_{-1} & \cdots & u_{1-n} \\ u_1 & u_0 & & u_{2-n} \\ \vdots & & & \vdots \\ u_{N-1} & u_{N-2} & \cdots & u_{N-n} \end{bmatrix} \tag{6.3.7}$$

From Eq. (6.3.6) the average information matrix defined by

$$\overline{M} = M/N \tag{6.3.8}$$

is

$$\overline{M} = \begin{bmatrix} \hat{\Gamma}/\Sigma & 0 \\ 0 & 1/2\Sigma^2 \end{bmatrix} \tag{6.3.9}$$

where

$$\hat{\Gamma} = \frac{1}{N} \sum_{t=1}^{N} \begin{bmatrix} u_{t-1}^2 & \cdots & u_{t-1}u_{t-n} \\ \vdots & & \vdots \\ u_{t-1}u_{t-n} & \cdots & u_{t-n}^2 \end{bmatrix} \tag{6.3.10}$$

Now, the experimental design problem is to choose $\{u_t: t = 1 - n, \ldots, N - 1\}$ so that the design criterion is optimized. Here we shall employ the standard design criterion

$$J = -\log \det \overline{M} \tag{6.3.11}$$

As mentioned in Section 6.1, the nature of the constraints must be specified. We consider the following input power constraint

$$\frac{1}{N} \sum_{t=1}^{N} u_{t-i}^2 = 1, \qquad i = 1, \ldots, n \tag{6.3.12}$$

Result 6.3.1   Subject to the constraint (6.3.12), the criterion $J = -\log \det \overline{M}$ is minimized when the off-diagonal elements of $\hat{\Gamma}$ are zero.

Proof   From Eq. (6.3.9)

$$-\log \det \overline{M} = -\log \det \hat{\Gamma} + \log 2\Sigma^{n+2} \tag{6.3.13}$$

The constraint implies that the diagonal elements of $\hat{\Gamma}$ are equal to 1. Applying Result E.2.3 establishes the result.   ▽

The only remaining problem is to find $\{u_t: t = 1 - n, \ldots, N - 1\}$ to yield

$$\frac{1}{N} \sum_{t=1}^{N} u_{t-i} u_{t-j} = \begin{cases} 1, & i = j \\ 0, & i \neq j \end{cases} \tag{6.3.14}$$

For large N, (6.3.14) can be achieved by use of a white noise sequence. For finite N, pseudorandom binary sequences can be found which very nearly satisfy (6.3.14). Furthermore, for certain values of N, there exist binary sequences which have the desired exact property [202]. The loss in efficiency by use of a special class of pseudorandom binary sequences, known as m-sequences [203], is discussed in problem (6.3). A further point worth noting is that, if the constraint is on the input amplitude, say $(-1 \leq u_t \leq 1, \ ^-t)$, then an uncorrelated binary sequence maximizes $J = -\log \det M$. This follows immediately from the above discussion (see Problem 6.4).

Although the above results were obtained for the criterion $J = -\log \det M$, the results are in fact stronger as we now show.

Result 6.3.2   Subject to the constraint (6.3.12), the solution $\hat{\Gamma} = I$, (i) simultaneously minimizes all the diagonal elements of $(\overline{M})^{-1}$. Furthermore, (ii) the design minimizes the maximum eigenvalue of $(\overline{M})^{-1}$.

Proof (i) We partition $\overline{M}$ as follows:

$$\overline{M} = \begin{bmatrix} \overline{M}_{11} & \overline{M}_{21}^{\mathsf{T}} & 0 \\ \overline{M}_{21} & \overline{M}_{22} & 0 \\ 0 & 0 & 1/2\Sigma^2 \end{bmatrix} \tag{6.3.15}$$

where the scalar M_{11} corresponds to b_1.

The $(1, 1)$ element of $\overline{M}^{-1}$ is given by

$$\operatorname{cov} \hat{b}_1 = [\overline{M}_{11} - \overline{M}_{21}^{\mathsf{T}}\overline{M}_{22}^{-1}\overline{M}_{21}]^{-1}$$

$$\geq \overline{M}_{11}^{-1} \qquad \text{with equality iff} \quad M_{21} \equiv 0 \tag{6.3.16}$$

Now $\overline{M}_{11}^{-1}$ is a constant since $\overline{M}_{11}$ is fixed. Thus $\operatorname{cov} \hat{b}_1$ is minimized when $M_{21} \equiv 0$, i.e., when $\hat{\Gamma}$ is diagonal. Similar arguments apply for $\operatorname{cov} \hat{b}_i$, $i = 2, \ldots, n$.

(ii) Follows immediately, since (i) gives $\operatorname{trace}[\overline{M}]^{-1} \geq n$ with equality for the optimal design but $n\lambda_{\max}[\overline{M}]^{-1} \geq \operatorname{trace}[\overline{M}]^{-1}$. Therefore

$$\lambda_{\max}[\overline{M}]^{-1} \geq 1$$

with equality for the optimal design. $\triangledown$

6.3.2 *General model* We return to the general model of Eq. (6.3.1) but for simplicity restrict attention to a single-input, single-output case

$$y_t = G_1(z)u_t + G_2(z)\varepsilon_t \tag{6.3.17}$$

with $G_2(\infty) = 1$.

We recall that the log likelihood function is given by (cf. Section 5.4.3).

$$\log p(Y|\beta) = -\frac{N}{2}\log 2\pi - \frac{N}{2}\log \Sigma - \frac{1}{2\Sigma}\sum_{t=1}^{N}\omega_t^2 \tag{6.3.18}$$

where β is the vector of parameters in G_1, G_2, and Σ and $\{\omega_k\}$ is the residual sequence given by

$$\omega_t = G_2^{-1}(z)[y_t - G_1(z)u_t] \tag{6.3.19}$$

Fisher's information matrix is given by

$$M = E_{Y|\beta}\left[\left(\frac{\partial \log p(Y|\beta)}{\partial \beta}\right)^{\mathsf{T}}\left(\frac{\partial \log p(Y|\beta)}{\partial \beta}\right) \right] \tag{6.3.20}$$

$$\frac{\partial \log p(Y|\beta)}{\partial \beta} = -\frac{1}{\Sigma}\sum_{t=1}^{N}\omega_t\frac{\partial \omega_t}{\partial \beta} - \frac{1}{2\Sigma}\frac{\partial \Sigma}{\partial \beta}\left[N - \frac{1}{\Sigma}\sum_{t=1}^{N}\omega_t^2\right] \tag{6.3.21}$$

where

$$\partial\omega_t/\partial\beta = -G_2^{-1}(z)[(\partial G_2(z)/\partial\beta)\omega_t + (\partial G_1(z)/\partial\beta)u_t] \tag{6.3.22}$$

(see Problem 6.5). Substituting (6.3.22) and (6.3.21) into (6.3.20) yields

$$M = E_{Y|\beta}\left\{\left[-\frac{1}{\Sigma}\sum_{t=1}^{N}\omega_t\frac{\partial\omega_t}{\partial\beta} - \frac{1}{2\Sigma}\frac{\partial\Sigma}{\partial\beta}\left(N - \frac{1}{\Sigma}\sum_{t=1}^{N}\omega_t^2\right)\right]^{\mathrm{T}}\right.$$
$$\left.\times\left[-\frac{1}{\Sigma}\sum_{s=1}^{N}\omega_s\frac{\partial\omega_s}{\partial\beta} - \frac{1}{2\Sigma}\frac{\partial\Sigma}{\partial\beta}\left(N - \frac{1}{\Sigma}\sum_{s=1}^{N}\omega_s^2\right)\right]\right\}$$

Now we note that $E_{Y|\beta}\,\omega_t = 0$, $E_{Y|\beta}\,\omega_t\,\omega_s = \Sigma$ for $t = s$, and $E_{Y|\beta}\,\omega_t\,\omega_s = 0$ for $t \neq s$.

Using the fact that $\partial\omega_t/\partial\beta$ depends upon past values of ω_t, we can readily establish (see Problem 6.7) that

$$M = E_{Y|\beta}\left\{\frac{1}{\Sigma}\sum_{t=1}^{N}\left(\frac{\partial\omega_t}{\partial\beta}\right)^{\mathrm{T}}\left(\frac{\partial\omega_t}{\partial\beta}\right)\right\} + \frac{N}{2\Sigma^2}\left(\frac{\partial\Sigma}{\partial\beta}\right)^{\mathrm{T}}\left(\frac{\partial\Sigma}{\partial\beta}\right) \qquad (6.3.23)$$

Using superposition in (6.3.22) yields

$$\partial\omega_t/\partial\beta = (\partial\overline{\omega}_t/\partial\beta) + (\partial\tilde{\omega}_t/\partial\beta) \qquad (6.3.24)$$

where

$$\partial\overline{\omega}_t/\partial\beta = -G_2^{-1}(z)(\partial G_1(z)/\partial\beta)u_t \qquad (6.3.25)$$

$$\partial\tilde{\omega}_t/\partial\beta = -G_2^{-1}(z)(\partial G_2(z)/\partial\beta)\omega_t \qquad (6.3.26)$$

We assume, for the moment, that $\{u_t\}$ and $\{\varepsilon_t\}$ are independent. We treat the more general case where $\{u_t\}$ can be generated partially or wholly by feedback in a later section. Substituting (6.3.25) and (6.3.26) into (6.3.23) yields

$$M = \frac{1}{\Sigma}\sum_{t=1}^{N}\left(\frac{\partial\overline{\omega}_t}{\partial\beta}\right)^{\mathrm{T}}\left(\frac{\partial\overline{\omega}_t}{\partial\beta}\right) + E_{Y|\beta}\left[\frac{1}{\Sigma}\sum_{t=1}^{N}\left(\frac{\partial\tilde{\omega}_t}{\partial\beta}\right)^{\mathrm{T}}\left(\frac{\partial\tilde{\omega}_t}{\partial\beta}\right)\right] + \frac{N}{2\Sigma^2}\left(\frac{\partial\Sigma}{\partial\beta}\right)^{\mathrm{T}}\left(\frac{\partial\Sigma}{\partial\beta}\right)$$
$$(6.3.27)$$

From Eq. (C.1.2),

$$E_{Y|\beta}\left[\frac{1}{\Sigma}\sum_{t=1}^{N}\left(\frac{\partial\tilde{\omega}_t}{\partial\beta}\right)^{\mathrm{T}}\left(\frac{\partial\tilde{\omega}_t}{\partial\beta}\right)\right] = \frac{N}{2\pi\Sigma}\int_{-\pi}^{\pi}\Phi(e^{j\omega})\,d\omega \qquad (6.3.28)$$

where $\Phi(e^{j\omega})$ is given via Eq. (C.1.8) as

$$\Phi(e^{j\omega}) = \left[\frac{\partial G_2(e^{j\omega})}{\partial\beta}\right]^{\mathrm{T}}G_2^{-1}(e^{j\omega})\Sigma G_2^{-1}(e^{-j\omega})\left[\frac{\partial G_2(e^{-j\omega})}{\partial\beta}\right] \qquad (6.3.29)$$

Finally, Eq. (6.3.27) is of the form

$$M = \frac{1}{\Sigma}\sum_{t=1}^{N}(\partial\overline{\omega}_t/\partial\beta)^{\mathrm{T}}(\partial\overline{\omega}_t/\partial\beta) + M_c \qquad (6.3.30)$$

where

$$\partial \bar{\omega}_t / \partial \beta = -G_2^{-1}(z)(\partial G_1(z)/\partial \beta)u_t \qquad (6.3.31)$$

and M_c is a constant matrix which does not depend upon the choice of the input $\{u_t\}$.

The experiment design problem can now be stated as: choose $\{u_t\}$ subject to the given constraints so that a scalar function of M is optimized where M is related to $\{u_t\}$ via (6.3.30) and the dynamic Eq. (6.3.31). This is a standard nonlinear optimal control problem and can be solved, at least in principle, by the usual optimization procedures. For example, if a gradient algorithm is used, then the gradient of the design criterion with respect to the input can be readily computed using the usual adjoint system approach (see [164] and [165] for details). For example, if the design criterion is

$$J = -\log \det M \qquad (6.3.32)$$

then the gradient $\partial J/\partial u_t$ is given by

$$\partial J/\partial u_t = \lambda_t \qquad (6.3.33)$$

where $\{\lambda_t\}$ is calculated in reverse time from

$$\lambda_t = -G_2^{-1}(z^{-1})(\partial G_1(z^{-1})/\partial \beta)M^{-1}(\partial \omega_t/\partial \beta) \qquad (6.3.34)$$

with $\partial \omega_t / \partial \beta = 0, t > N; \lambda_t = 0, t > N$.

The above approach can be readily extended to nonlinear systems with amplitude constraints on inputs, outputs, or internal variables. The details are problem dependent and thus we shall not pursue the topic further. However, the interested reader is referred to [163]–[165], [170], [171], [174]–[176], [180]–[187] for typical design examples.

Example 6.3.1   Consider the following single-input, single-output system [51]:

$$y_t = \frac{b_1 z^{-1} + b_2 z^{-2}}{1 + a_1 z^{-1} + a_2 z^{-1}} u_t + \frac{1}{1 + c_1 z^{-1}} \varepsilon_t$$

where $a_1 = -1.3$, $a_2 = 0.6$, $b_1 = 0.15$, $b_2 = 0.15$, and $c_1 = -0.95$.

The input is constrained to lie in the interval $[-\alpha, \alpha]$ for some α. The design criterion $J = -\log \det M$ is optimized using the adjoint state approach outlined above. The optimal input is shown in Fig. 6.3.1b. For comparison, a pseudorandom binary noise test signal is shown in Fig. 6.3.1a. The improvement in the parameter variances achieved by use of the optimal signal is approximately 1.4 to 1 compared with the use of the pseudorandom binary signal. ∇

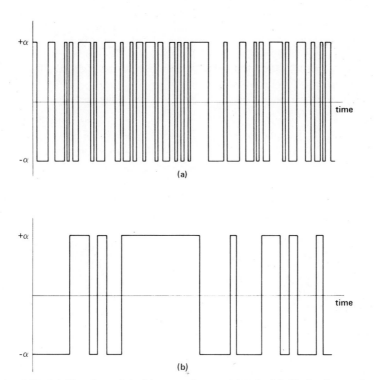

Figure 6.3.1. (a) Pseudorandom binary noise test signal. (b) Optimal test signal for Example 6.3.1.

The time domain approach outlined above is applicable to a wide class of problems, but unfortunately offers little insight. We are thus motivated in the next section, to make further simplifying assumptions.

6.4. FREQUENCY DOMAIN DESIGN OF INPUT SIGNALS

6.4.1 *Input power constraints* We refer back to Eq. (6.3.30) for the information matrix, i.e.,

$$M = \frac{1}{\Sigma} \sum_{t=1}^{N} \left(\frac{\partial \overline{\omega}_t}{\partial \beta}\right)^{\mathrm{T}} \left(\frac{\partial \overline{\omega}_t}{\partial \beta}\right) + M_c \tag{6.4.1}$$

where

$$\partial \overline{\omega}_t / \partial \beta = -G_2^{-1}(z)(\partial G_1(z)/\partial \beta)u_t \tag{6.4.2}$$

We now make the following simplifying assumptions:

(1)   The experiment time (i.e., N) is large.
(2)   The input $\{u_t\}$ is restricted to the class admitting a spectral representation.
(3)   The constraint is taken to be the allowable input power.

Since N is large, it is more convenient to work with the average information matrix defined as follows:

$$\bar{M} = \lim_{N \to \infty} \frac{1}{N} M \qquad (6.4.3)$$

If the spectral distribution function of the input is $F(\omega)$, $\omega \in [-\pi, \pi]$ (see Section C.3), then substituting (6.4.1) and (6.4.2) into (6.4.3) yields

$$\bar{M} = \frac{1}{2\pi\Sigma} \int_{-\pi}^{\pi} \left[\frac{\partial G_1(e^{j\omega})}{\partial \beta}\right]^T G_2^{-1}(e^{j\omega}) G_2^{-1}(e^{-j\omega}) \left[\frac{\partial G_1(e^{-j\omega})}{\partial \beta}\right] dF(\omega) + \bar{M}_c$$

$$(6.4.4)$$

where, from (6.3.27) to (6.3.29)

$$\bar{M}_c = \frac{1}{2\pi} \int_{-\pi}^{\pi} \left[\frac{\partial G_2(e^{j\omega})}{\partial \beta}\right]^T G_2^{-1}(e^{j\omega}) G_2^{-1}(e^{-j\omega}) \left[\frac{\partial G_2(e^{-j\omega})}{\partial \beta}\right] d\omega$$

$$+ \frac{1}{2\Sigma^2} \left(\frac{\partial \Sigma}{\partial \beta}\right)^T \left(\frac{\partial \Sigma}{\partial \beta}\right) \qquad (6.4.5)$$

Equation (6.4.4) can be written more succinctly as

$$\bar{M} = \frac{1}{\pi} \int_0^{\pi} \tilde{M}(\omega) \, d\xi(\omega) + \bar{M}_c \qquad (6.4.6)$$

where

$$d\xi(\omega) = \tfrac{1}{2} dF(\omega), \qquad \omega = 0 \quad \text{or} \quad \pi$$
$$= dF(\omega), \qquad \omega \in (0, \pi) \qquad (6.4.7)$$

and

$$\tilde{M}(\omega) = \text{real part} \left\{\frac{1}{\Sigma} \left[\frac{\partial G_1(e^{j\omega})}{\partial \beta}\right]^T G_2^{-1}(e^{j\omega}) G_2^{-1}(e^{-j\omega}) \left[\frac{\partial G_1(e^{-j\omega})}{\partial \beta}\right]\right\} \qquad (6.4.8)$$

We note from Eq. (C.3.1) that the input power is given by

$$P_u = \frac{1}{\pi} \int_0^{\pi} d\xi(\omega) \qquad (6.4.9)$$

Hence, the input power constraint can be expressed as

$$\frac{1}{\pi} \int_0^\pi d\xi(\omega) = 1 \tag{6.4.10}$$

We introduce the following notation:

(i) $\mathcal{K}$ is the set of all measures $\xi(\omega)$ satisfying (6.4.10). We shall call an input design belonging to $\mathcal{K}$ an *input power constrained design*.

(ii) We denote by $\overline{M}(\xi)$ the average information matrix corresponding to the measure $\xi(\omega)$.

(iii) $\mathcal{M}$ is the set of all average information matrices corresponding to input power constrained designs, i.e., $\mathcal{M} = \{\overline{M}(\xi) : \xi(\omega) \in \mathcal{K}\}$.

(iv) $\mathcal{F}$ is the set of all $\xi(\omega) \in \mathcal{K}$ which are piecewise constant with at most one discontinuity on $[0, \pi]$, i.e., $\xi(\omega) \in \mathcal{F}$ implies that the input is a single sinusoid with power $= 1$ and frequency given by the point of discontinuity (zero frequency if no discontinuity) (see [16, Chapters 7 and 8]).

Remark 6.4.1 If $\xi(\omega) \in \mathcal{F}$ with discontinuity at $\omega = \omega_1$, then $\overline{M}(\xi(\omega_1)) = \tilde{M}(\omega_1) + \overline{M}_c$, i.e., $\tilde{M}(\omega_1) + \overline{M}_c$ is the average information matrix corresponding to an input comprising a single sinusoid of frequency ω_1. ∇

Theorem 6.4.1 The set $\mathcal{M}$ is the convex hull of the set of all average information matrices corresponding to single frequency designs (i.e., $\xi(\omega) \in \mathcal{F}$).

Proof Equation (6.4.6) may be rewritten as

$$\overline{M} = \frac{1}{\pi} \int_0^\pi (\tilde{M}(\omega) + \overline{M}_c) \, d\xi(\omega) \tag{6.4.11}$$

since $\xi(\omega)$ is power constrained, i.e.,

$$\frac{1}{\pi} \int_0^\pi d\xi(\omega) = 1 \tag{6.4.12}$$

(6.4.11) and (6.4.12) define a convex hull [210] of the set of average information matrices $\{\overline{M}(\omega) = \tilde{M}(\omega) + \overline{M}_c, \omega \in [0, \pi]\}$. With Remark 6.4.1 in mind, we have thus established the theorem. ∇

We illustrate the above theorem with the following simple example.

Example 6.4.1 Consider the moving average model discussed in Section 6.3.1. For simplicity we assume that $G_1(z)$ is third order. This yields

$$G_1(z) = b_1 z^{-1} + b_2 z^{-2} + b_3 z^{-3}$$
$$G_2(z) = 1, \qquad \beta^\mathsf{T} = (b_1, b_2, b_3, \Sigma)$$

From Eq. (6.4.5),

$$\overline{M}_c = \begin{bmatrix} 0 & 0 & 0 & 0 \\ 0 & 0 & 0 & 0 \\ 0 & 0 & 0 & 0 \\ 0 & 0 & 0 & \frac{1}{2}\Sigma^{-2} \end{bmatrix} \qquad (6.4.13)$$

From Eq. (6.4.8),

$$\tilde{M}(\omega) = \frac{1}{\Sigma} \begin{bmatrix} 1 & \cos\omega & \cos 2\omega & 0 \\ \cos\omega & 1 & \cos\omega & 0 \\ \cos 2\omega & \cos\omega & 1 & 0 \\ 0 & 0 & 0 & 0 \end{bmatrix} \qquad (6.4.14)$$

Inspection of Eqs. (6.4.13) and (6.4.14) indicates that $\overline{M}$ corresponding to an input power constrained design is of the form

$$\overline{M} = \frac{1}{\Sigma} \begin{bmatrix} 1 & x_1 & x_2 & 0 \\ x_1 & 1 & x_1 & 0 \\ x_2 & x_1 & 1 & 0 \\ 0 & 0 & 0 & \frac{1}{2}\Sigma^{-1} \end{bmatrix} \qquad (6.4.15)$$

where

$$x_1(\xi) = \frac{1}{\pi} \int_0^\pi x_1(\omega)\, d\xi(\omega) \qquad (6.4.16)$$

$$x_2(\xi) = \frac{1}{\pi} \int_0^\pi x_2(\omega)\, d\xi(\omega) \qquad (6.4.17)$$

where $x_1(\omega) = \cos\omega$, $x_2(\omega) = \cos(2\omega)$.

Now it follows from Theorem 6.4.1 that the set of all average information matrices corresponding to input power constrained designs is the convex hull of information matrices of the form

$$\overline{M}(\omega) = \frac{1}{\Sigma} \begin{bmatrix} 1 & x_1(\omega) & x_2(\omega) & 0 \\ x_1(\omega) & 1 & x_1(\omega) & 0 \\ x_2(\omega) & x_1(\omega) & 1 & 0 \\ 0 & 0 & 0 & \frac{1}{2}\Sigma^{-1} \end{bmatrix} \qquad (6.4.18)$$

The convex hull is shown diagrammatically in Fig. 6.4.1.
It follows from Eq. (6.4.15) that

$$\det \overline{M} = \frac{1}{2}\Sigma^{-4}[1 + 2x_1^2(x_2 - 1) - x_2^2] \qquad (6.4.19)$$

It can be seen from Fig. 6.4.1 that $x_2 \leq 1$. Hence from (6.4.19), det $\overline{M}$ achieves its maximum value when $x_1 = 0$, $x_2 = 0$. One way that this result can be realized is to use an input having equal power at all frequencies

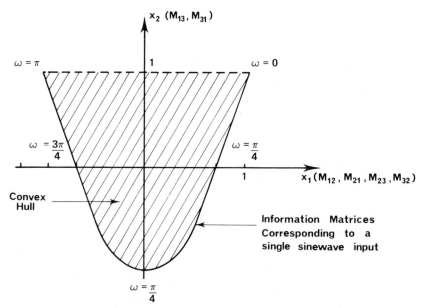

Figure 6.4.1. Convex hull for Example 6.4.1.

between 0 and π since this yields the origin in Fig. 6.4.1. This corresponds
to the white noise input found to be optimal for this example in Section
6.3.1. From Fig. 6.4.1 it is clear that the optimal solution is not unique. For
example, another optimal input is the sum of two sinusoids with frequencies
$\pi/4$ and $3\pi/4$. This gives the optimal input as

$$
\begin{aligned}
u_t = \cos(\pi/4)t + \cos(3\pi/4)t &= \quad 2, &&\text{for } t = 0 \quad \text{modulo } 8 \\
&= -2, &&\text{for } t = 4 \quad \text{modulo } 8 \\
&= \quad 0, &&\text{otherwise}
\end{aligned}
$$

We shall prove below that it is always possible to find an optimal input
comprising a finite number of sinusoids. ∇

Theorem 6.4.2   For any $\xi_1(\omega) \in \mathcal{K}$ with corresponding $(p \times p)$ average
information matrix $\overline{M}(\xi_1)$, there always exists a $\xi_2(\omega) \in \mathcal{K}$ which is piecewise constant with at most $(p(p + 1)/2) + 1$ discontinuities and $\overline{M}(\xi_2) =
\overline{M}(\xi_1)$.

Proof   From Theorem 6.4.1 we know that $\mathcal{M}$ is the convex hull of
average information matrices $\tilde{M}(\omega) + \overline{M}_c \cdot \omega \in [0, \pi]$. Hence, from Caratheodory's theorem (Theorem E.3.1), it follows that

$$
\overline{M}(\xi) = \frac{1}{\Sigma} \sum_{k=1}^{l} \mu_k (\tilde{M}(\omega_k) + \overline{M}_c) \tag{6.4.20}
$$

where each $\mu_k \geq 0$, $l = (p(p+1)/2) + 1$ and

$$\sum_{k=1}^{l} \mu_k = 1 \qquad (6.4.21)$$

Defining

$$\xi_2(\omega) = \sum_{k=1}^{l} \mu_k h(\omega - \omega_k) \qquad (6.4.22)$$

where

$$h(x) = 1 \qquad \text{if} \quad \omega \geq \omega_k$$
$$= 0 \qquad \text{otherwise} \qquad (6.4.23)$$

yields the required result. ∇

The implication of Theorem 6.4.1 is that *any* information matrix can be obtained by applying an input which is simply the sum of a finite number of sinusoids of various amplitudes. In particular, it suffices to search for an optimal design of this form. We can do better however, as we show below.

Theorem 6.4.3 For the design criteria $J = -\log \det \overline{M}$, optimal designs exist comprising not more than $p(p+1)/2$ sinusoidal components (i.e., one less than predicted by Theorem 6.4.2).

Proof It is readily shown that the information matrix corresponding to an optimal design is a boundary point of the convex hull. (If M is not on the boundary, then $\exists \alpha > 0$ such that $(1 + \alpha)M \in \mathcal{M}$; but $\det(1 + \alpha)M > \det M$.) The theorem then follows from Caratheodory's theorem (Theorem E.3.1). ∇

Theorem 6.4.4 Consider the system of Eq. (6.3.1) where

$$G_1(z) = \frac{B(z)}{A(z)} = \frac{b_1 z^{-1} + \cdots + b_n z^{-n}}{1 + a_1 z^{-1} + \cdots + a_n z^{-n}} \qquad (6.4.24)$$

$$G_2(z) = \frac{D(z)}{C(z)} = \frac{1 + d_1 z^{-1} + \cdots + d_q z^{-q}}{1 + c_1 z^{-1} + \cdots + c_q z^{-q}} \qquad (6.4.25)$$

and

$$\beta^{\mathrm{T}} = (b_1, \ldots, b_n, a_1, \ldots, a_n, d_1, \ldots, d_q, c_1, \ldots, c_q, \Sigma) \qquad (6.4.26)$$

Then an optimal power constrained input minimizing $J = -\log \det \overline{M}$ exists comprising not more than $2n$ sinusoids.

Proof From Eq. (6.4.8) it follows that (see Problem 6.8),

$$\tilde{M}(\omega) = \frac{1}{\Sigma} \begin{vmatrix} S\Gamma_X S^{\mathrm{T}} & \vdots & 0 \\ \cdots & & \cdots \\ 0 & \vdots & 0 \end{vmatrix} \tag{6.4.27}$$

where

$$S = \begin{bmatrix} 1 & a_1 \cdots & -a_n & 0 \cdots 0 \\ 0 & 1 & a_1 & & 0 \\ 0 & \cdots 0 & 1 & a_1 \cdots -a_n \\ 0 & -b_1 \cdots & -b_n & 0 \cdots 0 \\ 0 & \cdots 0 & b_1 \cdots & b_n \end{bmatrix} \tag{6.4.28}$$

$$\Gamma_X = H(\omega) \begin{bmatrix} 1 & \cos(\omega) & \cdots & \cos(\overline{2n-1}\omega) \\ \cos(\omega) & & \ddots & \vdots \\ \vdots & \ddots & \ddots & \cos(\omega) \\ \cos(\overline{2n-1}\omega) & \cdots & \cos(\omega) & 1 \end{bmatrix}$$

where

$$H(\omega) = \frac{C(e^{j\omega})C(e^{-j\omega})}{D(e^{j\omega})A^2(e^{j\omega})A^2(e^{-j\omega})D(e^{-j\omega})} \tag{6.4.29}$$

Hence

$$\overline{M}(\xi) = \overline{M}_c + \sum_{k=1}^{2n} \alpha_k(\xi) \begin{vmatrix} SD_k S^{\mathrm{T}} & 0 \\ 0 & 0 \end{vmatrix} \tag{6.4.30}$$

where

$$\alpha_k(\xi) = \frac{1}{\cdot \Sigma \pi} \int_0^{\pi} H(\omega) \cos(\overline{k-1}\omega) \, d\xi(\omega) \tag{6.4.31}$$

D_k has ijth element $\delta_{|i-j|-k+1}$, i.e., $\overline{M}(\xi)$ lies in a $2n$-dimensional linear variety [210] and we can again apply Caratheodory's theorem (Theorem E.3.1) to obtain the desired result. ▽

It is interesting to compare the number of frequency components required for optimality with the number required for nonsingularity of $\overline{M}$ (this is required for consistency of parameter estimation, see part (c) of Theorem 5.5.4 and the asymptotic covariance given in Theorem 5.5.2).

Theorem 6.4.5 Consider the system of Theorem 6.4.4 with (A, B) and (C, D) coprime. (i) A sufficient condition for the average information matrix $(\overline{M})$ to be nonsingular is that the input comprises n distinct sinusoidal components in $(0, \pi)$. Furthermore, (ii) $\overline{M}$ is singular for any input spectral distribution comprising fewer than n points of nonzero measure in $[0, \pi]$.

Proof (i) Assume that the n input frequencies are $\omega_1, \omega_2, \ldots, \omega_n$ and that λ_i, $i = 1, \ldots, n$ represents the fraction of the total input power at the ith frequency. By assumption, we have $0 < \omega_1 < \omega_2 \cdots < \omega_n < \pi$, $\lambda_i \neq 0$, $i = 1, \ldots, n$.

We first note that, for this system, the information matrix has the following structure (cf. Eqs. 6.4.5 and 6.4.8)

$$\overline{M} = \begin{bmatrix} \overline{M}_u' & 0 \\ 0 & \overline{M}_c' \end{bmatrix}$$

where $\overline{M}_c'$ is a nonsingular constant matrix (independent of the input) and $\overline{M}_u'$ is given by

$$\overline{M}_u' = \frac{1}{\Sigma} S \left[\sum_{i=1}^n \lambda_i \Gamma_X(\omega_i) \right] S^T$$

where

$$\Gamma_X(\omega_i) = H(\omega_i) \begin{bmatrix} 1 & \cos(\omega_i) & \cdots & \cos(\overline{2n-1}\omega_i) \\ \cos(\omega_i) & 1 & & \vdots \\ \vdots & & \ddots & \\ \cos(\overline{2n-1}\omega_i) & & \cdots & 1 \end{bmatrix}$$

and $H(\omega_i)$ is given by Eq. (6.4.29).

It can be shown by induction [211] that the determinant of the information matrix is given by

$$[\det \overline{M}] = \frac{4^{n(n-1)}}{\Sigma^{2n}} [\det \overline{M}_c'][\det S]^2 \prod_{i=1}^n \{\lambda_i H(\omega_i) \sin(\omega_i)\}^2$$

$$\times \prod_{1 \leq s < t \leq n} (\cos \omega_s - \cos \omega_t)^4 \tag{6.4.32}$$

Now, we note that S is the Sylvester matrix [212] for the polynomial pair $[A(z), B(z)]$. Hence S is nonsingular provided A and B are coprime. Hence $\det \overline{M}$ is nonsingular since $\lambda_i \neq 0$, $\sin(\omega_i) \neq 0$, $\cos(\omega_i) \neq \cos(\omega_j)$, $i \neq j$, $i = 1, \ldots, n; j = 1, \ldots, n$.

(ii)   This follows immediately since the rank of $\Gamma_X(\omega_i)$ for $\omega_i \in [0, \pi]$ is at most 2. ∇

The above theorem places a lower bound on the number of input frequencies that can be used for identification. If it can be shown, that the optimal information matrix can be achieved with this number of frequencies, then the following theorem applies.

Theorem 6.4.6   Consider the system of Theorem 6.4.4. If the optimal information matrix can be shown to be achievable with n sinusoids, then the optimal partitioning of the input power is

$$\lambda_i = \frac{1}{n}, \qquad i = 1, \ldots, n$$

Proof   From Eq. (6.4.32), det $\overline{M}$ can be expressed in the form

$$[\det \overline{M}] = \prod_{i=1}^{n} \lambda_i^2 \alpha(\omega_i)$$

where $\alpha(\omega_i)$, $i = 1, \ldots, n$ does not depend upon λ_i, $i = 1, \ldots, n$.

Hence maximizing $[\det \overline{M}]$ with respect to λ_i, $i = 1, \ldots, n$ subject to the constraint $\sum_{i=1}^{n} \lambda_i = 1$, yields $\lambda_i = 1/n$; $i = 1, \ldots, n$.   ▽

We illustrate the above theorems by a number of examples.

Example 6.4.2   Consider the systems of (6.4.24) and (6.4.25) with $n = 1$ and $q = 0$, i.e.,

$$\beta^{\mathrm{T}} = (b_1, a_1, \Sigma)$$

From Eq. (6.4.31),

$$\overline{M}(\xi) = \begin{vmatrix} M_{ab} & 0 \\ 0 & \tfrac{1}{2}\Sigma^{-2} \end{vmatrix}$$

where

$$M_{ab} = \alpha_1(\xi) \begin{bmatrix} b_1^2 & a_1 b_1 \\ a_1 b_1 & 1 + a_1^2 \end{bmatrix} + \alpha_2(\xi) \begin{bmatrix} 0 & b_1 \\ b_1 & 2a_1 \end{bmatrix}$$

and

$$\alpha_1(\xi) = \frac{1}{\pi\Sigma} \int_0^{\pi} \frac{1}{(1 + a_1^2 + 2a_1 \cos \omega)^2} \, d\xi(\omega)$$

$$\alpha_2(\xi) = \frac{1}{\pi\Sigma} \int_0^{\pi} \frac{\cos \omega}{(1 + a_1^2 + 2a_1 \cos \omega)^2} \, d\xi(\omega)$$

Figure 6.4.2 shows the convex hull of all average information matrices corresponding to input power constrained designs together with the contours of constant determinant. It can be seen that a single sinusoid suffices for optimality in this case.   ▽

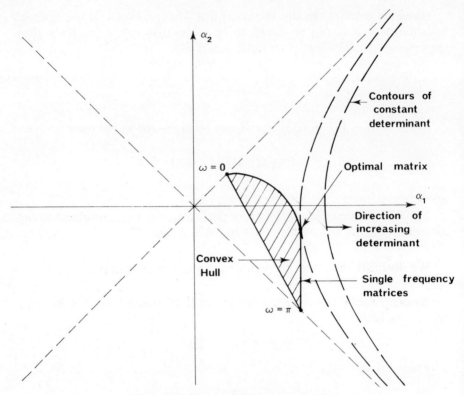

Figure 6.4.2. Convex hull for Example 6.4.2.

The theory that we have presented above for experiment design applies, with minor modifications, to continuous time systems. We have presented the theory for the discrete case since in most applications the data will be available in sampled data form. However, we present below two simple continuous time design problems to illustrate the modifications needed for continuous problems.

For the continuous case the changes required are simply

(i)   replace summations over time by integration;
(ii) replace integration over the frequency range $[0, \pi]$ by integrations over $[0, \infty]$;
(iii) replace $z = e^{j\omega}$ by $s = j\omega$.

Example 6.4.3   Consider the following first-order continuous time system

$$y = [1/(\tau s + 1)]u + n \qquad (6.4.33)$$

where u and y are the input and output, respectively, and n denotes colored measurement noise having spectral density (cf. Problems 4.3 and 4.4)

$$\Psi(\omega) = 1/(a^2\omega^2 + 1) \qquad (6.4.34)$$

Here the only parameter of interest is τ, the system time constant. In the light of Theorem 6.4.4, it suffices to consider single sinusoids as candidates for optimal inputs. The corresponding average information matrix (per unit time) is

$$\overline{M} = 2\omega^2(a^2\omega^2 + 1)/(\tau^2\omega^2 + 1)^2 \qquad (6.4.35)$$

Hence it follows that, for continuous observations, the input which minimizes $-\log \det \overline{M}$ has frequency $\overline{\omega}$ given by

$$\overline{\omega} = (\tau^2 - 2a^2)^{-1/2} \quad \text{if} \quad 2a^2 < \tau^2 \qquad (6.4.36)$$

$$= \infty \quad \text{if} \quad 2a^2 \geq \tau^2 \qquad (6.4.37)$$

That is, if the noise is wide band $(a \to 0)$, then the following appealing result is obtained:

$$\overline{\omega} = 1/\tau \qquad (6.4.38)$$

Equation (6.4.38) implies that we should place the test frequency at the system $3dB$ point. This is intuitively reasonable since for low frequencies the response is constant independent of τ, and for high frequencies the response is lost in the noise. ∇

Example 6.4.4   We now consider a model, closely related to the one in Example 6.4.3, but containing an additional gain parameter, i.e.,

$$y = [K/(\tau s + 1)]u + n \qquad (6.4.39)$$

The parameter vector is $\beta = [K, \tau]$. The average information matrix for a sinewave input is

$$\overline{M} = 2(a^2\omega^2 + 1)\begin{bmatrix} 1/(\tau^2\omega^2 + 1) & -K\tau\omega^2/(\tau^2\omega^2 + 1)^2 \\ -K\tau\omega^2/(\tau^2\omega^2 + 1)^2 & K^2\omega^2/(\tau^2\omega^2 + 1)^2 \end{bmatrix} \qquad (6.4.40)$$

The convex hull of all input power constrained information matrices is illustrated in Fig. 6.4.3 (for the case $a = 0$).

Geometric arguments show that the optimal information matrix is obtained with a single sinewave input. For the case of wide band noise $(a \to 0)$, the optimal input frequency is

$$\overline{\omega} = 1/\tau\sqrt{3} \qquad (6.4.41)$$

The variation of the determinant of the information matrix with input frequency is shown in Fig. 6.4.4.

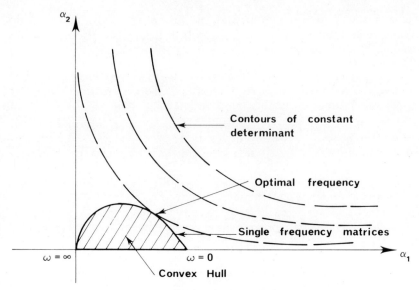

Figure 6.4.3. Convex hull for Example 6.4.4.

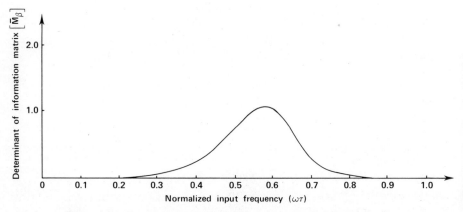

Figure 6.4.4. Determinant of information matrix versus input frequency.

Example 6.4.5 Consider the following second-order system

$$y = \frac{\omega_0{}^2}{s^2 + 2\xi\omega_0 s + \omega_0{}^2} u + n \qquad (6.4.42)$$

where u, y, and n denote the input, output, and noise, respectively. For simplicity we assume that the noise is wide band.

The parameter vector is $\beta = [\omega_0, \xi]$. The average information matrix is

$$\overline{M} = \frac{2}{\left\{\left[1 - \left(\dfrac{\omega}{\omega_0}\right)^2\right]^2 + 4\xi^2 \left|\dfrac{\omega}{\omega_0}\right|^2\right\}} \begin{bmatrix} 4\left(\dfrac{\omega^4}{\omega_0^6}\right) + 4\xi^2\left(\dfrac{\omega^2}{\omega_0^4}\right) & -4\xi\left[\dfrac{\omega^2}{\omega_0^3}\right] \\ -4\xi\left[\dfrac{\omega^2}{\omega_0^3}\right] & 4\left[\dfrac{\omega^2}{\omega_0^2}\right] \end{bmatrix}$$

$$(6.4.43)$$

For this problem it can again be shown that the optimal information matrix is achievable with a single sinusoid. The optimal input frequency is

$$\bar{\omega} = (\omega_0/\sqrt{5})\{(1 - 2\xi^2) + [(1 - 2\xi^2)^2 + 15]^{1/2}\}^{1/2} \qquad (6.4.44)$$

The variation of optimal input frequency with damping ratio is shown in Fig. 6.4.5. It can be seen that for $\xi \to 0$, the optimal input frequency is the undamped natural frequency of the system. The optimal input frequency varies inversely with the damping ratio.

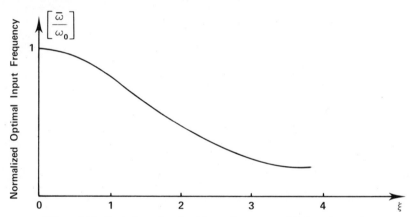

Figure 6.4.5. Variation of optimal input frequency with damping ratio.

In the above simple examples it has been possible to determine analytically the optimal design. However, for general systems it is usually the case that some iterative design procedure will be required. We proceed to establish a theorem that is useful in the checking for optimality of designs. We first introduce some notation and some preliminary results.

Definition 6.4.1 The *response dispersion* $v(\xi, \omega)$ of a design ξ at frequency ω is defined to be

$$v(\xi, \omega) = \text{trace}([\overline{M}(\xi)]^{-1}[\overline{M}(\omega)]) \qquad (6.4.45)$$

where $\overline{M}(\xi)$ and $\overline{M}(\omega)$ are the information resulting from the design ξ and a single frequency input, respectively. ∇

We remark that $v(\xi, \omega)$ can in certain circumstances be given the interpretation of the ratio of the variance of the system frequency response to the noise power at frequency ω. To show this, consider

$$y_t = G_1(z)u_t + G_2(z)\varepsilon_t$$

The frequency transfer function is $G_1(e^{j\omega})$. The change in the transfer function due to a small change in parameters is approximately given by

$$\Delta G_1(e^{j\omega}) \simeq [\partial G_1(e^{j\omega})/\partial \beta]\, \Delta \beta \tag{6.4.46}$$

Therefore the variance in the transfer function is

$$V(\omega) = E[\Delta G_1(e^{j\omega})\, \Delta G_1(e^{-j\omega})] \simeq E\left|\frac{\partial G_1(e^{j\omega})}{\partial \beta}\, \Delta \beta\right|\left|\frac{\partial G_1(e^{-j\omega})}{\partial \beta}\, \Delta \beta\right|$$

$$= \left(\frac{\partial G_1(e^{j\omega})}{\partial \beta}\right) \operatorname{cov}\hat\beta \left(\frac{\partial G_1(e^{-j\omega})}{\partial \beta}\right)^{\mathrm{T}}$$

$$= \operatorname{trace}\operatorname{cov}\hat\beta \left|\frac{\partial G_1(e^{-j\omega})}{\partial \beta}\right|^{\mathrm{T}}\left|\frac{\partial G_1(e^{j\omega})}{\partial \beta}\right|$$

$$= \frac{1}{N}\operatorname{trace}\left([\overline{M}]^{-1}\left|\frac{\partial G_1(e^{-j\omega})}{\partial \beta}\right|^{\mathrm{T}}\left|\frac{\partial G_1(e^{j\omega})}{\partial \beta}\right|\right)$$

$$\left|\frac{V(\omega)}{G_2(e^{j\omega})G_2(e^{-j\omega})}\right| = \frac{1}{N}\operatorname{trace}\left([\overline{M}]^{-1}\left|\frac{\partial G_1(e^{-j\omega})}{\partial \beta}\right|^{\mathrm{T}} G_2^{-1}(e^{-j\omega})G_2^{-1}(e^{j\omega})\right.$$

$$\times \left|\frac{\partial G_1(e^{j\omega})}{\partial \beta}\right|$$

$$= \frac{\Sigma}{N}\operatorname{trace}([\overline{M}]^{-1}\tilde{M}(\omega)) \tag{6.4.47}$$

The last equality follows from (6.4.8) since the real part and trace operators commute.

We note that (6.4.47) is related to (6.4.45). Equality is achieved when G_1 and G_2 are independently parameterized (cf. the system of Theorem 6.4.4).

Result 6.4.1 Consider an experiment with input distribution function ξ and response dispersion $v(\xi, \omega)$, then

(1)
$$\frac{1}{\pi}\int_0^\pi v(\xi, \omega)\, d\xi(\omega) = p \tag{6.4.48}$$

where p is the number of parameters

(2)
$$\max_{\omega \in [0,\pi]} v(\xi, \omega) \geq p \tag{6.4.49}$$

Proof

(1) $\dfrac{1}{\pi} \displaystyle\int_0^\pi \text{trace}([\overline{M}(\xi)]^{-1}[\overline{M}(\omega)]\,d\xi(\omega))$

$$= \text{trace}\left([\overline{M}(\xi)]^{-1} \cdot \frac{1}{\pi}\int_0^\pi \overline{M}(\omega)\,d\xi(\omega)\right)$$

$$= \text{trace}([\overline{M}(\xi)]^{-1}[\overline{M}(\xi)]) \qquad \text{from} \quad (6.4.11)$$

$$= p$$

(2) Use contradiction, assume $\max_{\omega \in [0,\,\pi]} v(\xi,\omega) < p$, then

$$\frac{1}{\pi}\int_0^\pi v(\xi,\omega)\,d\xi(\omega) \le \frac{1}{\pi}\int_0^\pi \max_\omega v(\xi,\omega)\,d\xi(\omega)$$

$$< \frac{p}{\pi}\int_0^\pi d\xi(\omega)$$

$$= p \qquad \text{from (6.4.12)}$$

This contradicts (6.4.48) thus establishing (6.4.49). ∇

Theorem 6.4.7 The following characterizations of an optimal energy constrained design ($\xi^* \in \mathcal{K}$) are equivalent:

(1) the design ξ^* minimizes $-\log \det \overline{M}(\xi)$,
(2) the design ξ^* minimizes $\max_{\omega \in [0,\,\pi]} v(\xi,\omega)$, and
(3) $\max_{\omega \in [0,\,\pi]} v(\xi^*,\omega) = p$,

where p is the number of parameters.

Proof We proceed $(1) \Rightarrow (3)$, $(3) \Rightarrow (2)$, and $(3) \Rightarrow (1)$. We first show that $(1) \Rightarrow (3)$. Let

$$\tilde{\xi} = (1-\alpha)\xi^* + \alpha\xi$$

for any $\xi \in \mathcal{F}$ (i.e., a sinusoidal input). From Eq. (6.4.6),

$$\overline{M}(\tilde{\xi}) = (1-\alpha)\overline{M}(\xi^*) + \alpha\overline{M}(\xi)$$

Using Result E.2.1,

$$\frac{d}{d\alpha}[-\log \det \overline{M}(\tilde{\xi})]\bigg|_{\alpha=0} = -\text{trace}([\overline{M}(\tilde{\xi})]^{-1}[\overline{M}(\xi) - \overline{M}(\xi^*)])\bigg|_{\alpha=0}$$

$$= \text{trace}([\overline{M}(\xi^*)]^{-1}[\overline{M}(\xi^*) - \overline{M}(\xi)]$$

$$= p - \text{trace}([\overline{M}(\xi^*)]^{-1}\overline{M}(\xi))$$

$$\ge 0 \qquad \text{by hypothesis that (1) is true}$$

Now $\xi \in \mathcal{F}$ implies $\overline{M}(\xi) = \overline{M}(\omega)$ for some ω, thus

$$\text{trace}([\overline{M}(\xi^*)]^{-1}\overline{M}(\omega)) \leq p$$

i.e.,

$$v(\xi^*, \omega) \leq p \qquad\qquad (6.4.50)$$

Equations (6.4.50) and (6.4.49) establish (3). (3) $\Rightarrow$ (2) immediately follows from (6.4.49). Also, (2) $\Rightarrow$ (3) since (1) $\Rightarrow$ (3) ensures the existence of an ξ^* satisfying (3) and then (6.4.49) guarantees that (2) is true.

To show (3) $\Rightarrow$ (1), we assume (3) is true but that (1) is false, i.e., $\exists\, \tilde{\xi} \in \mathcal{K}$ such that

$$-\log \det \overline{M}(\tilde{\xi}) < -\log \det \overline{M}(\xi^*) \qquad\qquad (6.4.51)$$

Therefore, consider the design $\xi = (1 - \alpha)\xi^* + \alpha\tilde{\xi}$. Then,

$$\overline{M}(\xi) = (1 - \alpha)\overline{M}(\xi^*) + \alpha\overline{M}(\tilde{\xi})$$

It follows from Result E.4.1 that for $0 < \alpha < 1$,

$$-\log \det \overline{M}(\xi) < -(1 - \alpha)\log \det \overline{M}(\xi^*) - \alpha \log \det \overline{M}(\tilde{\xi})$$

$$< -\log \det \overline{M}(\xi^*) \qquad \text{from (6.4.51)}$$

It follows that $(\partial/\partial\alpha)[-\log \det \overline{M}(\xi)]|_{\alpha=0} < 0$, i.e., from Result E.2.1,

$$-\text{trace}([\overline{M}(\xi)]^{-1}[\overline{M}(\tilde{\xi}) - \overline{M}(\xi^*)])\Big|_{\alpha=0} < 0$$

i.e.,

$$\text{trace}([\overline{M}(\xi^*)]^{-1}[\overline{M}(\tilde{\xi})]) > p$$

$$\frac{1}{\pi} \int_0^\pi v(\xi^*, \omega)\, d\tilde{\xi}(\omega) > p \qquad\qquad (6.4.52)$$

However, (3) implies that

$$\frac{1}{\pi} \int_0^\pi v(\xi^*, \omega)\, d\tilde{\xi}(\omega) \leq p \qquad\qquad (6.4.53)$$

This contradiction means that (1) must be true. ∇

Corollary 6.4.5   For an optimal design ξ^* comprising a finite number of sinusoids (the existence of which is guaranteed by Theorem 6.4.2), the response dispersion $v(\xi^*, \omega)$ achieves its maximum value of p at the frequencies of the design.

Proof We assume the contrary, i.e., $\exists\ \omega_k$ in the design such that $v(\xi^*, \omega_k) < p$. Then

$$\frac{1}{\pi} \sum_{k=1}^{l} \lambda_k^* v(\xi^*, \omega_k) < p \qquad \text{since} \qquad \frac{1}{\pi} \sum_{k=1}^{l} \lambda_k^* = 1$$

However, this contradicts (6.4.48) thus establishing the result. ∇

The implications of Theorem 6.4.5 and Corollary 6.4.5 are diagrammatically represented in Fig. 6.4.6.

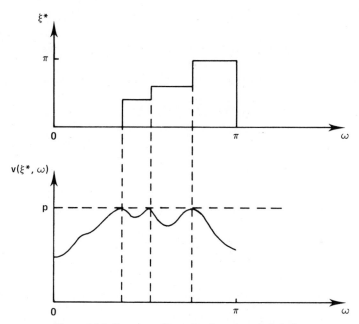

Figure 6.4.6. Response dispersion for an optical design.

The above results give us tests for the optimality (or otherwise) of input spectra. In particular,

(1) If the response dispersion $v(\xi, \omega)$ is less than or equal to p for all ω, then ξ is optimal.

(2) If the response dispersion $v(\xi, \omega)$ is not equal to p at the frequencies of ξ, then ξ is not optimal.

We can also use the results to develop sequential design algorithms. Suppose we have an arbitrary (nonoptimal) design ξ_k with response dispersion $v(\xi_k, \omega)$. Further, let $v(\xi_k, \omega)$ attain its maximum (necessarily $> p$)

at $\omega = \omega_k{}^0$. Then, the design

$$\xi_{k+1} = (1 - \alpha_k)\xi_k + \alpha_k \xi_k{}^0 \qquad (6.4.54)$$

(where $\xi_k{}^0 \in \mathcal{F}$ has a single step at $\omega_k{}^0$) leads to a decrease in the value of $-\log \det \overline{M}$ for suitably small α_k. This follows since the gradient with respect to α_k is negative, i.e.,

$$\frac{\partial}{\partial \alpha_k} [-\log \det \overline{M}(\xi_{k+1})] \Big|_{\alpha_k = 0} = p - v(\xi_k, \omega_k{}^0) < 0$$

It can be shown [156, 158] that the algorithm given in (6.4.54) converges to an optimal design provided $\{\alpha_k\}$ is suitably chosen. For example, any sequence satisfying $\sum_{k=1}^{\infty} \alpha_k = \infty$, $\lim_{k \to \infty} \alpha_k = 0$ will yield convergence.

Theorem 6.4.2 can also be used directly to obtain optimal designs by optimizing in a $p(p + 1)$-dimensional space.

We have used the single-input, single-output case as a vehicle for presenting the above results on experiment design. The reason for this has been to simplify the notation. However, the results can be readily extended to the multiple-input, multiple-output case [190, 193]. For the multivariable case, Eqs. (6.4.4)–(6.4.10) for the information matrix should be replaced by the following (see also Problem 6.9):

$$[\overline{M}]_{ij} = [\overline{M}_u]_{ij} + [\overline{M}_c]_{ij} \qquad (6.4.55)$$

where $[\overline{M}]_{ij}$ is the ijth element of $\overline{M}$ and

$$[\overline{M}_u]_{ij} = \frac{1}{2\pi} \int_{-\pi}^{\pi} \operatorname{trace}\{B_{ij}(\omega)\,dF(\omega)\} \qquad (6.4.56)$$

$$B_{ij}(\omega) = \frac{\partial G_1{}^{\mathrm{T}}(e^{-j\omega})}{\partial \beta_i} G_2^{-\mathrm{T}}(e^{-j\omega})\Sigma^{-1}G_2^{-1}(e^{j\omega})\frac{\partial G_1(e^{j\omega})}{\partial \beta_j} \qquad (6.4.57)$$

$F(\omega)$ is the multivariable input spectral distribution function

$$[\overline{M}_c]_{ij} = \frac{1}{2\pi} \operatorname{trace}\left\{\int_{-\pi}^{\pi} \frac{\partial G_2{}^{\mathrm{T}}(e^{j\omega})}{\partial \beta_i} G_2^{-\mathrm{T}}(e^{-j\omega})\Sigma^{-1}G_2^{-1}(e^{j\omega})\frac{\partial G_2(e^{j\omega})}{\partial \beta_j} \Sigma\,d\omega\right\}$$

$$+ \tfrac{1}{2} \operatorname{trace}\{\Sigma^{-1}\,\partial\Sigma/\partial\beta_i\} \operatorname{trace}\{\Sigma^{-1}\,\partial\Sigma/\partial\beta_j\} \qquad (6.4.58)$$

The reader can verify that the above expressions reduce to those given earlier for the single-input, single-output case.

For the multivariable case, the analog of Theorem 6.4.2 holds for input power constraints of the form

$$\operatorname{trace} F(\pi) = \int_{-\pi}^{\pi} \operatorname{trace} dF(\omega) = 1 \qquad (6.4.59)$$

The result is that the optimal information matrix can be achieved with a design comprising not more than $p(p + 1)/2$ sinusoids [193], where p is the number of parameters. In the multiple-input case, however, it is also necessary to design $2r$ quantities expressing the relative magnitudes and phases of the r inputs at each frequency.

An alternative interpretation of the optimal spectra obtained in this section is possible. Each sinusoid in the optimal input can be applied separately, each with amplitude $\sqrt{2}$ (1 for $\omega = 0$ or π) and for a duration proportional to the height of the corresponding line in the optimal spectrum (i.e., the magnitude of the jump in $\xi(\omega)$ at the frequency of the sinusoid). Thus we can apply $p(p + 1)/2$ sinusoids of equal magnitudes sequentially for times proportional to the heights of the lines in the optimal spectrum.

6.4.2 *Feedback and output power constraints* The experiment design problem statement made in Section 6.4.1 assumed that the input would be generated exogenously. However, a more general statement allows the possibility of generating the input either wholly or partially by feedback. As pointed out in Section 5.6 feedback may be an important ingredient in practical identification problems and we therefore proceed to study the corresponding experiment design problem.

Consider the multivariable feedback system depicted in Fig. 6.4.7. In Figure 6.4.7 we have

$s \in R^r$: externally applied set point perturbation
$u \in R^r$: system input
$y \in R^m$: system output
$\varepsilon \in R^m$: white noise (assumed Gaussian with covariance Σ).

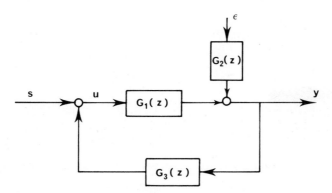

Figure 6.4.7. Linear system with feedback.

G_1, G_2, and G_3 are transfer functions of appropriate dimensions. β is the vector of parameters in G_1, G_2, and Σ. G_3 is known and is available for adjustment by the experimenter, i.e., G_3 is a design variable.

We shall be concerned with the following categories of experimental conditions:

(A) open Loop (i.e., $G_3 \equiv 0$);
(B) a single feedback ($G_3 \neq 0$) together with set point perturbations;
(C) a concentration of l subexperiments each from group (B).

The likelihood function for the data is

$$p(Y|\beta) = [(2\pi)^r \det \Sigma]^{-N/2} \exp\left\{-\frac{1}{2}\sum_{t=1}^{N} \omega_t^T \Sigma^{-1} \omega_t\right\} \qquad (6.4.60)$$

where $\{\omega_t\}$ is the residual sequence given by

$$\omega_t = G_2^{-1}[y_t - G_1 u_t] \qquad (6.4.61)$$

$$u_t = s_t + G_3 y_t \qquad (6.4.62)$$

An expression for the information matrix is now developed.

Result 6.4.2   For the system depicted in Fig. 6.4.7 the ikth element of the average information matrix is

$$
\begin{aligned}
[\overline{M}]_{ik} &= \frac{1}{2} \operatorname{trace}\left\{\Sigma^{-1}\frac{\partial \Sigma}{\partial \beta_i}\right\} \operatorname{trace}\left\{\Sigma^{-1}\frac{\partial \Sigma}{\partial \beta_k}\right\} \\
&\quad + \frac{1}{2\pi}\operatorname{trace}\left\{\int_{-\pi}^{\pi}\left[\frac{\partial G_1}{\partial \beta_i}G_3(I-G_1G_3)^{-1}G_2 + \frac{\partial G_2}{\partial \beta_i}\right]^* G_2^{-*}\Sigma^{-1}\right. \\
&\quad \times G_2^{-1}\left[\frac{\partial G_1}{\partial \beta_k}G_3(I-G_1G_3)^{-1}G_2 + \frac{\partial G_2}{\partial \beta_k}\right]\Sigma \, d\omega\Big\} \\
&\quad + \frac{1}{2\pi}\operatorname{trace}\left\{\int_{-\pi}^{\pi}[I-G_3G_1]^{-*}\frac{\partial G_1^*}{\partial \beta_i}G_2^{-*}\Sigma^{-1}\right. \\
&\quad \times G_2^{-1}\frac{\partial G_1}{\partial \beta_k}[I-G_3G_1]^{-1}\,dF_s(\omega)\Big\}
\end{aligned}
\qquad (6.4.63)
$$

where $F_s(\omega)$ is the spectral distribution function of the set point sequence $\{s_t\}$. The notations $G \equiv G(e^{j\omega})$, $G^* \equiv G^T(e^{-j\omega})$, and $G^{-*} = [G^*]^{-1}$ have been used.

Proof   From Fig. 6.4.7, we have the following equivalent open loop model

$$y_t = Ls_t + N\varepsilon_t \qquad (6.4.64)$$

where

$$L = (I - G_1 G_3)^{-1} G_1 \qquad (6.4.65)$$

$$N = (I - G_1 G_3)^{-1} G_2 \qquad (6.4.66)$$

Treating L and N as the G_1 and G_2 of Eqs. (6.4.55)–(6.4.58) yields the following expressions for the ikth element of the information matrix

$$[\overline{M}]_{ik} = \frac{1}{2} \operatorname{trace}\left\{\Sigma^{-1}\frac{\partial\Sigma}{\partial\beta_i}\right\}\operatorname{trace}\left\{\Sigma^{-1}\frac{\partial\Sigma}{\partial\beta_k}\right\}$$

$$+ \frac{1}{2\pi}\operatorname{trace}\left\{\int_{-\pi}^{\pi}\frac{\partial N^*}{\partial\beta_i}N^{-*}\Sigma^{-1}N^{-1}\frac{\partial N}{\partial\beta_k}\Sigma\,d\omega\right\}$$

$$+ \frac{1}{2\pi}\operatorname{trace}\left\{\int_{-\pi}^{\pi}\frac{\partial L^*}{\partial\beta_i}N^{-*}\Sigma^{-1}N^{-1}\frac{\partial L}{\partial\beta_k}dF_s(\omega)\right\} \qquad (6.4.67)$$

where from Eqs. (6.4.65) and (6.4.66),

$$\partial N/\partial\beta_i = (I - G_1 G_3)^{-1}(\partial G_1/\partial\beta_i)G_3(I - G_1 G_3)^{-1}G_2 + (I - G_1 G_3)^{-1}$$
$$\times (\partial G_2/\partial\beta_i) \qquad (6.4.68)$$

$$\partial L/\partial\beta_i = (I - G_1 G_3)^{-1}(\partial G_1/\partial\beta_i)G_3(I - G_1 G_3)^{-1}G_1 + (I - G_1 G_3)^{-1}$$
$$\times (\partial G_1/\partial\beta_i) \qquad (6.4.69)$$

Substituting (6.4.68) and (6.4.69) into (6.4.67) yields the required result.   ▽

It is obvious that the set of information matrices obtainable with experiments from group (B) (closed loop) includes the set of information matrices obtainable with experiments from group (A). However, as we shall show in the next theorem, for the usual parameterization of G_1, G_2, and Σ it is the case that the optimal information matrices is obtainable by a group (A) experiment.

Theorem 6.4.8 Consider a system in which G_1 has no parameters in common with G_2 and Σ. If a constraint is placed on the total input power, then the design criterion $-\log\det\overline{M}$ is minimized by an experiment from group (A) (open loop).

Proof We define $\beta^T = (\theta^T, \gamma^T)$, where θ corresponds to the parameters in G_1 and γ corresponds to the parameters in G_2 and Σ. The corresponding partitions in the information matrix are given by

$$\overline{M} = \begin{bmatrix} \overline{M}_{\theta\theta} & \overline{M}_{\theta\gamma} \\ \overline{M}_{\theta\gamma}^T & \overline{M}_{\gamma\gamma} \end{bmatrix} \qquad (6.4.70)$$

Noting that $\partial G_1/\partial \gamma$ and $\partial \Sigma/\partial \gamma$ are zero in Eq. (6.4.63) yields that $\overline{M}_{\gamma\gamma}$ is a constant matrix independent of G_3 and $F_s(\omega)$. Also from (6.4.63),

$$
\begin{aligned}
[\overline{M}_{\theta\theta}]_{ik} = \frac{1}{2\pi} \int_{-\pi}^{\pi} \text{trace} & \left\{ \left(\frac{\partial G_1^*}{\partial \theta_i} G_2^{-*}\Sigma^{-1}G_2^{-1} \frac{\partial G_1}{\partial \theta_k} \right) \right. \\
& \times \left([I - G_3 G_1]^{-1} dF_s[I - G_3 G_1]^{-*} \right. \\
& \left. + G_3(I - G_1 G_3)^{-1}G_2 \Sigma G_2^*(I - G_1 G_3)^{-*}G_3^* \, d\omega \right\}
\end{aligned}
\tag{6.4.71}
$$

Now from Eqs. (6.4.62) and (6.4.64), it follows that the spectral distribution function for the system input u is given by

$$
\begin{aligned}
dF_u = & (I - G_3 G_1)^{-1} \, dF_s(I - G_3 G_1)^{-*} \\
& + (I - G_3 G_1)^{-1}G_3 G_2 \Sigma G_2^*G_3^*(I - G_3 G_1)^{-*} \, d\omega
\end{aligned}
\tag{6.4.72}
$$

Substituting (6.4.72) into (6.4.71) and noting that $(I - G_3 G_1)^{-1}G_3 = G_3(I - G_1 G_3)^{-1}$ (Result E.1.4) yields

$$
[\overline{M}_{\theta\theta}]_{ik} = \frac{1}{2\pi} \int_{-\pi}^{\pi} \text{trace} \left\{ \left(\frac{\partial G_1^*}{\partial \theta_i} G_2^{-*}\Sigma^{-1}G_2^{-1} \frac{\partial G_1}{\partial \theta_k} \right) dF_u(\omega) \right\}
\tag{6.4.73}
$$

Thus $\overline{M}_{\theta\theta}$ depends only on G_3 and dF_s via F_u, i.e., it is only the spectral distribution of u that affects $\overline{M}_{\theta\theta}$; not the input generating mechanism. Hence Eq. (6.4.70) can be written as

$$
\overline{M} = \left[\begin{array}{c|c} \overline{M}_{\theta\theta}(F_u) & \overline{M}_{\theta\gamma}(G_3) \\ \hline \overline{M}_{\theta\gamma}^T(G_3) & C \end{array} \right]
\tag{6.4.74}
$$

where C is a constant matrix.

Now from (6.4.74) and Result E.1.2

$$
\det \overline{M} = [\det \overline{M}_{\theta\theta}][\det(C - \overline{M}_{\theta\gamma}^T[\overline{M}_{\theta\theta}]^{-1}\overline{M}_{\theta\gamma})]
\tag{6.4.75}
$$

For fixed F_u, it is obvious that $\det \overline{M}$ achieves its maximum value when $\overline{M}_{\theta\gamma} = 0$, i.e., when $G_3 \equiv 0$. Let F_u^* maximize $\det \overline{M}_{\theta\theta}$ subject to the given input power constraint. The resulting maximum value of $\det \overline{M}_{\theta\theta}$ is independent of G_3 since $\overline{M}_{\theta\theta}$ does not depend on G_3. It follows that

$$
\det \overline{M} \le [\det \overline{M}_{\theta\theta}(F_u^*)][\det C]
\tag{6.4.76}
$$

with equality if $G_3 \equiv 0$ and $F_s = F_u^*$. This completes the proof. ∇

The implication of the above theorem is that, with an input power constraint, an open loop experiment is preferable to a closed loop experiment. Feedback will, in general, deteriorate the achievable accuracy. However, there are cases when feedback is beneficial. The importance of feedback in experiment design was first recognized by Soderstrom, Ljung, and Gustavsson [249, 237]. An important case when feedback is desirable is when the constraint is on the available output power. We illustrate this below for a special model structure.

Theorem 6.4.9 Consider the following autoregressive model

$$y_t = a_1 y_{t-1} + \cdots + a_n y_{t-n} + b u_{t-1} + \varepsilon_t \tag{6.4.77}$$

where $\{\varepsilon_t\}$ is a white noise sequence having a Gaussian distribution with variance Σ. The constraint is on the available output power. Then the design criterion $-\log \det \overline{M}$ is minimized when the system input is generated by a minimum variance feedback control together with a white set point sequence.

Proof Using Eq. (6.3.23) the information matrix is given by

$$\overline{M} = E_{Y|\beta} \left\{ \frac{1}{N\Sigma} \sum_{t=1}^{N} \left(\frac{\partial \omega_t}{\partial \beta} \right)^{\mathsf{T}} \left(\frac{\partial \omega_t}{\partial \beta} \right) + \frac{1}{2\Sigma^2} \left(\frac{\partial \Sigma}{\partial \beta} \right)^{\mathsf{T}} \left(\frac{\partial \Sigma}{\partial \beta} \right) \right\} \tag{6.4.78}$$

where $\beta = (a_1, \ldots, a_n, b, \Sigma)^{\mathsf{T}} = (\alpha^{\mathsf{T}}, b, \Sigma)^{\mathsf{T}}$ and where ω_t is the residual sequence given by

$$\omega_t = y_t - a_1 y_{t-1}, \ldots, -a_n y_{t-n} - b u_{t-1} \tag{6.4.79}$$

Thus

$$\partial \omega_t / \partial a_i = -y_{t-i} \tag{6.4.80}$$

$$\partial \omega_t / \partial b = -u_{t-1} \tag{6.4.81}$$

Substituting (6.4.80) and (6.4.81) into (6.4.78) yields

$$\overline{M} = E \begin{bmatrix} y_{t-1}y_{t-1} & \cdots & y_{t-1}y_{t-n} & y_{t-1}u_{t-1} & 0 \\ \vdots & & \vdots & \vdots & \vdots \\ y_{t-n}y_{t-1} & \cdots & y_{t-n}^2 & y_{t-n}u_{t-1} & 0 \\ \hline u_{t-1}y_{t-1} & \cdots & u_{t-1}y_{t-n} & u_{t-1}^2 & 0 \\ \hline 0 & \cdots & 0 & 0 & 1/2\Sigma^2 \end{bmatrix} \tag{6.4.82}$$

$$= \begin{bmatrix} A & B & 0 \\ \hline B^{\mathsf{T}} & C & 0 \\ \hline 0 & 0 & 1/2\Sigma^2 \end{bmatrix} \tag{6.4.83}$$

where the partitions correspond to the partitioning of θ into $(\alpha^{\mathsf{T}}, b, \Sigma)^{\mathsf{T}}$. We now define $\{\rho_i\}$ as

$$E\{y_{t-j} y_{t-k}\} \triangleq \rho_{|j-k|} \tag{6.4.84}$$

Substituting (6.4.84) into (6.4.82) and using (6.4.77) yields

$$A = \begin{bmatrix} \rho_0 & \rho_1 & \cdots & \rho_{n-1} \\ \rho_1 & \rho_0 & & \\ \vdots & & & \vdots \\ \rho_{n-1} & & \cdots & \rho_0 \end{bmatrix} \tag{6.4.85}$$

$$B = (1/b)(V - A\alpha) \tag{6.4.86}$$

$$C = (1/b^2)(\rho_0 - 2\alpha^T V + \alpha^T A\alpha - \Sigma) \tag{6.4.87}$$

where V is given by

$$V^T = [\rho_1, \rho_2, \ldots, \rho_n] \tag{6.4.88}$$

From Eq. (6.4.83) we have

$$-\log \det \overline{M} = \log(2\Sigma^2) - \log \det(A) - \log(C - B^T A^{-1} B) \tag{6.4.89}$$

Using (6.4.86) to (6.4.88) yields

$$-\log \det \overline{M} = \log(2\Sigma^2) - \log \det(A) - \log(\rho_0 - \Sigma - V^T A^{-1} V) + \log b^2 \tag{6.4.90}$$

If the output power is constrained, then ρ_0 is fixed (necessarily $\geq \Sigma$), and hence the diagonal elements of A are fixed. Result E.2.3 then yields that $-\log \det(A)$ is minimized when A is diagonal. From Eq. (6.4.85), A is diagonal when $\rho_i = 0$, $i > 0$. Moreover, $\rho_i = 0$, $i > 0$ yields $V = 0$ from Eq. (6.4.88). Thus the term $-\log(\rho_0 - \Sigma - V^T A^{-1} V)$ simultaneously achieves its minimum value of $(\rho_0 - \Sigma)$.

Thus we have established that $-\log \det \overline{M}$ achieves its minimum value of $\log(2\Sigma^2 b^2/\rho_0{}^n[\rho_0 - \Sigma])$ when $\rho_i = 0$, $i > 0$.

Finally, we note that $\rho_i = 0$, $i > 0$ is achieved when $\{y_t\}$ is an uncorrelated sequence and this is true if the input $\{u_t\}$ is chosen to satisfy

$$u_t = -(1/b)(a_1 y_t + a_2 y_{t-1} + \cdots + a_n y_{t-n+1}) + \eta_t \tag{6.4.91}$$

where $\{\eta_t\}$ is a white set point perturbation with variance $(1/b^2)(\rho_0 - \Sigma)$. Equation (6.4.91) will be recognized as the minimum variance control law [14] for the model of Eq. (6.4.77). ∇

The optimal experiment described in Theorem 6.4.7 is depicted in Fig. 6.4.8.

The above theorem has demonstrated that feedback is, in general, advantageous when an experiment is to be designed to satisfy an output power constraint. It is obvious that experiments from group (C) (a concatenation of l subexperiments each with a single feedback setting together

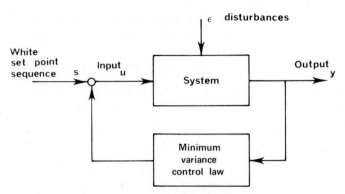

Figure 6.4.8. Optimal experiment for autoregresive model and constrained output powers.

with set point perturbations) yield a set of information matrices including the sets obtainable from group (A) or group (B) designs. Actually, it can be readily shown using Caratheodory's theorem (Theorem E.3.1) that an optimal power constrained experiment is achievable with a finite number of subexperiments each having a single sinewave input and single feedback. Other studies, e.g., [121], have indicated that a good, if not optimal, experiment design procedure with constrained output variations is to use a tight control law, e.g., the minimum variance controller, and then to design the set point perturbations by standard open loop methods.

6.5. SAMPLING STRATEGY DESIGN

Here we consider the design of experiments in which it is possible to adjust not only the input but also the sampling instants and presampling filters.

6.5.1 *Nonuniform sampling* In this section the following problem is considered:

(a) The parameters in a continuous time linear system are to be estimated from N samples.

(b) The samples are taken at times $t_1, t_2, \ldots, t_N$ where the sampling intervals

$$\Delta_k = t_{k+1} - t_k, \qquad k = 0, \ldots, N-1 \qquad (6.5.1)$$

are not necessarily equal.

(c)   The experimental conditions available for adjustment are
(1)   system input (assumed constant between sampling instants),
(2)   the sampling instants $t_1, \ldots, t_N$, and
(3)   the presampling filter (assumed linear time invariant).

It was shown in Section 4.5.3 that a general PEM model for this situation is given by

$$y_k = \bar{y}_k + v_k \tag{6.5.2}$$

where

$$\bar{y}_k = C\bar{x}_k \tag{6.5.3}$$

$$\bar{x}_{k+1} = (\Phi_k - \Gamma_k C)\bar{x}_k + \Psi_k u_k + \Gamma_k y_k \tag{6.5.4}$$

$\{u_k\}$, $\{y_k\}$, and $\{v_k\}$ are the input, output, and innovations sequences, respectively. y_k denotes $y(t_k)$, $k = 1, \ldots, N$ and so on.

The innovations sequence is nonstationary and has covariance at time t_k given by S_k.

The matrices $C, \Phi_k, \Gamma_k, \Psi_k, S_k, k = 1, 2, \ldots, N$ are, in general, nonlinear time varying functions of

(1)   the system parameters β,
(2)   the presampling filter, and
(3)   the sampling intervals, $\Delta_1, \Delta_2, \ldots, \Delta_N$ (the form of the dependence is given in Section 4.5.3).

The likelihood function for the data is given by (cf. Eq. (5.4.15))

$$p(Y|\beta) = \left([2\pi]^{mN} \prod_{k=1}^{N} \det S_k\right)^{-1/2} \exp\left\{-\frac{1}{2}\sum_{k=1}^{N}(y_k - \bar{y}_k)^{\mathsf{T}}S_k^{-1}(y_k - \bar{y}_k)\right\}$$

$$\tag{6.5.5}$$

This leads to the following expression for the ijth element of the information matrix

$$[M]_{ij} = E_{Y|\beta}\left\{\sum_{k=1}^{N}\left(\frac{\partial\bar{y}_k}{\partial\beta_i}\right)^{\mathsf{T}}S_k^{-1}\left(\frac{\partial\bar{y}_k}{\partial\beta_j}\right)\right\} + \frac{1}{2}\sum_{k=1}^{N}\operatorname{trace}\left\{S_k^{-1}\frac{\partial S_k}{\partial\beta_i}S_k^{-1}\frac{\partial S_k}{\partial\beta_j}\right\} \tag{6.5.6}$$

The proof of Eq. (6.5.6) proceeds along the same lines as those presented in Section 6.4 and is therefore left as an exercise (see Problem 6.9).

The quantities $\partial\bar{y}_k/\partial\beta_i$ in Eq. (6.5.6) can be evaluated from the following sensitivity equations:

$$\frac{\partial\bar{y}_k}{\partial\beta_i} = -\frac{\partial C}{\partial\beta_i}\bar{x}_k - C\frac{\partial\bar{x}_k}{\partial\beta_i} \tag{6.5.7}$$

$$\frac{\partial\bar{x}_{k+1}}{\partial\beta_i} = (\phi_k - \Gamma_k C)\frac{\partial\bar{x}_k}{\partial\beta_i} + \left(\frac{\partial\phi_k}{\partial\beta_i} - \Gamma_k\frac{\partial C}{\partial\beta_i}\right)\bar{x}_k + \frac{\partial\psi_k}{\partial\beta_i}u_k + \frac{\partial\Gamma_k}{\partial\beta_i}(y_k - \bar{y}_k)$$

$$\tag{6.5.8}$$

Equation (6.5.8) can be combined with Eq. (6.5.4) to give

$$\tilde{x}_{k+1} = F_k \tilde{x}_k + G_k u_k + H_k(y_k - \bar{y}_k) \qquad (6.5.9)$$

where

$$\tilde{x}_k = [\bar{x}_k^T, \partial \bar{x}_k^T / \partial \beta_1, \ldots, \partial \bar{x}_k^T / \partial \beta_p]^T \qquad (6.5.10)$$

and the matrices F_k, G_k, and H_k follow from (6.5.8) and (6.5.4).

Also, from Eq. (6.5.7) $\partial \bar{y}_k / \partial \beta_i$ can be expressed as a linear function of the composite state $\tilde{x}_k$ as

$$\partial \bar{y}_k / \partial \beta_i = \Omega_i \tilde{x}_k \qquad (6.5.11)$$

where

$$\Omega_i = \left| \frac{\partial C}{\partial \beta_i}, 0, \ldots, 0, C, 0, \ldots, 0 \right| \qquad (6.5.12)$$

Now substituting (6.5.11) into (6.5.6) leads to the following expression for the ijth element of the information matrix:

$$M_{ij} = E_{Y|\beta} \left| \sum_{k=1}^{N} \tilde{x}_k^T \Omega_i^T S_k^{-1} \omega_j \tilde{x}_k \right| + \frac{1}{2} \sum_{k=1}^{N} \mathrm{trace} \left| S_k^{-1} \frac{\partial S_k}{\partial \beta_i} \right| \mathrm{trace} \left| S_k^{-1} \frac{\partial S_k}{\partial \beta_j} \right|$$

$$(6.5.13)$$

Performing the expectation in Eq. (6.5.13) yields

$$M_{ij} = \sum_{k=1}^{N} \{ \hat{x}_k^T \Omega_i^T S_k^{-1} \Omega_j \hat{x}_k \}$$

$$+ \sum_{k=1}^{N} \mathrm{trace} \{ \Omega_i^T S_k^{-1} \Omega_j T_k + \tfrac{1}{2} \mathrm{trace} \left| S_k^{-1} \frac{\partial S_k}{\partial \beta_i} \right| \mathrm{trace} \left| S_k^{-1} \frac{\partial S_k}{\partial \beta_j} \right| \qquad (6.5.14)$$

where $\hat{x}_k$ and T_k are the mean and covariance, respectively, of $\tilde{x}_k$ and are given by

$$\hat{x}_{k+1} = F_k \hat{x}_k + G_k u_k, \qquad\qquad \hat{x}_0 = 0 \qquad (6.5.15)$$

$$T_{k+1} = F_k T_k F_k^T + H_k S_k H_k^T, \qquad T_0 = 0 \qquad (6.5.16)$$

Summing up, Eq. (6.5.14) provides a means for computing the information matrix before the experiment is performed. It is clear that M depends on the experimental conditions, namely, the input sequence $(u_0, \ldots, u_{N-1})$, the sampling intervals $(\Delta_0, \ldots, \Delta_{N-1})$, and the presampling filter. Hence, it would be possible, at least in principle, to minimize $-\log \det M$ with respect to these design variables.

Suitable algorithms have been developed for this type of optimization [200] but are extremely complex except for simple situations and unfortunately do not give a great deal of insight into the problem.

In Chapter 7 a suboptimal on-line design scheme will be described which allows the theory developed in this section to be applied to practical systems. In the next subsection it will be shown that, with additional constraints on the design, alternative simple design methods exist.

These alternative methods give further insight into the effect of different sampling strategies on the achievable accuracy.

6.5.2 *Uniform Sampling Interval Design*   Consider the following model of a constant coefficient linear single-input, single-output system with colored measurement noise:

$$\dot{x}(t) = Ax(t) + Bu(t) \tag{6.5.17}$$

$$y(t) = Cx(t) + Du(t) + w(t) \tag{6.5.18}$$

where $t \in T = [0, t_f]$, $x: T \rightarrow R^n$, is the state vector, u the input, y the output, and w the stationary measurement noise, assumed to have a zero mean Gaussian distribution with rational power density spectrum $\psi(f)$ for all $f \in (-\infty, \infty)$.

For large experiment time t_f, it is convenient to transform the above model to the frequency domain

$$y(s) = T(s)u(s) + w(s), \qquad s = j2\pi f \tag{6.5.19}$$

where

$$T(s) = (C(sI - A)^{-1}B + D) \tag{6.5.20}$$

Fisher's information for the parameters, β, in A, B, C, and D (assuming for the moment continuous observations over $[0, t_f]$) is given by

$$M_\beta = t_f \overline{M}_\beta \tag{6.5.21}$$

where $\overline{M}_\beta$ is the average information matrix per unit time with ikth element given by

$$(\overline{M}_\beta)_{ik} = \int_{-\infty}^{\infty} \frac{\partial T^T(j2\pi f)}{\partial \beta_i} \psi^{-1}(f) \frac{\partial T(j2\pi f)}{\partial \beta_k} d\xi(f) \tag{6.5.22}$$

where $\xi(f)$ is the input spectral distribution function satisfying the usual input power constraint:

$$1 = \int_{-\infty}^{\infty} d\xi(f) \tag{6.5.23}$$

Now, defining Q by

$$Q(s) = \partial T(s)/\partial \beta \tag{6.5.24}$$

the average information matrix is

$$\overline{M}_\beta = \int_{-\infty}^{\infty} Q^{\mathrm{T}}(-j2\pi f)\psi^{-1}(f)Q(j2\pi f)\,d\xi(f) \qquad (6.5.25)$$

Assuming that the test signal spectrum is band limited to $[-f_h, f_h]$, Eq. (6.5.25) may equivalently be written as

$$\overline{M}_\beta = \int_{-f_h}^{f_h} Q^{\mathrm{T}}(-j2\pi f)\psi^{-1}(f)Q(j2\pi f)\,d\xi(f) \qquad (6.5.26)$$

The effect of sampling on the average information matrix is now investigated. Suppose that the output $y(t)$, $t \in [0, t_f]$ is sampled at greater than the Nyquist rate for f_h (i.e., the sampling frequency f_s, is greater than $2f_h$) [207]. This form of sampler does not distort that part of the spectrum of y arising from the input. The part of the output spectrum arising from the noise will, however, be distorted due to aliasing. This distortion will result in an "information" loss as indicated by the following expression for the average information matrix (*per unit time*) from the *sampled* observations (obtained from (6.5.26) with ψ replaced by the distorted spectrum ψ_s):

$$\overline{M}_\beta{}^S = \int_{-f_h}^{f_h} Q^{\mathrm{T}}(-j2\pi f)\psi_s^{-1}(f)Q(j2\pi f)\,d\xi(f) \qquad (6.5.27)$$

where $\psi_s(f)$ is the aliased noise spectrum [28] given by

$$\psi_s(f) = \psi(f) + \sum_{k=1}^{\infty} \{\psi(kf_s + f) + \psi(kf_s - f)\}, \qquad f \in \left(-\frac{f_s}{2}, \frac{f_s}{2}\right) \qquad (6.5.28)$$

The fact that $\overline{M} - \overline{M}^S$ is nonnegative definite follows from Eq. (6.5.28), which indicates that $\psi_s(f) - \psi(f)$ is positive definite and hence that

$$\overline{M}_\beta - \overline{M}_\beta{}^S = \int_{-f_h}^{f_h} Q^{\mathrm{T}}(-j2\pi f)\{\psi^{-1}(f) - \psi_s^{-1}(f)\}Q(j2\pi f)\,d\xi(f) \quad (6.5.29)$$

is nonnegative definite.

It follows that $-\log \det \overline{M}_\beta{}^S \geq -\log \det \overline{M}_\beta$, i.e., the experiment with sampled data cannot be better than the corresponding experiment with continuous observations. This result is, of course, quite expected. However, it will now be shown that, by the inclusion of a suitable presampling filter, equality can be achieved.

Theorem 6.5.1   For the design criterion $-\log \det \overline{M}_\beta$, and sampling rate f_s satisfying $f_s > 2f_h$, then a presampling filter having transfer function F with the property

$$|F(j2\pi f)| = 0 \qquad \text{for all } f \text{ not contained in } (-f_s/2, f_s/2) \quad (6.5.30)$$

and invertible otherwise, gives the same value for the design criterion using the sampled data as with continuous observations.

Proof   The inclusion of a presampling filter having transfer function $F(s)$ modifies the expression for the average information matrix (6.5.27) to

$$\overline{M}_\beta^{FS} = \int_{-f_h}^{f_h} Q^T(-j2\pi f)F^T(-j2\pi f)\psi_{FS}^{-1}(f)F(j2\pi f)Q(j2\pi f)\, d\xi(f) \quad (6.5.31)$$

where the filtered and sampled noise spectrum $\psi_{FS}(f)$ is

$$\psi_{FS}(f) = \psi_F(f) + \sum_{k=1}^{\infty} \{\psi_F(kf_s + f) + \psi_F(kf_s - f)\}$$

and $\psi_F(f)$ is the filtered noise spectrum given by

$$\psi_F(f) = F(j2\pi f)\psi(f)F^T(-j2\pi f) \quad (6.5.32)$$

Now, for a presampling filter satisfying (6.5.30), $\psi_{FS}(f)$ reduces to $\psi_F(f)$ as is readily verified. Furthermore, substituting (6.5.32) into (6.5.31) yields

$$\overline{M}_\beta^{FS} = \int_{-f_h}^{f_h} Q^T(-j2\pi f)\psi^{-1}(f)Q(j2\pi f)\, d\xi(f) \quad (6.5.33)$$

and comparison of (6.5.33) with (6.5.26) yields

$$\overline{M}_\beta^{FS} = \overline{M}_\beta \quad (6.5.34)$$

This completes the proof of the theorem. ∇

The significance of (6.5.34) is that the specified presampling filter F eliminates the information loss due to sampling. This form of filter is often called an aliasing filter [28].

So far, only the average information matrix per unit *time* has been considered. However, if the total number of samples is constrained to be N, say, then it is more appropriate to consider the average information matrix per *sample*. If the sampling rate is f_s samples per second, then the average information matrix per sample is given by

$$\tilde{M}_\beta = \overline{M}_\beta^S/f_s \quad (6.5.35)$$

where $\overline{M}_\beta^S$ is the average information matrix per unit time corresponding to an input spectrum having highest frequency component f_h and samples collected at $f_s > 2f_h$. Furthermore, inclusion of an aliasing filter ensures that $\overline{M}_\beta{}^S$ in (6.5.35) can be replaced by $\overline{M}_\beta$:

$$\tilde{M}_\beta = \overline{M}_\beta / f_s \qquad (6.5.36)$$

We wish to choose the sampling rate f_s and the input spectral distribution $\xi(f)$, $f \in (-f_s/2, f_s/2)$ so that $J = -\log \det \tilde{M}_\beta$ is minimized. We first note that for fixed f_s, J can be minimized by choosing the input spectral distribution to minimize $J' = -\log \det \overline{M}_\beta$ (using any technique of Section 6.4). Denote the minimizing $\overline{M}_\beta$ by $\overline{M}_\beta{}^*(f_s)$, where the dependence on f_s arises from the fact that the input spectrum is constrained to lie in $(-f_s/2, f_s/2)$. Finally, we simply determine f_s by minimizing $J = -\log \det \overline{M}_\beta{}^*(f_s)/f_s$ with respect to f_s.

The above joint optimal design with constrained number of samples, in general, leads to a compressed optimal input spectrum compared with the continuous observation case. The above design procedure is illustrated in Fig. 6.5.1.

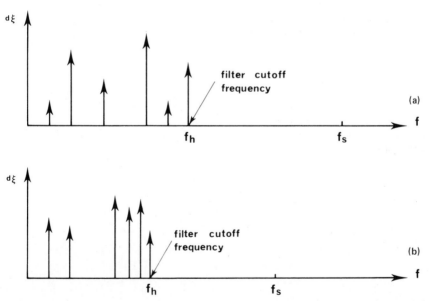

Figure 6.5.1. (a) Optimal spectrum for fixed experiment time. (b) Optimal spectrum for fixed number of samples.

The above result is heuristically reasonable since the compression of the input spectrum leads to a lower sampling rate and hence increased experiment time, which is potentially capable of yielding more information per sample. We illustrate this by means of two simple examples.

Example 6.5.1   Consider the first-order model of Example 6.4.3

$$y = [1/(1 + s\tau)]u + n$$

where n is wide band noise. It was shown in Example 6.4.3 that the average information matrix *per unit time* is

$$\overline{M}(\omega) = 2\omega^2/(1 + \omega^2\tau^2)^2 \tag{6.5.37}$$

The optimal input spectrum for fixed experiment time was shown to correspond to a single sinusoid of frequency $\bar{\omega} = 1/\tau$.

Assuming that an aliasing filter is used, the average information matrix *per sample* is given by

$$\tilde{M} = \overline{M}/f_s$$

Since we are sampling at the Nyquist rate, we have $f_s = \omega/\pi$. Hence

$$\tilde{M} = 2\pi\omega/(1 + \omega^2\tau^2)^2$$

and the optimal input frequency is thus

$$\tilde{\omega} = 1/\tau\sqrt{3}$$

together with optimal sampling rate

$$f_s = 1/\tau\pi\sqrt{3}$$

Example 6.5.2   Consider the model of Example 6.4.5

$$y = \frac{\omega_0^2}{s^2 + 2\xi\omega_0 s + \omega_0^2} u + n$$

where n is wide band noise. It was shown in Example 6.4.5 that the average information matrix *per unit time* is

$$\overline{M}(\omega) = \frac{2}{\left\{\left|1 - \left(\dfrac{\omega}{\omega_0}\right)^2\right|^2 + 4\xi^2\left(\dfrac{\omega}{\omega_0}\right)^2\right\}^2} \begin{bmatrix} 4\dfrac{\omega^4}{\omega_0^6} + 4\xi^2\dfrac{\omega}{\omega_0^4} & -4\xi\dfrac{\omega^2}{\omega_0^3} \\[2ex] -4\xi\dfrac{\omega^2}{\omega_0^3} & 4\dfrac{\omega^2}{\omega_0^2} \end{bmatrix}$$

The optimal input for fixed experiment time was seen to be a single sinusoid of frequency $\bar{\omega} = (\omega_0/\sqrt{5})\{(1 - 2\xi^2) + [(1 - 2\xi^2)^2 + 15]^{1/2}\}^{1/2}$.

Assuming the inclusion of an aliasing filter, the average information matrix *per sample* is given by

$$\tilde{M}(\omega) = \pi\overline{M}(\omega)/\omega$$

and hence the optimal input for N samples is a single sinusoid of frequency

$$\tilde{\omega} = (\omega_0/\sqrt{3})\{(1 - 2\xi^2) + [(1 - 2\xi^2)^2 + 3]^{1/2}\}^{1/2}$$

with optimal sampling rate $= \tilde{\omega}/\pi$. The variation of $\tilde{\omega}$ and $\bar{\omega}$ versus damping ratio ξ is shown in Fig. 6.5.2. ∇

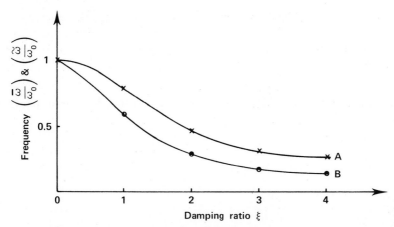

Figure 6.5.2. Variation of optimal input frequency versus damping. Ratio for second-order system with "white" output noise. A—continuous observations $(\overline{\omega}/\omega_0)$; B—sampled observations $(\tilde{\omega}/\omega_0)$.

6.5.3 *Suboptimal design of nonuniform sampling*   The most general sampling interval design problem arises when the samples can be arbitrarily located. This has been studied in Section 6.5.1. However, the resultant design algorithm was complex and offered little insight. For this reason, attention was focused on the fixed sampling rate problem in Section 6.2.5. For this case, we have seen that the design can be carried out in a straight-forward fashion in the frequency domain. However, it is sometimes the case that a single sampling rate cannot cover the whole frequency range of interest. We are thus led to consider a suboptimal approach to the design of nonuniform sampling schemes in which a number of subexperiments is considered, each having a fixed sampling rate.

It will be clear from the comments of Section 6.4.2 that the optimal strategy will be to apply a single sinusoid in each subexperiment and to sample at the appropriate Nyquist rate. It is also clear that a separate aliasing filter should be used in each subexperiment. This arises from the fact that it is better to match the sampling rate to each frequency component of the input spectrum in turn, rather than being forced to use a uniform high sampling rate because of the existence of high frequency components.

The only change required for design purposes is to replace $\tilde{M}_\beta$ given in (6.5.36) by the following expression for the average information matrix per sample:

$$\tilde{M}_\beta = \sum_{k=1}^{l} \frac{\pi \lambda_k}{\omega_k} \overline{M}(\omega_k) \tag{6.5.38}$$

where $l = p(p + 1)/2$ (cf. Theorem 6.4.2) and λ_k denotes the fraction of the total number of samples collected in the kth subexperiment (comprising a single sinusoid of amplitude $\sqrt{2}$ and frequency ω_k together with sampling rate ω_k/π).

Example 6.5.3   Consider the following third-order three parameter model

$$y = \frac{1}{(as + 1)(bs + 1)(cs + 1)} u + n \qquad (6.5.39)$$

where n is wide band noise.

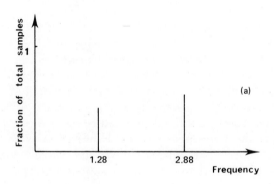

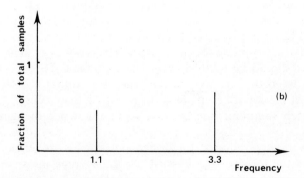

Figure 6.5.3. Optimal designs for example 6.5.3 (with $a = 0.1$, $b = 0.2$, and $c = 0.5$). (a) Optimal uniform sampling strategy. (b) Optimal subexperiment design.

The average information matrix per unit time for the parameters $\beta = (a, b, c)^{\mathrm{T}}$ is

$$
M = \begin{bmatrix}
\dfrac{\omega^2}{(a^2\omega^2 + 1)^2(b^2\omega^2 + 1)} & \dfrac{\omega^2(ab\omega^2 + 1)}{(a^2\omega^2 + 1)^2(b^2\omega^2 + 1)^2} & \dfrac{\omega^2(ac\omega^2 + 1)}{(a^2\omega^2 + 1)^2(b^2\omega^2 + 1)} \\[2pt]
\times (c^2\omega^2 + 1) & \times (c^2\omega^2 + 1) & \times (c^2\omega^2 + 1) \\[8pt]
\dfrac{\omega^2(ab\omega^2 + 1)}{(a^2\omega^2 + 1)^2(b^2\omega^2 + 1)^2} & \dfrac{\omega^2}{(a^2\omega^2 + 1)(b^2\omega^2 + 1)^2} & \dfrac{\omega^2(bc\omega^2 + 1)}{(a^2\omega^2 + 1)(b^2\omega^2 + 1)^2} \\[2pt]
\times (c^2\omega^2 + 1) & \times (c^2\omega^2 + 1) & \times (c^2\omega^2 + 1) \\[8pt]
\dfrac{\omega^2(ac\omega^2 + 1)}{(a^2\omega^2 + 1)^2(b^2\omega^2 + 1)} & \dfrac{\omega^2(bc\omega^2 + 1)}{(a^2\omega^2 + 1)(b^2\omega^2 + 1)^2} & \dfrac{\omega^2}{(a^2\omega^2 + 1)(b^2\omega^2 + 1)} \\[2pt]
\times (c^2\omega^2 + 1)^2 & \times (c^2\omega^2 + 1)^2 & \times (c^2\omega^2 + 1)^2
\end{bmatrix}
$$

It turns out that the optimal information matrix can be achieved with a two-line input spectrum. Figure 6.5.3 shows the optimal design for a uniform sampling rate together with the optimal design for two subexperiments, each with a different sampling rate and aliasing filter. The improvement in det $\overline{M}$ for the two subexperiment case as compared with the single uniform sampling rate experiment was about $3:1$. This represents only about 20% improvement in parameter accuracy, but for larger spreads in time constants the improvement can be substantial, e.g., $200:1$ for a $50:1$ spread in time constants [204].

6.6.   DESIGN FOR STRUCTURE DISCRIMINATION

The previous sections of this chapter have been concerned with the design of experiments for accurate parameter estimation within a model of specified structure. However, it is sometimes the case that there are two or more rival models and the purpose of the experiment may be to determine which, if any, of the models are adequate.

The most difficult aspect of the structure discrimination problem is to obtain a meaningful criterion. A nice discussion of some of the difficulties associated with this problem is given in [138]. The key problem is that we often prefer a simple model even though a more complex model is bound to give a better fit to the data. This preference is tied up with some vague notion of model cost effectiveness, but this may be difficult to quantify in practical situations. The problem is compounded by the fact that it will seldom, if ever, be the case that the true data generating mechanism corresponds to one of the models under study. Structure discrimination is currently an area of active research and a number of different criteria for structure discrimination have been proposed based on hypotheses testing,

Bayesian methods, and information theory. The interested reader is referred to [64], [65], [144]–[146], and [219] for further details.

Here we shall adopt a classical hypothesis testing approach and use the likelihood ratio test as a means of discriminating between competing model structures. Our aim shall be to adjust the experimental conditions so that the power of the test is maximized.

It was shown in Section 3.6 that the power of the likelihood ratio test is related to the noncentrality parameter of the noncentral χ^2 distribution. As in Section 3.6, we assume that the parameter vector is divided into two sections as follows:

$$\theta = \begin{vmatrix} \alpha \\ -- \\ \beta \end{vmatrix} \tag{6.6.1}$$

We assume that θ corresponds to a general model and that the structure discrimination problem can be stated as a hypothesis testing problem with null hypothesis $H_0: \alpha = 0$. It has been shown in Section 3.6 that for a specific alternative hypothesis $H_A: \alpha = \alpha_A$, the noncentrality parameter is

$$h = N\alpha_A^T[\overline{M}_{\alpha\alpha} - \overline{M}_{\alpha\beta}\overline{M}_{\beta\beta}^{-1}\overline{M}_{\alpha\beta}^T]\alpha_A \tag{6.6.2}$$

where N is the number of data points and $\overline{M}_{\alpha\alpha}$, $\overline{M}_{\alpha\beta}$, and $\overline{M}_{\beta\beta}$ are the partitions of the average information matrix for θ, i.e.,

$$\overline{M}_\theta = \begin{vmatrix} \overline{M}_{\alpha\alpha} & \overline{M}_{\alpha\beta} \\ \overline{M}_{\alpha\beta}^T & \overline{M}_{\beta\beta} \end{vmatrix} \tag{6.6.3}$$

For fixed significance level (i.e., fixed probability of rejecting H_0 when it is true), the power (i.e., 1—the probability of accepting H_0 when H_A is true) of the test is made large by making the noncentrality parameter h large. It is clear from (6.6.2) that the experimental conditions affect h via the information matrix. Three experiment design strategies have been studied:

(i)   locally optimum designs which maximize h for a particular α_A of interest;

(ii)   minimax designs which maximize the minimum value of the noncentrality parameter over the sphere $\alpha_A^T\alpha_A = 1$ (equivalent to maximizing minimum eigenvalue);

(iii) D_s-optimum designs which maximize

$$J_s = \det[\overline{M}_{\alpha\alpha} - \overline{M}_{\alpha\beta}\overline{M}_{\beta\beta}^{-1}\overline{M}_{\alpha\beta}^T] \tag{6.6.4}$$

Here we shall concentrate on the D_s criterion as it does not depend on parameter scaling factors and because, at the beginning of an experiment, there is unlikely to be reliable information available on the particular value of α_A that it is desired to discriminate against.

Applying Result E.1.2 to Eq. (6.6.4) yields

$$J_s = \det \overline{M}_\theta / \det \overline{M}_{\beta\beta} \tag{6.6.5}$$

The design criterion given in (6.6.5) is a simple scalar function of the information matrix and hence Theorems 6.4.1–6.4.4 apply without modification to the current problem. Thus an optimal experiment can be found in which the input comprises a finite number ($2n$ in the single-input, single-output case) of sinusoidal components. Of course, there will be differences in the frequencies and powers between D and D_s optimal designs.

An analog for D_s optimality of Theorem 6.4.7 can be proved by defining the *structure dispersion* for design ξ and frequency ω as

$$v_s(\xi, \omega) = \text{trace}\{\overline{M}_\theta(\xi)^{-1}\overline{M}_\theta(\omega) - \overline{M}_{\beta\beta}(\xi)^{-1}\overline{M}_{\beta\beta}(\omega)\} \tag{6.6.6}$$

We then have the following theorem:

Theorem 6.6.1 The following characterizations of a D_s optimal power construed design ($\xi^* \in K$) are equivalent:

(1) the design ξ^* maximizes $\det \overline{M}_\theta / \det \overline{M}_{\beta\beta}$ (minimizes $\log \det \overline{M}_\theta + \log \det \overline{M}_{\beta\beta}$)

(2) the design ξ^* minimizes

$$\max_{\omega \in [0, \pi]} v_s(\xi, \omega)$$

(3) $\displaystyle\max_{\omega \in [0, \pi]} v_s(\xi^*, \omega) = \text{dimension } \alpha$

Proof See proof for Theorem 6.4.7. ▽

Sequential design procedures similar to those described in Section 6.4 can also be developed for the D_s optimal case. We illustrate the above ideas by a simple example.

Example 6.6.1 Consider the following linear single-input, single-output system

$$y_t = G_1(z)u_t + G_2(z)\varepsilon_t \tag{6.6.7}$$

where

$$G_1(z) = K(1 + b_1 z^{-1})/(1 + a_1 z^{-1})(1 + a_2 z^{-1}) \tag{6.6.8}$$

$$G_2(z) = 1 \tag{6.6.9}$$

$\{\varepsilon_t\}$ is a white noise sequence with $N(0, \sigma^2)$ distribution.

The parameter vector is $\phi = (K, a_1, a_2, b_1, \sigma^2)^T$ and the information matrix is given by $\int_0^\pi \overline{M}_\phi(\omega)\, d\xi(\omega)$, where

$$
\overline{M}_\phi(\omega) = \begin{bmatrix}
\overline{M}_{KK} & \overline{M}_{Ka_1} & \overline{M}_{Ka_2} & \overline{M}_{Kb_1} & 0 \\
\overline{M}_{Ka_1} & \overline{M}_{a_1a_1} & \overline{M}_{a_1a_2} & \overline{M}_{a_1b_1} & 0 \\
\overline{M}_{Ka_2} & \overline{M}_{a_1a_2} & \overline{M}_{a_2a_2} & \overline{M}_{a_2b_1} & 0 \\
\overline{M}_{Kb_1} & \overline{M}_{a_1b_1} & \overline{M}_{a_2b_1} & \overline{M}_{b_1b_1} & 0 \\
0 & 0 & 0 & 0 & 1/2\sigma^4
\end{bmatrix}
\tag{6.6.10}
$$

with

$$
\overline{M}_{KK} = \frac{1 + b_1{}^2 + 2b_1 \cos \omega}{(1 + a_1{}^2 + 2a_1 \cos \omega)(1 + a_2{}^2 + 2a_2 \cos \omega)}
\tag{6.6.11}
$$

$$
\overline{M}_{Ka_1} = \frac{-K(1 + b_1{}^2 + 2b_1 \cos \omega)(a_1 + \cos \omega)}{(1 + a_1{}^2 + 2a_1 \cos \omega)^2(1 + a_2 + 2a_2 \cos \omega)}
\tag{6.6.12}
$$

$$
\overline{M}_{Ka_2} = \frac{-K(1 + b_1{}^2 + 2b_1 \cos \omega)(a_2 + \cos \omega)}{(1 + a_1{}^2 + 2a_1 \cos \omega)(1 + a_2 + 2a_2 \cos \omega)^2}
\tag{6.6.13}
$$

$$
\overline{M}_{Kb_1} = \frac{Kb_1 + K \cos \omega}{(1 + a_1{}^2 + 2a_1 \cos \omega)(1 + a_2{}^2 + 2a_2 \cos \omega)}
\tag{6.6.14}
$$

$$
\overline{M}_{a_1a_1} = \frac{K^2(1 + b_1{}^2 + 2b_1 \cos \omega)}{(1 + a_1{}^2 + 2a_1 \cos \omega)^2(1 + a_2{}^2 + 2a_2 \cos \omega)}
\tag{6.6.15}
$$

$$
\overline{M}_{a_1a_2} = \frac{K^2(1 + b_1{}^2 + 2b_1 \cos \omega)(1 + a_1a_2 + (a_1 + a_2) \cos \omega)}{(1 + a_1{}^2 + 2a_1 \cos \omega)^2(1 + a_2{}^2 + 2a_2 \cos \omega)^2}
\tag{6.6.16}
$$

$$
\overline{M}_{a_1b_1} = \frac{-K^2(1 + a_1b_1 + (a_1 + b_1) \cos \omega)}{(1 + a_1{}^2 + 2a_1 \cos \omega)(1 + a_2{}^2 + 2a_2 \cos \omega)}
\tag{6.6.17}
$$

$$
\overline{M}_{a_{22}} = \frac{K^2(1 + b_1{}^2 + 2b_1 \cos \omega)}{(1 + a_1{}^2 + 2a_1 \cos \omega)(1 + a_2{}^2 + 2a_2 \cos \omega)^2}
\tag{6.6.18}
$$

$$
\overline{M}_{a_2b_1} = \frac{-K^2(1 + a_2b_1 + (a_2 + b_1) \cos \omega)}{(1 + a_1{}^2 + 2a_1 \cos \omega)(1 + a_2{}^2 + 2a_2 \cos \omega)^2}
\tag{6.6.19}
$$

$$
\overline{M}_{b_1b_1} = \frac{K^2}{(1 + a_1{}^2 + 2a_1 \cos \omega)(1 + a_2{}^2 + 2a_2 \cos \omega)}
\tag{6.6.20}
$$

The following nominal values of the parameters are assumed

$$
\begin{aligned}
K &= 1 \\
a_1 &= -0.5 \\
a_2 &= -0.8 \\
b_1 &= -0.82
\end{aligned}
\tag{6.6.21}
$$

We reparameterize the system as

$$\theta = L\phi \qquad (6.6.22)$$

with

$$L = \begin{bmatrix} 0 & 0 & 1 & -1 & 0 \\ 0 & 0 & 1 & 0 & 0 \\ 0 & 1 & 0 & 0 & 0 \\ 1 & 0 & 0 & 0 & 0 \\ 0 & 0 & 0 & 0 & 1 \end{bmatrix} \qquad (6.6.23)$$

This gives

$$\theta^{\mathrm{T}} = (a_2 - b_1, a_2, a_1, K, \sigma^2) \qquad (6.6.24)$$

and

$$\overline{M}_\theta = L^{\mathrm{T}} \overline{M}_\phi L \qquad (6.6.25)$$

Finally we wish to test the null hypothesis that the model order should be 1, i.e., H_0: $\alpha = 0$ where α is the first component of θ.

For this problem all of the design criteria mentioned earlier are equivalent since the dimension of α is 1. The optimal experiment design for the above problem was found [227] by the sequential design procedure and is shown in Fig. 6.6.1.

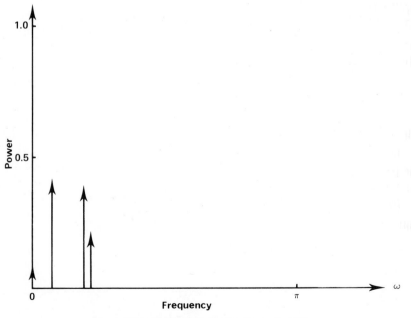

Figure 6.6.1. Optimal design for Example 6.6.1.

For the above example it turns out there is very little difference between the D and D_s optimal experiments. However, this result is example dependent. Application of the above ideas to more general problems is described in [227], [188], [169], [228], and [229].

6.7.   CONCLUSIONS

This chapter has considered the question of achievable accuracy in identification. It has been shown that the experimental conditions may be adjusted, subject to constraints, to make an experiment maximally informative. Simple experiment design procedures have been presented for a number of commonly occurring situations.

The insight into the identification problem gained by a knowledge of the factors affecting the information content of data gives an appreciation of the limitations of data even when no design is carried out.

PROBLEMS

6.1.   Show that the criterion $J = \log \det M$ (Eq. (6.2.7)) is independent of nonlinear changes in the parameters that are one to one. (Assume the Jacobian exists and is nonsingular.)

6.2.   For the power constraint (6.3.12), show that the input $u_t = 1$ for all t achieves the maximum of $J = \text{trace } M$. What would be wrong with this input from a pragmatic viewpoint?

6.3.   Consider the problem of estimating the pulse response of a linear system from observations corrupted by zero mean Gaussian white noise. The model for the process can be expressed as

$$y_k = \sum_{j=1}^{m} h_j u_{k-j} + \varepsilon_k$$

where $\{y_k\}$ is the output, $\{u_k\}$ the input, and $\{\varepsilon_k\}$ a white Gaussian sequence.

(a)   Write down an expression for Fisher's information matrix M, for the parameters $h_1, \ldots, h_m$ assuming that N observations of $\{y_k\}$ will be used for estimation purposes.

(b)   An "m-sequence" is used as the input to the process and it is known that an "m-sequence" has the property that

$$\sum_{k=1}^{N} u_{k-j} u_{k-j-i} = N \qquad \text{if} \quad i = 0$$
$$= -1 \qquad \text{if} \quad i \neq 0$$

Show that the variance of the BLUE estimator of each of the parameters $h_1, \ldots, h_m$ is

$$\text{var}(\hat{h}_i) = \left| \frac{N - m + 2}{N^2 - Nm + 2N - m + 1} \right| \sigma^2$$

(Hint: Express M in the form $D - vv^T$ where D is diagonal and v is a vector of 1's).

(c)   Compare the variances obtained in part (b) with the theoretical minimum variances obtained with the optimal input having the same power as the "m-sequence."

6.4.   For the system considered in Problem 6.3, show that a binary sequence (if it exists) yields minimum variance estimates if the input is constrained to lie between ± 1.

6.5.   Establish Eq. (6.3.22) by differentiating (6.3.19) with respect to β_i.

6.6.   Show that the fact that $\partial \omega_t / \partial \beta_i$ (Eq. 6.3.22)) is independent of ω_t is a consequence of the fact that $G_2(\infty) = 1$.

6.7.   Establish Eq. (6.3.23) from (6.3.22) using the fact that the third central moment of a normal distribution is zero and that the fourth central moment is $3\Sigma^2$ where Σ is the variance (cf. Problems 2.8 and 2.9).

6.8.   Establish Eq. (6.4.27).

6.9.   Establish Eq. (6.5.6).

6.10.   Consider the following model for a process $\{y_k\}$:

$$y_k = ay_{k-1} + u_k + \varepsilon_k$$

where $|a| < 1$, $\{u_k\}$ is an adjustable input and $\{\varepsilon_k\}$ is a Gaussian white noise sequence with zero mean and variance σ^2.

(a)   Develop an expression for Fisher's information for the parameter a in terms of the spectral density of $\{u_k\}$ and $\{\varepsilon_k\}$. (Let the number of observations be large.)

(b)   Assuming that an input of the form

$$u_k = \sqrt{2} \cos[kw], \quad k = 1, 2, \ldots, \quad w \in (0, \pi)$$

$$= \cos[kw], \qquad k = 1, 2, \ldots, \quad w = 0 \text{ or } \pi$$

will be used, show that the information about a is maximized if $w \Rightarrow 0$ when $a > 0$, and $w \Rightarrow \pi$ when $a < 0$.

(c)   Show that $w = 0$ gives a constant input whereas $w = \pi$ gives an input which switches alternately between $+1$ and -1.

(d)   Comment on the assumption made in part (b) in relation to the set of all possible information matrices that can be achieved with an input having unity power.

6.11.   Consider the following model for a stochastic process $\{y_k\}$:

$$y_k = -a_1 y_{k-1} - a_2 y_{k-2} + \varepsilon_k$$

where $\{\varepsilon_k\}$ is a white noise process with zero mean and variance v. (Assume the process is stable.)

It is desired to estimate the parameter vector $\beta = (a_1, a_2, v)^T$ from N equally spaced observations of $\{y_k\}$.

(a)   Assuming that $\{\varepsilon_k\}$ has a Gaussian distribution and that N is sufficiently large to be able to ignore initial condition effects, develop an expression for Fisher's information matrix for the parameter vector β.

(b)   Using the result of part (a), or otherwise, show that the information matrix is diagonal if the nominal value of the parameter a_1 is zero.

CHAPTER

7

Recursive Algorithms

7.1. INTRODUCTION

The algorithms discussed in this chapter have the common property that they can be computed in real time. Possible advantages of these algorithms are

(i)   the ability to track time varying parameters,
(ii)   small computational burden at each step thus enabling use of small computers for off- or on-line data analysis, and
(iii)   as an aid in model building where form of parameter time variation may suggest cause of a model inadequacy [109].

We shall consider three classes of algorithm, viz., parameter estimation, experiment design, and stochastic control. We devote most attention to the principles and properties of the parameter estimation and experiment design algorithms. We then show how these algorithms can be utilized in active or passive adaptive control schemes.

7.2.   RECURSIVE LEAST SQUARES

The least squares method was studied in Chapter 2. In its simplest form, the method is based on the minimization of functions of the form

$$S(\theta) = \sum_{t=1}^{N} (y_t - x_t^T\theta)^2 \tag{7.2.1}$$

where y_t is the scalar observation at time t and $\theta \in R^p$ is a parameter vector.

The least squares estimate of θ, based on observations $y_1, y_2, \ldots, y_N$ is given by (cf. Eq. (2.2.5))

$$\hat{\theta}_N = (X_N^T X_N)^{-1} X_N^T Y_N \tag{7.2.2}$$

where

$$X_N = \begin{bmatrix} x_1^T \\ x_2^T \\ \vdots \\ x_N^T \end{bmatrix}, \qquad Y_N = \begin{bmatrix} y_1 \\ y_2 \\ \vdots \\ y_N \end{bmatrix} \tag{7.2.3}$$

To establish the recursive least square algorithm, we consider the situation where we make just one extra observation, viz., y_{N+1}. Defining

$$P_N = (X_N^T X_N)^{-1} \tag{7.2.4}$$

we have

$$P_{N+1} = \left[\begin{bmatrix} X_N \\ x_{N+1}^T \end{bmatrix}^T \begin{bmatrix} X_N \\ x_{N+1}^T \end{bmatrix} \right]^{-1} \tag{7.2.5}$$

i.e.,

$$P_{N+1} = [X_N^T X_N + x_{N+1} x_{N+1}^T]^{-1} \tag{7.2.6}$$

The inversion indicated in (7.2.6) is facilitated by the matrix inversion lemma (Lemma E.1.1) which forms the basis of the recursive algorithm described in the following result:

Result 7.2.1   The least squares estimate $\hat{\theta}$ can be computed recursively from the following equations:

$$\hat{\theta}_{N+1} = \hat{\theta}_N + K_{N+1}(y_{N+1} - x_{N+1}^T\hat{\theta}_N) \tag{7.2.7}$$

where K_N is a time varying gain matrix satisfying

$$K_{N+1} = P_N x_{N+1}/(1 + x_{N+1}^T P_N x_{N+1}) \tag{7.2.8}$$

Furthermore, P_N, as defined in (7.2.4), may be computed recursively from

$$P_{N+1} = \left(I - P_N \frac{x_{N+1}x_{N+1}^T}{1 + x_{N+1}^T P_N x_{N+1}}\right)P_N \qquad (7.2.9)$$

provided a suitable boundary value, P_0 say, is known.

Proof Equation (7.2.9) follows immediately by applying Lemma E.1.1 to Eq. (7.2.6). Hence from Eq. (7.2.2),

$$\hat{\theta}_{N+1} = P_{N+1}X_{N+1}^T Y_{N+1} = P_{N+1}[X_N^T Y_N + x_{N+1}y_{N+1}]$$

$$= \left|I - P_N \frac{x_{N+1}x_{N+1}^T}{(1 + x_{N+1}^T P_N x_{N+1})}\right| P_N[X_N^T Y_N + x_{N+1}y_{N+1}]$$

$$= \hat{\theta}_N + P_N x_{N+1}y_{N+1} - P_N \frac{x_{N+1}x_{N+1}^T}{(1 + x_{N+1}^T P_N x_{N+1})} P_N x_{N+1}y_{N+1}$$

$$\quad - K_{N+1}x_{N+1}\hat{\theta}_N$$

$$= \hat{\theta}_N + \frac{P_N x_{N+1}(1 + x_{N+1}^T P_N x_{N+1} - x_{N+1}^T P_N x_{N+1})y_{N+1}}{(1 + x_{N+1}^T P_N x_{N+1})}$$

$$\quad - K_{N+1}x_{N+1}\hat{\theta}_N$$

$$= \hat{\theta}_N + K_{N+1}[y_{N+1} - x_{N+1}\hat{\theta}_N]$$

This completes the proof of the result. $\triangledown$

Remark 7.2.1 The starting value P_0 may be obtained either by evaluating $(X_0^T X_0)^{-1}$, where X_0 is obtained from an initial block of data, $-N_0$ to 0 say, or by simply letting $P_0 = I/\varepsilon$ where ε is small. For large N, the choice of ε is unimportant. $\triangledown$

Remark 7.2.2 The simplicity of the recursive Eqs. (7.2.7)–(7.2.9) arises because of the replacement of a matrix inversion at each step by a simple scalar division. $\triangledown$

Recursive equations of the form (7.2.7)–(7.2.9) have found wide application in parameter estimation dating from Gauss who used a similar technique for estimating the orbit of Ceres.

As discussed in Section 2.6, there may be numerical difficulties associated with the direct solution of Eq. (7.2.2). These difficulties may also be present in the recursive scheme outlined above. However, the numerically robust algorithm described in Section 2.6 also has a recursive counterpart described in the following result.

Result 7.2.2 The least squares estimate can be computed recursively
from

$$\hat{\theta}_{N+1} = R_{N+1}^{-1} \eta_{N+1} \qquad (7.2.10)$$

where R_{N+1} is a $p \times p$ upper triangular matrix and η_{N+1} is a p vector.
R_{N+1} and η_{N+1} can be computed recursively as follows:

$$\begin{bmatrix} R_{N+1} \\ 0 \end{bmatrix} = \Psi_N \begin{bmatrix} R_N \\ x_{N+1}^T \end{bmatrix} \qquad (7.2.11)$$

where Ψ_N is an orthonormal matrix expressible as

$$\Psi_N = Q_p Q_{p-1} Q_{p-2} \cdots Q_1 \qquad (7.2.12)$$

where

$$Q_i = \begin{bmatrix} I_{i-1} & & 0 & \\ & c_i & 0 & s_i \\ 0 & 0 & I_{p-i} & 0 \\ & s_i & 0 & c_i \end{bmatrix} \qquad (7.2.13)$$

where

$$c_i = (R_N)_{ii}/\lambda_i \qquad (7.2.14)$$

$$s_i = (x_{N+1})_i/\lambda_i \qquad (7.2.15)$$

$$\lambda_i = [(R_N)_{ii}^2 + (x_{N+1})_i^2]^{1/2} \qquad (7.2.16)$$

and

$$\begin{bmatrix} \eta_{N+1} \\ v_{N+1} \end{bmatrix} = \Psi_N \begin{bmatrix} \eta_N \\ y_{N+1} \end{bmatrix} \qquad (7.2.17)$$

Proof Consider the situation when N observations have been collected.
We assume that the Householder transformation procedure has been used
and, analogously to Eq. (2.6.27):

$$S_N(\theta) = (\eta_N - R_N \theta)^T (\eta_N - R_N \theta) + S_N(\hat{\theta}_N) \qquad (7.2.18)$$

When another observation has been collected, we have

$$S_{N+1}(\theta) = (\eta_N - R_N \theta)^T (\eta_N - R_N \theta) + (y_{N+1} - x_{N+1}^T \theta)^2 + S_N(\hat{\theta}_N) \qquad (7.2.19)$$

$$= \left\| \begin{bmatrix} \eta_N \\ y_{N+1} \end{bmatrix} - \begin{bmatrix} R_N \\ x_{N+1}^T \end{bmatrix} \theta \right\|^T \left\| \begin{bmatrix} \eta_N \\ y_{N+1} \end{bmatrix} - \begin{bmatrix} R_N \\ x_{N+1}^T \end{bmatrix} \theta \right\| + S_N(\hat{\theta}_N) \qquad (7.2.20)$$

An orthonormal transformation Ψ_N can be found such that

$$\Psi_N \begin{bmatrix} R_N \\ x_{N+1}^T \end{bmatrix} = \begin{bmatrix} R_{N+1} \\ Q \end{bmatrix} \qquad (7.2.21)$$

where R_{N+1} is upper triangular (Lemma 2.6.4). Ψ_N can be expressed as $Q_p Q_{p-1} \cdots Q_1$ where Q_i has the form (7.2.13) (see Problem 7.1).

Define η_{N+1}, v_{N+1} as follows:

$$\begin{vmatrix} \eta_{N+1} \\ v_{N+1} \end{vmatrix} = \Psi_N \begin{vmatrix} \eta_N \\ y_{N+1} \end{vmatrix} \tag{7.2.22}$$

Substituting (7.2.22) and (7.2.21) into (7.2.20) yields

$$S_{N+1}(\theta) = (\eta_{N+1} - R_{N+1}\theta)^{\mathrm{T}}(\eta_{N+1} - R_{N+1}\theta) + S_N(\hat{\theta}_N) + v_{N+1}^2 \tag{7.2.23}$$

Hence, the best estimate of θ at time $N+1$ becomes

$$\hat{\theta}_{N+1} = R_{N+1}^{-1} \eta_{N+1} \tag{7.2.24}$$

and this gives

$$S_{N+1}(\hat{\theta}_{N+1}) = S_N(\hat{\theta}) + v_{N+1}^2 \tag{7.2.25}$$

Induction on N completes the proof. ∇

Equation (7.2.10) must be solved on-line but this presents no difficulty since R_N is upper triangular and thus $\hat{\theta}_N$ can be found by back substitution. The sequence of transformations described in (7.2.13)–(7.2.16) must also be computed on-line. However these transformations are elementary coordinate rotations (in fact this is the motivation for using C_i and S_i since these are actually cos and sin terms). The algorithm described in Result 7.2.2 is, in general, a preferable alternative to the algorithm described in Result 7.2.1 both from the point of view of speed and numerical robustness [230]. The algorithm of Result 7.2.2 is frequently called a *square root estimator* since R_N is propagated rather than P_N $(P_N = R_N^{-1} R_N^{-\mathrm{T}})$.

It is also possible to exploit certain shift invariant properties to develop fast algorithms for recursive estimation. These fast algorithms are generalizations of Levinson's algorithm [233] for solving the normal equations. The interested reader is referred to [234] for further discussion of these algorithms and their applications to recursive identification.

7.3.   TIME VARYING PARAMETERS

The recursive methods described in the previous section are not directly applicable when the parameters vary with time since new data are swamped by past data. However, the algorithms can be modified to handle time varying parameters by discounting old data. We describe three basic ways in which this can be achieved.

7.3.1 *Exponential window* The idea of the exponential window approach is to emphasize artificially the effect of current data by exponentially weighting past data values. We consider the following cost function:

$$S'_{N+1}(\theta) = \alpha S_N'(\theta) + (y_{N+1} - x_{N+1}^T\theta)^2 \tag{7.3.1}$$

where $0 < \alpha < 1$. Note that $\alpha = 1$ gives the standard least squares algorithm. For $0 < \alpha < 1$, we have

Result 7.3.1 The exponentially weighted least squares estimator satisfies the following recursive equations:

$$\hat{\theta}_{N+1} = \hat{\theta}_N + K_{N+1}(y_{N+1} - x_{N+1}^T\hat{\theta}_N) \tag{7.3.2}$$

where

$$K_{N+1} = P_N x_{N+1}/(\alpha + x_{N+1}^T P_N x_{N+1}) \tag{7.3.3}$$

$$P_{N+1} = \frac{1}{\alpha}\left(I - P_N \frac{x_{N+1}x_{N+1}^T}{\alpha + x_{N+1}^T P_N x_{N+1}}\right)P_N \tag{7.3.4}$$

Proof We shall use induction and assume that $S_N'(\theta)$ can be expressed as a quadratic form as follows:

$$S_N'(\theta) = (\theta - \hat{\theta}_N)^T P_N^{-1}(\theta - \hat{\theta}_N) + \beta_N \tag{7.3.5}$$

Then

$$S'_{N+1}(\theta) = \alpha(\theta - \hat{\theta}_N)P_N^{-1}(\theta - \hat{\theta}_N) + (y_{N+1} - x_{N+1}^T\theta)^2 + \beta_N \tag{7.3.6}$$

Collecting terms gives

$$S'_{N+1}(\theta) = \theta^T(\alpha P_N^{-1} + x_{N+1}x_{N+1}^T)\theta - 2(\hat{\theta}_N^T\alpha P_N^{-1} + y_{N+1}x_{N+1}^T)\theta$$
$$+ (\hat{\theta}_N^T\alpha P_N^{-1}\hat{\theta}_N) + y_{N+1}^2 + \beta_N \tag{7.3.7}$$

Completing the square gives

$$S'_{N+1}(\theta) = (\theta - \hat{\theta}_{N+1})^T P_{N+1}^{-1}(\theta - \hat{\theta}_{N+1}) + \beta_{N+1} \tag{7.3.8}$$

where

$$P_{N+1}^{-1} = (\alpha P_N^{-1} + x_{N+1}x_{N+1}^T) \tag{7.3.9}$$

and

$$\hat{\theta}_{N+1} = P_{N+1}(\alpha P_N^{-1}\hat{\theta}_N + x_{N+1}y_{N+1}) \tag{7.3.10}$$

Application of the matrix inversion lemma (Lemma E.1.1) to (7.3.9) yields Eq. (7.3.4). Substituting (7.3.4) into Eq. (7.3.10) and proceeding analogously to the proof of Result 7.2.1 yields Eq. (7.3.2) and (7.3.3) (see Problem 7.2). All that is required to complete the proof is to show that $S_0'(\theta) = (\theta - \hat{\theta}_0)P_0^{-1}(\theta - \hat{\theta}_0) + \beta_0$ for some P_0, β_0. This is always the case if we define S_0 as the sum of the squares of the errors for some initial block of data, i.e.,

$$S_0(\theta) = (Y_0 - X_0\theta)^{\mathrm{T}}(Y_0 - X_0\theta)$$
$$= (\theta - \hat{\theta}_0)^{\mathrm{T}}(X_0{}^{\mathrm{T}}X_0)(\theta - \hat{\theta}_0) + (Y_0 - X_0\hat{\theta}_0)^{\mathrm{T}}(Y_0 - X_0\hat{\theta}_0) \quad (7.3.11)$$

where

$$\hat{\theta}_0 = (X_0{}^{\mathrm{T}}X_0)^{-1}X_0{}^{\mathrm{T}}Y_0 \quad (7.3.12)$$

i.e., choose

$$P_0 = (X_0{}^{\mathrm{T}}X_0)^{-1} \quad (7.3.13)$$

$$\beta_0 = (Y_0 - X_0\hat{\theta}_0)^{\mathrm{T}}(Y_0 - X_0\hat{\theta}_0) \quad (7.3.14)$$

Induction completes the proof. $\triangledown$

Further discussion on this algorithm may be found in [59], [62], and [112].

7.3.2 *Rectangular windows*   The essential feature of rectangular window algorithms is that the estimate at time t is based only on a finite number of past data. All old data are completely discarded.

We first consider a rectangular window of fixed length N. Each time a new data point is added, an old data point is discarded thus keeping the active number of points equal to N. Consider the situation at time $t = i + N$. We first accept the latest observation y_{i+N}. Then Eqs. (7.2.7)–(7.2.9) may be written (with an obvious extension of notation) as

$$P_{i+N, i} = \left(I - P_{i+N-1, i} \frac{x_{i+N}x_{i+N}^{\mathrm{T}}}{(1 + x_{i+N}^{\mathrm{T}}P_{i+N-1, i}x_{i+N})}\right)P_{i+N-1, i} \quad (7.3.15)$$

$$\hat{\theta}_{i+N, i} = \hat{\theta}_{i+N-1, i} + K_{i+N, i}[y_{i+N} - x_{i+N}^{\mathrm{T}}\hat{\theta}_{i+N-1, i}] \quad (7.3.16)$$

$$K_{i+N, i} = P_{i+N-1, i}\left|\frac{x_{i+N}}{1 + x_{i+N}^{\mathrm{T}}P_{i+N-1}x_{i+N}}\right| \quad (7.3.17)$$

where $\hat{\theta}_{i+N, i}$ denotes the estimate of θ based on observations between i and $i + N$, i.e., $y_i, y_{i+1}, \ldots, y_{i+N}$.

To maintain the data window at length N, we now discard the observation at i:

$$P_{i+N, i+1} = \left(I + P_{i+N, i} \frac{x_i x_i^{\mathrm{T}}}{1 - x_i^{\mathrm{T}}P_{i+N, i}x_i}\right)P_{i+N, i} \quad (7.3.18)$$

$$\hat{\theta}_{i+N, i+1} = \hat{\theta}_{i+N, i} - K_{i+N, i+1}[y_i - x_i^{\mathrm{T}}\hat{\theta}_{i+N, i}] \quad (7.3.19)$$

$$K_{i+N, i+1} = P_{i+N, i}\left|\frac{x_i}{1 - x_i^{\mathrm{T}}P_{i+N, i}x_i}\right| \quad (7.3.20)$$

where $\hat{\theta}_{i+N, i+1}$ denotes the estimate of θ based on observations between $i+1$ and $i+N$.

The proof of Eqs. (7.3.18)–(7.3.20) relies upon the matrix inversion lemma and is left as an exercise (Problem 7.3).

The above algorithm requires that the last N data points be stored. For large window sizes, it may be preferable to use an oscillating memory estimator, since this will usually result in a considerable decrease in the storage requirements [59]. The oscillating memory estimator discards old data in blocks of N. The effective memory varies between N and $2N$. When adding new data we simply use Eqs. (7.3.18)–(7.3.20) as for the fixed length rectangular window. When we have reached a window length of $2N$, we simply discard the first N as follows:

$$\hat{\theta}_{i+2N, i+N+1} = P_{i+2N, i+N+1}[P_{i+2N, i+1}^{-1}\hat{\theta}_{i+2N, i+1} - P_{i+N, i+1}^{-1}\hat{\theta}_{i+N, i+1}]$$

$$(7.3.21)$$

$$P_{i+2N, i+N+1}^{-1} = P_{i+2N, i+1}^{-1} - P_{i+N, i+1}^{-1} \tag{7.3.22}$$

Thus, to discard N points, all we need to store are the estimate $\hat{\theta}_{i+N, i+1}$ and the covariance $P_{i+N, i+1}$, rather than the points themselves. The proof of Eqs. (7.3.21)–(7.3.22) is left as an exercise (Problem 7.4) or see [59]. The oscillating memory estimator is illustrated in the following example.

Example 7.3.1[†]   We consider the problem of estimating the orbit of a satellite in approximately synchronous orbit with equation of motion

$$\ddot{r} = -C^2 r$$

where $C = 7.293 \times 10^{-5} \text{ sec}^{-1}$ and r is the vector position of the satellite measured in fixed axes with origin at the center of the earth. The above equation can be scaled, discretized, and written in state space form as $x_{k+1} = A_k x_k$, where

$$x(t_k) = x_k = \begin{bmatrix} Cr_1 \\ Cr_2 \\ Cr_3 \\ \dot{r}_1 \\ \dot{r}_2 \\ \dot{r}_3 \end{bmatrix}$$

$$A_k = \begin{bmatrix} \cos Ct_k & 0 & 0 & \sin Ct_k & 0 & 0 \\ 0 & \cos Ct_k & 0 & 0 & \sin Ct_k & 0 \\ 0 & 0 & \cos Ct_k & 0 & 0 & \sin Ct_k \\ \sin Ct_k & 0 & 0 & \cos Ct_k & 0 & 0 \\ 0 & \sin Ct_k & 0 & 0 & \cos Ct_k & 0 \\ 0 & 0 & \sin Ct_k & 0 & 0 & \cos Ct_k \end{bmatrix}$$

[†] Example cleared for publication by the British Aircraft Corporation.

We consider the problem of estimating the position of the satellite, nominally in a synchronous orbit but actually in an inclined elliptical orbit due to an apogee motor malfunction. It is also assumed that there is a radar station capable of measuring range, azimuth angle, and elevation angle and from these measurements it is possible to calculate x_1, x_2, and x_3 to within a standard deviation of 0.2.

A difficulty here is that due to the modeling inaccuracy the dynamic equations are slightly in error and hence, for this problem, estimates of the satellite position based on data from the distant past become less and less meaningful as time progresses. In the light of the discussion in this section, a possible solution is to discount old data so that the estimate of the satellite position is based on the most recent information.

The above problem is discussed in detail in [220]. Simulation studies were conducted on a typical communications satellite in the following orbit:

Apogee altitude: 35,760 km
Perigee altitude: 29,630 km
Eccentricity: 0.0784
Period: 21 hrs, 23 min
Inclination to equatorial plane: 1.1°

Observations were assumed to be available every 20 min and two types of estimator were compared, namely, a standard least squares estimator and an oscillating memory estimator with memory varying between 10 and 20 observations. The resulting error trace for the r_1 coordinate is shown in Fig. 7.3.1. Clearly the oscillating memory estimator is superior for this problem.

7.3.3   *The Kalman filter approach*   The algorithms described in Sections 7.3.1 and 7.3.2 treat time varying parameters in a direct fashion. The rate at which old data are discarded or discounted depends upon the nature of the time variations of the parameters. In this section we assume that the parameter variations can be modeled by stochastic difference equations, i.e.,

$$\theta_{N+1} = F_N \theta_N + \omega_N \tag{7.3.23}$$

The data are modeled as before:

$$y_N = x_N{}^T \theta_N + v_N \tag{7.3.24}$$

We assume that F_N is known for all N (known time variation) and that

$$E\{\omega_N \omega_M{}^T\} = Q_N \delta_{NM} \tag{7.3.25}$$

$$E\{v_N v_M\} = \Sigma_N \delta_{NM} \tag{7.3.26}$$

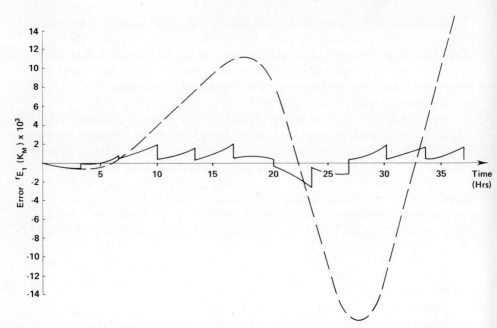

Figure 7.3.1. Error traces for oscillating memory and infinite memory estimators. ——, oscillating memory; - - - - -, infinite memory.

We note that (7.3.23) and (7.3.24) are the state equations of a linear system with time varying state transition matrix F_N, and time varying output matrix x_N^T. We can thus use the Kalman filter equations (Problem 1.8) to obtain the best linear unbiased estimate of the state θ_N based on past observations (maximum likelihood estimate of θ_N if ω_N, v_N are normal) [59]. The recursive equations are

$$\hat{\theta}_{N+1} = F_N \hat{\theta}_N + K_{N+1}(y_{N+1} - x_{N+1}^T \hat{\theta}_N) \qquad (7.3.27)$$

where

$$K_{N+1} = F_N P_N x_{N+1}/(\Sigma_{N+1} + x_{N+1}^T P_N x_{N+1}) \qquad (7.3.28)$$

and

$$P_{N+1} = F_N P_N F_N^T + Q_N - F_N P_N \frac{x_{N+1} x_{N+1}^T}{(\Sigma_{N+1} + x_{N+1}^T P_N x_{N+1})} P_N F_N^T \qquad (7.3.29)$$

Equations (7.3.27)–(7.3.29) bear a marked similarity to those derived earlier in this chapter. In particular we can obtain the exponential window estimator (Eqs. (7.3.2)–(7.3.4)) by making the following assignments:

$$Q_N = \left(\frac{1}{\alpha} - 1\right)[I - K_{N+1}x_{N+1}^{\mathrm{T}}]P_N \qquad (7.3.30)$$

$$\Sigma_N = \alpha \qquad (7.3.31)$$

$$F_N = I \qquad (7.3.32)$$

or we can obtain the least squares estimator (Eq. (7.2.7)–(7.2.9)) by making the assignments

$$Q_N = 0 \qquad (7.3.33)$$

$$\Sigma_N = 1 \qquad (7.3.34)$$

$$F_N = I \qquad (7.3.35)$$

For the case when the model is nonlinear in the parameters, recursive algorithms of the Kalman filter type can be easily derived by linearizing the model about the current estimate. The resulting recursive equations are identical to those presented above save for minor changes required to incorporate the local linearization. Algorithms of this type are known by the name *extended Kalman filter*. The convergence properties of the algorithm depend upon the type of nonlinearity (see [59]).

7.4. FURTHER RECURSIVE ESTIMATORS FOR DYNAMIC SYSTEMS

Since the least squares estimator of Section 7.2 is equivalent to the ordinary least squares estimator described in Chapters 2 and 5, it has the same asymptotic properties as these estimators. Thus, from Section 5.2, the recursive least squares estimator is a consistent estimator of the parameter vector θ in the following model (the so-called least squares structure):

$$A(z)y_t = B(z)u_t + \varepsilon_t \qquad (7.4.1)$$

where

$$A(z) = 1 + a_1 z^{-1} + \cdots + a_n z^{-n} \qquad (7.4.2)$$

$$B(z) = b_1 z^{-1} + \cdots + b_n z^{-n} \qquad (7.4.3)$$

$$\theta^{\mathrm{T}} = (b_1, b_2, \ldots, b_n, \ a_1, a_2, \ldots, a_n) \qquad (7.4.4)$$

and $\{u_t\}$, $\{y_t\}$, and $\{\varepsilon_t\}$ are the input, output, and wide sense stationary white noise sequences, respectively. The consistency result relied upon the uncorrelatedness of the sequence $\{\varepsilon_t\}$. If $\{\varepsilon_t\}$ is correlated, then the least squares estimator will generally be biased. In Chapter 5, we described three algorithms which overcome this bias problem, viz., instrumental variables, generalized least squares, and maximum likelihood. In this section we describe recursive algorithms inspired by these three methods.

7.4.1 *Recursive instrumental variables*   The off-line instrumental variable algorithm was described in Section 5.3.1 where the definition of an instrumental variable matrix was given. There are many ways in which an instrumental variable matrix Z may be obtained. One way is the auxiliary model method described in Chapter 5. This method suggests the following recursive algorithm [109], for obtaining an estimate $\tilde{\theta}_N$ of θ in the model (7.4.1)–(7.4.4) from the data $y_1, \ldots, y_N$:

$$\tilde{\theta}_{N+1} = \tilde{\theta}_N + K_{N+1}(y_{N+1} - x_{N+1}^T \tilde{\theta}_N) \tag{7.4.5}$$

$$K_{N+1} = P_N z_{N+1}/(1 + x_{N+1}^T P_N z_{N+1}) \tag{7.4.6}$$

$$P_{N+1} = \left(I - P_N \frac{z_{N+1} x_{N+1}^T}{(1 + x_{N+1}^T P_N z_{N+1})}\right) P_N \tag{7.4.7}$$

where

$$z_N^T = [u_{N-1}, u_{N-2}, \ldots, u_{N-n}, -\tilde{y}_{N-1}, -\tilde{y}_{N-2}, \ldots, -\tilde{y}_{N-n}] \tag{7.4.8}$$

$$x_N^T = [u_{N-1}, u_{N-2}, \ldots, u_{N-n}, -y_{N-1}, -y_{N-2}, \ldots, -y_{N-n}] \tag{7.4.9}$$

and

$$\tilde{y}_t = z_t^T \tilde{\theta}_t \tag{7.4.10}$$

Suitable starting values are (cf. Remark 7.2.1):

$$P_0 = I/\varepsilon, \qquad \varepsilon \text{ small} \tag{7.4.11}$$

$$\tilde{y}_0 = \tilde{y}_{-1} = \tilde{y}_{-2} = \cdots = 0 \tag{7.4.12}$$

Whereas recursive least squares and ordinary least squares are mathematically equivalent, the same is not true for the recursive and off-line versions of the instrumental variables estimator. Further properties of the above algorithm may be found in [125].

7.4.2 *Recursive generalized least squares*   As in Section 5.3.2, we seek estimates of the parameters in the following model:

$$A(z)y_t = B(z)u_t + \eta_t \tag{7.4.13}$$

$$F(z)\eta_t = \varepsilon_t \tag{7.4.14}$$

where $A(z)$, $B(z)$, and θ are as defined in Eqs. (7.4.2)–(7.4.4). Also,

$$F(z) = 1 + f_1 z^{-1} + \cdots + f_q z^{-q} \tag{7.4.15}$$

and

$$\gamma^T = (f_1, f_2, \ldots, f_q) \tag{7.4.16}$$

$$\beta = \begin{vmatrix} \theta \\ \gamma \end{vmatrix} \tag{7.4.17}$$

Thus, from data $y_1, y_2, \ldots, y_N$ we wish to obtain an estimate $\hat{\beta}_N$ of β. The following recursive algorithm [55] has an obvious relation to the off-line generalized least squares of Section 5.3.2:

$$\hat{\theta}_{N+1} = \hat{\theta}_N + K_{N+1}^1 (\bar{y}_{N+1} - \bar{x}_{N+1}^T \hat{\theta}_N) \tag{7.4.18}$$

$$K_{N+1}^1 = P_N^1 \bar{x}_{N+1} / (1 + \bar{x}_{N+1}^T P_N^1 \bar{x}_{N+1}) \tag{7.4.19}$$

$$P_{N+1}^1 = \left(I - P_N^1 \frac{\bar{x}_{N+1} \bar{x}_{N+1}^T}{(1 + \bar{x}_{N+1}^T P_N^1 \bar{x}_{N+1})} \right) P_N^1 \tag{7.4.20}$$

$$\hat{\gamma}_{N+1} = \hat{\gamma}_N + K_{N+1}^2 (\eta_{N+1} - \xi_{N+1}^T \hat{\gamma}_N) \tag{7.4.21}$$

$$K_{N+1}^2 = P_N^2 \xi_{N+1} / (1 + \xi_{N+1}^T P_N^2 \xi_{N+1}) \tag{7.4.22}$$

$$P_{N+1}^2 = \left(I - P_N^2 \frac{\xi_{N+1} \xi_{N+1}^T}{(1 + \xi_{N+1}^T P_N^2 \xi_{N+1})} \right) P_N^2 \tag{7.4.23}$$

where

$$\bar{y}_t = \hat{F}(z) y_t \tag{7.4.24}$$

$$\bar{u}_t = \hat{F}(z) u_t \tag{7.4.25}$$

$$\bar{x}_t^T = [\bar{u}_{t-1}, \bar{u}_{t-2}, \ldots, \bar{u}_{t-n}, -\bar{y}_{t-1}, -\bar{y}_{t-2}, \ldots, -\bar{y}_{t-n}] \tag{7.4.26}$$

where $\hat{F}$ is defined by (7.4.15) and (7.4.16) with γ replaced by $\hat{\gamma}_t$. Also,

$$\eta_t = \hat{A}(z) y_t - \hat{B}(z) u_t \tag{7.4.27}$$

$$\xi_t^T = [-\eta_{t-1}, -\eta_{t-2}, \ldots, -\eta_{t-q}] \tag{7.4.28}$$

where $\hat{A}$ and $\hat{B}$ are defined by (7.4.2) and (7.4.3) with θ replaced by $\hat{\theta}_t$.

As for instrumental variables, the above algorithm is not equivalent to the off-line version.

7.4.3 *Extended matrix method* Here we consider the following model structure:

$$A(z) y_t = B(z) u_t + C(z) \varepsilon_t \tag{7.4.29}$$

where

$$A(z) = 1 + a_1 z^{-1} + \cdots + a_n z^{-n} \tag{7.4.30}$$

$$B(z) = b_1 z^{-1} + \cdots + b_n z^{-n} \tag{7.4.31}$$

$$C(z) = 1 + c_1 z^{-1} + \cdots + c_n z^{-n} \tag{7.4.32}$$

and $\{u_t\}$, $\{y_t\}$, and $\{\varepsilon_t\}$ are the input, output, and innovations sequence, respectively.

If we assume, for the moment, that $\{\varepsilon_t\}$ is measurable, then we can write the model of Eq. (7.4.29) in the following form:

$$y_t = x_t^T \theta + \varepsilon_t \tag{7.4.33}$$

where

$$x_t^T = [-y_{t-1}, \ldots, -y_{t-n}, u_{t-1}, \ldots, u_{t-n}, \varepsilon_{t-1}, \ldots, \varepsilon_{t-n}] \tag{7.4.34}$$

$$\theta^T = [a_1, \ldots, a_n, b_1, \ldots, b_n, c_1, \ldots, c_n] \tag{7.4.35}$$

The above set of equations immediately suggests use of the recursive least squares estimator described in Result 7.2.1, i.e., $\hat{\theta}_N$ is given by Eqs. (7.2.7)–(7.2.9).

Of course, the sequence $\{\varepsilon_t\}$ is not known. However, a natural way to proceed is to replace x_t^T in Eq. (7.4.34) by $\hat{x}_t^T$ defined below

$$\hat{x}_t^T = [-y_{t-1}, \ldots, -y_{t-n}, u_{t-1}, \ldots, u_{t-n}, \hat{\varepsilon}_{t-1}, \ldots, \hat{\varepsilon}_{t-n}] \tag{7.4.36}$$

where $\{\hat{\varepsilon}_t\}$ is the residual sequence given by

$$\hat{\varepsilon}_t = y_t - \hat{x}_t^T \hat{\theta}_{t-1} \tag{7.4.37}$$

This leads to the following recursive algorithm:

$$\hat{\theta}_{N+1} = \hat{\theta}_N + K_{N+1} \hat{\varepsilon}_N \tag{7.4.38}$$

$$K_{N+1} = \frac{P_N \hat{x}_{N+1}}{(1 + \hat{x}_{N+1}^T P_N \hat{x}_{N+1})} \tag{7.4.39}$$

$$P_{N+1} = \left(I - P_N \frac{\hat{x}_{N+1} \hat{x}_{N+1}^T}{(1 + \hat{x}_{N+1}^T P_N \hat{x}_{N+1})} \right) P_N \tag{7.4.40}$$

The above algorithm has been extensively used in practical applications (see [240]–[248]). We shall use the name *extended matrix method* [247], for the algorithm but it has also been called *Panuska's method* [240], *approximate maximum likelihood* [243], and *recursive maximum likelihood* version 1 [125]. The method has been found to have good convergence properties in almost all cases studied. However, by using a newly developed ordinary differential equation for studying convergence of recursive algorithms (see Section 7.6),

Ljung *et al.* [236] have produced a counterexample to general convergence. In practice, one could have a fair degree of confidence in the algorithm based on the extensive number of cases where it has been found to work satisfactorily. A modification to the algorithm which overcomes the convergence problem is discussed in the next subsection.

7.4.4 *Approximate maximum likelihood* As in Section 7.4.3, we consider the following model structure:

$$A(z)y_t = B(z)u_t + C(z)\varepsilon_t \qquad (7.4.41)$$

where $A(z)$, $B(z)$, and $C(z)$ are as defined in Eq. (7.4.30)–(7.4.32).

Prediction error estimates (or maximum likelihood estimates when the noise is assumed to have a Gaussian distribution) are obtained by minimizing

$$J_N(\theta) = \sum_{t=1}^{N} w_t(\theta)^2 \qquad (7.4.42)$$

where $\{w_t(\theta)\}$ satisfies

$$w_t(\theta) = [C(z)]^{-1}[A(z)y_t - B(z)u_t] \qquad (7.4.43)$$

If w_t were to be a linear function of θ, then a recursive least squares estimator could be developed using the principles described in Section 7.2. This motivates an approximate maximum likelihood algorithm in which a local quadratic approximation is made to J. Using a Taylor series expansion, we obtain

$$w_t(\theta) \simeq \hat{w}_t + (\partial \hat{w}_t/\partial\theta)(\theta - \hat{\theta}) \qquad (7.4.44)$$

where $\hat{w}_t = w_t(\hat{\theta})$ and $\partial \hat{w}_t/\partial\theta$ satisfies

$$\partial \hat{w}_t/\partial a_i = [\hat{C}(z)]^{-1}y_{t-i} \qquad (7.4.45)$$

$$\partial \hat{w}_t/\partial b_i = -[\hat{C}(z)]^{-1}u_{t-i} \qquad (7.4.46)$$

$$\partial \hat{w}_t/\partial c_i = -[\hat{C}(z)]^{-1}\hat{w}_{t-i} \qquad (7.4.47)$$

$\hat{C}(z)$ is evaluated at $\hat{\theta}$.

We now approximate $J_N(\theta)$ by a quadratic function of θ, i.e., we assume existence of $\hat{\theta}_N$, P_N, and B_N such that

$$J_N(\theta) \simeq (\theta - \hat{\theta}_N)^{\mathrm{T}} P_N^{-1}(\theta - \hat{\theta}_N) + \beta_N \qquad (7.4.48)$$

We shall use (7.4.44) and (7.4.48) to develop an approximate recursive maximum likelihood estimate. We first note from (7.4.44) and (7.4.48) that

$$J_{N+1}(\theta) \simeq \Delta^{\mathrm{T}} P_N^{-1} \Delta + \beta_N + (\hat{w}_{N+1} + \phi_{N+1}^{\mathrm{T}}\Delta)^2 \qquad (7.4.49)$$

where

$$\phi_{N+1}^{\mathrm{T}} = \partial \hat{w}_{N+1} / \partial \theta \qquad (7.4.50)$$

$$\Delta = \theta - \hat{\theta}_N \qquad (7.4.51)$$

Substituting (7.4.50) and (7.4.51) into (7.4.49) gives

$$J_{N+1}(\theta) \simeq \Delta^{\mathrm{T}}(P_N^{-1} + \phi_{N+1}\phi_{N+1}^{\mathrm{T}}) \Delta + 2\hat{w}_{N+1}\phi_{N+1}^{\mathrm{T}} \Delta + \beta_N \qquad (7.4.52)$$

Completing the square in (7.4.52) yields

$$J_{N+1}(\theta) \simeq (\Delta - \gamma_{N+1})^{\mathrm{T}} P_{N+1}^{-1} (\Delta - \gamma_{N+1}) + \beta_{N+1} \qquad (7.4.53)$$

where

$$P_{N+1}^{-1} = P_N^{-1} + \phi_{N+1}\phi_{N+1}^{\mathrm{T}} \qquad (7.4.54)$$

$$\gamma_{N+1} = P_{N+1}\phi_{N+1}\hat{w}_{N+1} \qquad (7.4.55)$$

Using the matrix inversion lemma (Lemma E.1.1) in Eq. (7.4.54) and substituting the result into (7.4.55) gives

$$P_{N+1} = P_N\left(I - \frac{\phi_{N+1}\phi_{N+1}^{\mathrm{T}}P_N}{(1 + \phi_{N+1}^{\mathrm{T}}P_N \phi_{N+1})}\right) \qquad (7.4.56)$$

$$\gamma_{N+1} = \frac{P_N \phi_{N+1}}{(1 + \phi_{N+1}^{\mathrm{T}}P_N \phi_{N+1})} \hat{w}_{N+1} \qquad (7.4.57)$$

Now the minimum of (7.4.53) is clearly achieved for Δ given by

$$\hat{\Delta}_{N+1} = \gamma_{N+1} \qquad (7.4.58)$$

Hence combining (7.4.51), (7.4.55), and (7.4.58)

$$\hat{\theta}_{N+1} = \hat{\theta}_N + K_{N+1}\hat{w}_{N+1} \qquad (7.4.59)$$

where

$$K_{N+1} = P_N \phi_{N+1}/(1 + \phi_{N+1}^{\mathrm{T}}P_N \phi_{N+1}) \qquad (7.4.60)$$

As in the previous section, $\{\hat{w}_t\}$ can be approximated by $\{\hat{\varepsilon}_t\}$ where $\hat{\varepsilon}_t$ satisfies the following recursion (cf., Eq. (7.4.37)):

$$\hat{w}_t \simeq \hat{\varepsilon}_t = y_t - \hat{x}_t^{\mathrm{T}}\hat{\theta}_t \qquad (7.4.61)$$

where

$$\hat{x}_t^{\mathrm{T}} = [-y_{t-1}, \ldots, -y_{t-n}, u_{t-1}, \ldots, u_{t-n}, \hat{\varepsilon}_{t-1}, \ldots, \hat{\varepsilon}_{t-n}] \qquad (7.4.62)$$

Finally, substituting (7.4.45) thru (7.4.47) into (7.4.50) yields the following recursion for ϕ_t:

$$\phi_{t+1} = \begin{bmatrix} -\hat{c}_1 & \cdots & -\hat{c}_n & & & & & \\ 1 & & & & & & & \\ & \ddots & & & & & & \\ & & 1 & 0 & & & & \\ & & & & -\hat{c}_1 & \cdots & -\hat{c}_n & \\ & & & & 1 & & & \\ & & & & & \ddots & & \\ & & & & & & 1 & 0 \\ & & & & & & & -\hat{c}_1 & \cdots & -\hat{c}_n \\ & & & & & & & 1 \\ & & & & & & & & \ddots \\ & & & & & & & & & 1 & 0 \end{bmatrix} \phi_t + \begin{bmatrix} y_t \\ 0 \\ \vdots \\ -u_t \\ 0 \\ \vdots \\ -\hat{\varepsilon}_t \\ 0 \\ \vdots \end{bmatrix}$$

(7.4.63)

We shall call the recursive algorithm given in Eqs. (7.4.56) and (7.4.59) (7.4.63) the *approximate maximum likelihood* method. The algorithm was originally described in [56] and [126]. Recent studies [236, 125] have shown that the approximate maximum likelihood method has superior convergence properties to the extended matrix method described in the previous subsection. Moreover, it has been shown in [236] that the method converges with probability one to a local minimum of the estimation criterion. This is the best one can hope for since the method has as good convergence properties as off-line prediction error methods.

7.5.   STOCHASTIC APPROXIMATION

Stochastic approximation [205] is the title given to a general class of recursive algorithms for solving equations of the form

$$g(\theta) = E_y\{f(y, \theta)\} = 0 \tag{7.5.1}$$

where f is a known function of the data y and θ is an unknown parameter.

The original stochastic approximation scheme, developed by Robbins and Monro [208], uses the following adjustment law to solve for θ given successive observations of y denoted by $y_1, y_2, \ldots, y_N, y_{N+1}$

$$\hat{\theta}_{N+1} = \hat{\theta}_N + \gamma_N f(y_{N+1}, \hat{\theta}_N) \tag{7.5.2}$$

where y_N is the observation made at time N and $\{\gamma_N\}$ is a suitably chosen sequence of scalars. It can be shown [213] that the above algorithm converges to the true parameter value θ, provided certain conditions are satisfied by $\{\gamma_N\}$, f and the noise variance.

The usual conditions on $\{\gamma_N\}$ are

$$\sum_{N=1}^{\infty} \gamma_N = \infty \tag{7.5.3}$$

$$\sum_{N=1}^{\infty} \gamma_N^{\,p} < \infty \qquad \text{for some} \quad p > 1 \tag{7.5.4}$$

A common choice for the sequence $\{\gamma_N\}$ is $\gamma_N = 1/N$. This choice will result in a.s. convergence of $\hat{\theta}_N$ to the true value of θ [127]. However, it is also pointed out in [127] that a more slowly decreasing sequences $\{\gamma_N\}$ can lead to faster convergence.

In identification problems, we are frequently interested in minimizing a least squares cost function of the form

$$J(\theta) = E\{(y_t - x_t^{\mathrm{T}}\theta)^2\} \tag{7.5.5}$$

This function can be minimized by taking the derivative with respect to θ to yield

$$g(\theta) = \partial J/\partial \theta = E\{2x_t(y_t - x_t^{\mathrm{T}}\theta)^2\} \tag{7.5.6}$$

The Robbins–Monro scheme applied to (7.5.6) then yields

$$\hat{\theta}_{N+1} = \hat{\theta}_N + \gamma_N\{x_{N+1}(y_{N+1} - x_{N+1}^{\mathrm{T}}\hat{\theta}_N)\} \tag{7.5.7}$$

Further discussion of the above algorithm may be found in [214].

For the above special case, it was an easy matter to compute the derivative $\partial J/\partial \theta$. In problems where the derivative is difficult to calculate, it seems reasonable to replace the derivative with a difference approximation. An algorithm of this type was suggested by Kiefer and Wolfowitz [215].

Details of the application of the above stochastic approximate algorithms to identification problems may be found in [127] and [58].

The convergence properties of a general class of stochastic approximation algorithms is studied in [127]. The class includes the Robbins–Monro and Kiefer–Wolfowitz procedures as special cases. It is shown that the convergence properties can be studied using an ordinary differential equation. We shall briefly discuss this equation in the next section.

7.6.   CONVERGENCE OF RECURSIVE ESTIMATORS

The recursive algorithms are time varying, nonlinear stochastic difference equations. This makes it relatively difficult to analyze their convergence properties directly. However, Ljung [127] has recently shown that an associated time invariant deterministic ordinary differential equation can be used to study convergence properties of recursive algorithms. Here we shall give a heuristic description of this ordinary differential equation and we shall quote some of the key convergence results. A rigorous development of these properties would be too great a diversion in this book but the interested reader is referred to [127], [125], and [236] for full details.

All of the recursive algorithms described in the previous sections have the following general form:

$$\hat{\theta}_{N+1} = \hat{\theta}_N + F_{N+1} x_{N+1} w_{N+1} \tag{7.6.1}$$

where $\{w_N\}$ is the residual sequence, x_N the regression vector, and F_N the adjustment factor.

For example, in the least squares procedure applied to the model of Eq. (7.4.1), we have

$$w_N = y_N - x_N^T \hat{\theta}_{N-1} \tag{7.6.2}$$

$$x_N^T = (-y_{N-1}, \ldots, -y_{N-n}, u_{N-1}, \ldots, u_{N-n}) \tag{7.6.3}$$

$$F_N = P_{N-1}/(1 + x_N^T P_{N-1} x_N) \tag{7.6.4}$$

For the extended matrix method described in Section 7.4.3;

$$w_N = \hat{\varepsilon}_N = y_N - \hat{x}_N^T \hat{\theta}_{N-1} \tag{7.6.5}$$

$$x_N^T = \hat{x}_N^T = [-y_{N-1}, \ldots, -y_{N-n}, u_{N-1}, \ldots, u_{N-n}, \hat{\varepsilon}_{N-1}, \ldots, \hat{\varepsilon}_{N-n}] \tag{7.6.6}$$

$$F_N = P_{N-1}/(1 + \hat{x}_N^T P_{N-1} \hat{x}_N) \tag{7.6.7}$$

For the simple stochastic approximation scheme described in Section 7.5, w_N is given by (7.6.2) or (7.6.5), x_N is given by (7.6.3) or (7.6.6), and F_N is given by

$$F_N = 1/N \tag{7.6.8}$$

We shall begin by studying a stochastic approximation version of the extended matrix method, i.e.,

$$\hat{\theta}_{N+1} = \hat{\theta}_N + \frac{1}{N+1} \hat{x}_{N+1} \hat{\varepsilon}_{N+1} \tag{7.6.9}$$

where $\{\hat{\varepsilon}_t\}$ and $\{\hat{x}_N\}$ are as defined in Eqs. (7.4.36) and (7.4.37).

We note that $\hat{x}_{N+1}$ and $\hat{\varepsilon}_{N+1}$ are functions of all previous estimates, i.e., of $\{\hat{\theta}_N, \hat{\theta}_{N-1}, \hat{\theta}_{N-2}, \hat{\theta}_{N-3}, \ldots\}$. However, the dependence on past estimates is a rapidly decreasing function of time since Eqs. (7.6.5) and (7.6.6) can be interpreted as stable time varying filters. Moreover, the adjustment factor decreases as time increases and therefore $\hat{\theta}_N$ will be close to $\hat{\theta}_{N-1}, \hat{\theta}_{N-2}$, etc.

Thus it is heuristically reasonable to replace $\hat{x}_{N+1}$ and $\hat{\varepsilon}_{N+1}$ in (7.6.9) by the stationary processes $\bar{x}_{N+1}(\hat{\theta}_N)$ and $\bar{\varepsilon}_{N+1}(\hat{\theta}_N)$ defined as follows:

$$\bar{\varepsilon}_N(\theta) = y_N - \bar{x}_N(\theta)^{\mathrm{T}}\theta \tag{7.6.10}$$

$$\bar{x}_N(\theta)^{\mathrm{T}} = (-y_{N-1}, \ldots, -y_{N-n}, u_{N-1}, \ldots, u_{N-n}, \bar{\varepsilon}_{N-1}(\theta), \ldots, \bar{\varepsilon}_{N-n}(\theta) \tag{7.6.11}$$

Hence Eq. (7.6.9) becomes

$$\hat{\theta}_{N+1} \simeq \hat{\theta}_N + \frac{1}{N+1}\,\bar{x}_{N+1}(\hat{\theta}_N)\bar{\varepsilon}_{N+1}(\hat{\theta}_N) \tag{7.6.12}$$

or

$$\hat{\theta}_{N+M} = \hat{\theta}_N + \sum_{k=N+1}^{N+M}\frac{1}{k}\,\bar{x}_k(\hat{\theta}_k)\bar{\varepsilon}_k(\hat{\theta}_k)$$

$$\simeq \hat{\theta}_N + E\{\bar{x}_N(\hat{\theta}_N)\bar{\varepsilon}_N(\hat{\theta}_N)\}\sum_{k=N+1}^{N+M}\frac{1}{k} \tag{7.6.13}$$

where the last step follows from the near constant nature of $\hat{\theta}_k$ over $k = N + 1$ to $N + M$.

Now for large N, the summation $\sum_{k=N+1}^{N+M} 1/k$ becomes small. Hence, if we denote this by Δt, then (7.6.13) can be thought of as a Euler solution to the following differential equation:

$$\dot{\theta} = f(\theta) \tag{7.6.14}$$

where

$$f(\theta) = E\{\bar{x}_N(\theta)\bar{\varepsilon}_N(\theta)\} \tag{7.6.15}$$

For the more general case where F_N satisfies an equation of the form (7.6.4), we introduce $R_N = P_N^{-1}/N$. Then from the usual matrix Riccati equation for P_N (cf. Eq. (7.4.40), in the extended matrix method), we have

$$P_{N+1}^{-1} = P_N^{-1} + \hat{x}_{N+1}\hat{x}_{N+1}^{\mathrm{T}} \tag{7.6.16}$$

or

$$R_{N+1} = R_N + [1/(N+1)](\hat{x}_{N+1}\hat{x}_{N+1}^{\mathrm{T}} - R_N) \tag{7.6.17}$$

Following arguments similar to those used in the derivation of (7.6.14), R_N can be thought of as the solution of the following ordinary differential equation

$$\dot{R} = G(\theta) - R \tag{7.6.18}$$

where
$$G(\theta) = E\{\bar{x}_N(\theta)\bar{x}_N(\theta)^{\mathsf{T}}\} \tag{7.6.19}$$

For the general case, $\hat{\theta}_N$ satisfies (7.6.1) with F_{N+1} given by

$$F_{N+1} = \frac{P_N}{1 + x_{N+1}^{\mathsf{T}} P_N x_N} = \frac{N^{-1} R_N^{-1}}{1 + N^{-1} x_{N+1}^{\mathsf{T}} R_N^{-1} x_{N+1}}$$

$$\simeq \frac{1}{N} R_N^{-1} \qquad \text{for large} \quad N \tag{7.6.20}$$

Hence Eq. (7.6.14) should be modified to give

$$\dot{\theta} = R^{-1} f(\theta) \tag{7.6.21}$$

where $\dot{R}$ satisfies (7.6.18), i.e.,

$$\dot{R} = G(\theta) - R \tag{7.6.22}$$

The above equations describe the expected asymptotic paths of the algorithm. The heuristic argument given here can be made rigorous [125] and can be applied to a wide class of recursive stochastic algorithms including those described in this chapter.

In the general case, we have the following theorem.

Theorem 7.6.1   Consider the ordinary differential equation:

$$\dot{\theta} = R^{-1} f(\theta) \tag{7.6.23}$$

$$\dot{R} = G(\theta) - R \tag{7.6.24}$$

where $f(\theta) = E\{\bar{x}_N(\theta)\bar{\varepsilon}_N(\theta)\}$ and $G(\theta) = E\{\bar{x}_N(\theta)\bar{x}_N(\theta)^{\mathsf{T}}\}$.

(a)   If $(\theta, R) = (\theta^*, G(\theta^*))$ is a globally asymptotically stable stationary point of the above ordinary differential equation, then

$$\hat{\theta}_N \xrightarrow{\text{a.s.}} \theta^* \tag{7.6.25}$$

(b)   The set of possible convergence points of the recursive algorithm is a subset of

$$D_s = \{\theta \mid f(\theta) = 0\} \tag{7.6.26}$$

(c)   A necessary condition for convergence to θ^* is that the ordinary differential equation is locally stable in the vicinity of θ^*, i.e., all eigenvalues of $[G(\theta^*)]^{-1}[df(\theta)/d\theta]_{\theta=\theta^*}$ have nonpositive real parts.

Proof   See [125].
Parts (a) and (b) follow from the observation that the expected asymptotic paths follow the solutions of the ordinary differential equations. To establish

part (c), we consider linearization of (7.6.23) and (7.6.24) at $\theta = \theta^*, R = G(\theta^*)$

$$\frac{d}{dt}\begin{bmatrix}\Delta\theta \\ \Delta R\end{bmatrix} = \begin{bmatrix}R^{-1}\,\partial f/\partial\theta & \vdots & -R^{-2}f \\ \hline \partial G/\partial\theta & \vdots & -I\end{bmatrix}_{\theta=\theta^*}\begin{bmatrix}\Delta\theta \\ \Delta R\end{bmatrix} \qquad (7.6.27)$$

$$= \begin{bmatrix}G^{-1}\partial f/\partial\theta & \vdots & 0 \\ \hline \partial G/\partial\theta & \vdots & -I\end{bmatrix}\begin{bmatrix}\Delta\theta \\ \Delta R\end{bmatrix} \qquad (7.6.28)$$

since $f(\theta^*) = 0$, $R(\theta^*) = G(\theta^*)$.

It follows from (7.6.28) that local stability is governed by the eigenvalues of $[G(\theta^*)]^{-1}[df(\theta)/d\theta]_{\theta=\theta^*}$. ∇

The following convergence properties can be established using the above result [125].

(i)   The recursive least squares method always converges to the true value. (This follows from the fact that it is mathematical equivalent to the off-line method.)

(ii)   The recursive instrumental variable method always converges to true value.

(iii)   The recursive generalized least squares method will converge to the true parameter value provided the signal-to-noise ratio is high. If the signal-to-noise ratio is low, then convergence to incorrect values can occur.

(iv)   The extended matrix method converges to the true parameter values for moving average processes and for first-order processes of autoregressive moving average type. However, counterexamples exist to convergence in general.

(v)   The approximate maximum likelihood method always converges to the true values for autoregressive-moving average processes. For systems with an input signal, convergence is achieved provided the signal-to-noise ratio is high. In all cases, the method will converge to a local minimum of the likelihood function. (Note $f(\theta)$ is the gradient of the expected value of the log likelihood function for this problem.)

Perhaps the most important aspect of Theorem 7.6.1 is its potential application in establishing convergence results for future algorithms. This last point has been our main motivation for summarizing the ordinary differential equation result here.

7.7.   RECURSIVE EXPERIMENT DESIGN

In this section, we turn to methods for recursively designing inputs and/or sampling instants. These algorithms may be used either off-line or on-line. Their main advantage is simplicity. A possible disadvantage is that no proofs of optimality, in the sense of Chapter 6, exist. They are optimal in the one-step-ahead sense, however. In Section 7.7.2, we describe an algorithm which is suitable for the design of either input, sampling instants or both concurrently. First, however, we illustrate the principles involved by describing a simple on-line input design algorithm for a model of the least squares structure.

7.7.1 *Input design*   Consider the least squares structure model

$$A(z)y_t = B(z)u_t + \varepsilon_t \qquad (7.7.1)$$

where $\{u_t\}$, $\{y_t\}$, and $\{\varepsilon_t\}$ are the input, output, and white noise sequences, respectively, and the parameter vector θ is defined by (7.4.2)–(7.4.4). The recursive least squares algorithm is given by Eqs. (7.2.7)–(7.2.9). In particular, the parameter covariance matrix P_N, satisfies the following recursion

$$P_{N+1} = \left(I - P_N \frac{x_{N+1}x_{N+1}^T}{1 + x_{N+1}^T P_N x_{N+1}}\right)P_N \qquad (7.7.2)$$

where

$$x_t^T = [u_{t-1}, u_{t-2}, \ldots, u_{t-n}, -y_{t-1}, -y_{t-2}, \ldots, -y_{t-n}] \qquad (7.7.3)$$

The design problem that we shall solve can be stated as follows. At time N choose u_N, subject to constraints, so that a scalar function of the parameter covariance matrix P_{N+1} is minimized. Suitable scalar functions were described in Chapter 6. We consider only one

$$J_1 = \log \det P_{N+1} \qquad (7.7.4)$$

Throughout this section we assume an amplitude constraint on the input, i.e.,

$$-1 \le u_t \le 1 \qquad \forall t \qquad (7.7.5)$$

Result 7.7.1   The value of u_N satisfying the amplitude constraint (7.7.5) and minimizing $J_1 = \log \det P_{N+1}$ is given by

$$u_N = u_N^{\,0} = \text{sign}(P_{12}^N \xi_{N+1}/P_{11}^N) \qquad (7.7.6)$$

where

$$\text{sign}(\alpha) = \begin{cases} +1 & \text{if } \alpha \ge 0 \\ -1 & \text{if } \alpha < 0 \end{cases} \qquad (7.7.7)$$

$$\xi_t^T = [u_{t-2}, \ldots, u_{t-n}, -y_{t-1}, -y_{t-2}, \ldots, -y_{t-n}] \qquad (7.7.8)$$

and P_{11}, P_{12} are 1×1 and $1 \times (2n-1)$ partitions of P as follows:

$$P = \left[\begin{array}{c|c} P_{11} & P_{12} \\ \hline P_{21} & P_{22} \end{array}\right] \qquad (7.7.9)$$

Proof From Eq. (7.7.2),

$$\log \det P_{N+1} = \log \det\left(I - P_N \frac{x_{N+1}x_{N+1}^T}{1 + x_{N+1}^T P_N x_{N+1}}\right) + \log \det P_N \quad (7.7.10)$$

Since only the first term on the right-hand side of (7.7.10) depends upon u_N, minimization of $J_1 = \log \det P_{N+1}$ is equivalent to minimization of

$$J' = \log \det\left(I - P_N \frac{x_{N+1}x_{N+1}^T}{1 + x_{N+1}^T P_N x_{N+1}}\right) \quad (7.7.11)$$

But, from Result E.1.3,

$$J' = \log\left(1 - \frac{x_{N+1}^T P_N x_{N+1}}{1 + x_{N+1}^T P_N x_{N+1}}\right) = \log(1/(1 + x_{N+1}^T P_N x_{N+1})) \quad (7.7.12)$$

Therefore, minimization of J' is equivalent to maximization of

$$J'' = x_{N+1}^T P_N x_{N+1} \quad (7.7.13)$$

$$= P_{11}u_N^2 + 2P_{12}\xi_{N+1}u_N + \xi_{N+1}^T P_{22}\xi_{N+1} \quad (7.7.14)$$

where (7.7.8) and (7.7.9) have been used. Thus J'' is quadratic in u_N with minimum given by

$$u_N^* = -P_{12}\xi_{N+1}/P_{11} \quad (7.7.15)$$

If u_N^* is negative, then J'' is maximized by choosing $u_N = +1$ and if u_N^* is negative, then J'' is maximized by choosing $u_N = -1$. Thus we have established (7.7.6), and hence the result. ∇

Applications of the above algorithm have been described in [166], [178], and [179]. Generalizations of the procedure are described in the next section.

In the above approach the input is chosen at each time instant to minimize the one-step-ahead cost function. Another approach is to perform an off-line design to find an optimal input spectrum (and hence autocorrelation), and then attempt to realize recursively an input with this spectrum. The off-line frequency domain procedures of the last chapter yield optimal power constrained input spectra. If in addition we have an amplitude constraint, then it is readily seen that if we can find a binary signal with a spectrum that closely approximates the optimal power constrained spectrum, then this binary signal is also optimal in the amplitude constrained case. We now describe a procedure for recursively realizing specified spectra (autocorrelations) with binary signals.

Consider a sequence $\{u_t\}$ where $|u_t| = 1$. The sample correlation is given by

$$\Gamma_N(\tau) = \frac{1}{N}\sum_{t=1}^{N-\tau} u_t u_{t+\tau}, \qquad \tau = 0, 1, \dots \quad (7.7.16)$$

The problem is to choose u_t, $t = 1, 2, \ldots$, so that the sample correlation approaches some desired value of the correlation function $\Gamma^*(\tau)$, $\tau = 0, 1, \ldots, s$. The recursive algorithm then chooses at each time N, the value of u_N (either $+1$ or -1) that minimizes the following cost function:

$$V_N = \sum_{\tau=1}^{s} [\Gamma_N(\tau) - \Gamma^*(\tau)]^2$$

where

$$\Gamma_N(\tau) = (1/N)[(N - 1)\Gamma_{N-1}(\tau) + u_{N-\tau}u_N]$$

This algorithm, though convergence is unproven, has been found to work extremely well in practice [217]. Typical results for 100 data points are illustrated in Fig. 7.7.1. The sequence obtained is shown in Fig. 7.7.2.

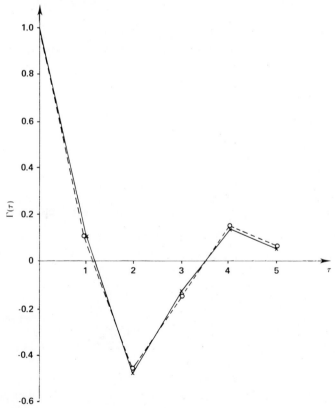

Figure 7.7.1. Desired and sample autocorrelation functions. ———, desired autocorrelation; -----, sample autocorrelation obtained using sequential design algorithm with 100 points.

Figure 7.7.2. Binary realization of autocorrelation shown in Fig. 7.7.1.

Other approaches to the problem of realizing binary signals with desired spectra (or more generally, signals with specified amplitude distribution and power spectrum) may be found in [191], [173], and [216].

7.7.2 *Recursive input and sampling interval design* This section describes a design procedure for sequential determination of the input and/or sampling intervals. We consider the same problem set-up as in Section 6.5.1.

It follows from Eq. (6.5.14) that the information matrix is given by the following recursion:

$$M_{k+1} = M_k + \mathscr{I}_{k+1}, \quad M_0 = 0 \qquad (7.7.17)$$

where $\mathscr{I}_k$ is the information increment resulting from the $(k+1)$th sample and has ijth element

$$[\mathscr{I}_{k+1}]_{ij} = \hat{x}_{k+1}^{\mathsf{T}} \Omega_i^{\mathsf{T}} S_{k+1}^{-1} \Omega_j \hat{x}_{k+1} + \text{trace}\{\Omega_i^{\mathsf{T}} S_{k+1}^{-1} \Omega_j T_{k+1}\}$$
$$+ \tfrac{1}{2}\,\text{trace}\{S_{k+1}^{-1}\,\partial S_{k+1}/\partial\beta_i\}\,\text{trace}\{S_{k+1}^{-1}\,\partial S_{k+1}/\partial\beta_j\} \qquad (7.7.18)$$

Using (6.5.15), Eq. (7.7.18) becomes

$$[\mathscr{I}_{k+1}]_{ij} = (F_k \hat{x}_k + G_k u_k)^{\mathsf{T}} \Omega_i^{\mathsf{T}} S_{k+1}^{-1} \Omega_j (F_k \hat{x}_k + G_k u_k) + \text{trace}\{\Omega_i^{\mathsf{T}} S_{k+1}^{-1} \Omega_j T_{k+1}\}$$
$$+ \tfrac{1}{2}\,\text{trace}\{S_{k+1}^{-1}\,\partial S_{k+1}/\partial\beta_i\}\,\text{trace}\{S_{k+1}^{-1}\,\partial S_{k+1}/\partial\beta_j\} \qquad (7.7.19)$$

Considering the situation at time k, if it is assumed that the past inputs and sampling intervals have been specified, then it follows from Eq. (7.7.17) and (7.7.18) that M_{k+1} is a function of only the two variables u_k and Δ_k. Hence it is a simple optimization problem to choose u_k and Δ_k to minimize $-\log \det M_{k+1}$. Then moving on to time $k+1$, the procedure can be repeated to choose u_{k+1} and Δ_{k+1}, etc. There is, however, no proof of global optimality in the sense that $-\log \det \overline{M}$ is minimized with respect to $u_0, u_1, \ldots, u_{N-1}$ and $\Delta_0, \Delta_1, \ldots, \Delta_{N-1}$.

The procedure described above is a generalization to the sampling problem of the sequential test signal design procedure described in Section 7.6.1.

The sequential design algorithm is now illustrated by a simple example.

Example 7.7.1   Consider the first-order system

$$\dot{x} = ax + bu \tag{7.7.20}$$

$$y = x + n \tag{7.7.21}$$

where the noise, $n(t)$ is assumed to be wide band. Equation (7.7.20) can be discretized as

$$x_{k+1} = \alpha_k x_k + \beta_k u_k \tag{7.7.22}$$

$$y_k = x_k + n_k \tag{7.7.23}$$

where $\alpha_k = \exp(a\Delta_k)$, $\beta_k = (\alpha_k - 1)b/a$ and n_k has zero mean and approximately constant variance. (This assumption will be valid provided the smallest interval allowed is much greater than the reciprocal of the noise bandwidth.)

Equation (7.7.17) becomes

$$M_{k+1} = M_k + \Omega \hat{x}_{k+1} \hat{x}_{k+1}^{\mathrm{T}} \Omega^{\mathrm{T}} \tag{7.7.24}$$

where

$$\Omega = \begin{bmatrix} 0 & 1 & 0 \\ 0 & 0 & 1 \end{bmatrix} \tag{7.7.25}$$

and

$$\hat{x}_{k+1} = \begin{bmatrix} \alpha_k & 0 & 0 \\ \alpha_k \Delta_k & \alpha_k & 0 \\ 0 & 0 & \alpha_k \end{bmatrix} \hat{x}_k + \begin{bmatrix} \beta_k \\ b(a\Delta_k \alpha_k - \alpha_k + 1)/a^2 \\ (\alpha_k - 1)/a \end{bmatrix} u_k \tag{7.7.26}$$

$$= A_k \hat{x}_k + B_k u_k \tag{7.7.27}$$

As usual, the optimality criterion is taken to be $-\log \det M$ or equivalently $-\det(M)$. Hence using Eq. (7.7.24),

$$\begin{aligned}
-\det(M_{k+1}) &= -\det(M_k) \det[I + M_k^{-1} \Omega \hat{x}_{k+1} \hat{x}_{k+1}^{\mathrm{T}} \Omega^{\mathrm{T}}] \\
&= -\det(M_k)(1 + \hat{x}_{k+1}^{\mathrm{T}} \Omega^{\mathrm{T}} M_k^{-1} \Omega \hat{x}_{k+1}) \quad \text{(from Result E.1.3)} \\
&= -\det(M_k)(1 + (A_k \hat{x}_k + B_k u_k)^{\mathrm{T}} \Omega^{\mathrm{T}} M_k^{-1} \Omega (A_k \hat{x}_k + B_k u_k))
\end{aligned} \tag{7.7.28}$$

Now assuming that u_k is constrained in amplitude and lies in the interval $-1 \le u_k \le 1$, it can be seen from Eq. (7.7.28) that the optimality criterion is minimized with respect to u_k if u_k is chosen as

$$u_k^* = \mathrm{sign}(B_k^{\mathrm{T}} \Omega^{\mathrm{T}} M_k^{-1} \Omega A_k \hat{x}_k) \tag{7.7.29}$$

Substituting (7.7.29) back into (7.7.28) yields

$$-\det M_{k+1} = -\det M_k[1 + \hat{x}_k^T A_k \Omega^T M_k^{-1} \Omega A_k \hat{x}_k + 2|B_k \Omega^T M_k^{-1} \Omega A_k \hat{x}_k|$$
$$+ B_k^T \Omega^T M_k^{-1} \Omega B_k] \quad (7.7.30)$$

A one-dimensional optimization can now be used to find the best value for Δ_k. This value can then be substituted into (7.7.29) to give u_k^*. Results of this optimization are shown in Fig. 7.7.3.

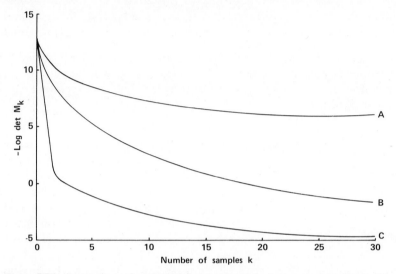

Figure 7.7.3. Comparison of sequential designs. Legend: A—square wave input, $\Delta_k = 0.2$; B—sequentially optimized input $\Delta_k = 0.2$; C—sequentially optimized input and sampling intervals.

It can be seen from Fig. 7.7.3 that, for this example, a significant improvement in information occurs if the input is sequentially optimized with fixing sampling intervals. However, a further substantial improvement is achieved by a coupled design of both the input and sampling intervals. However, the results are not as spectacular as a naïve interpretation of Fig. 7.7.3 would indicate since the samples have been assumed to be taken without any form of presample filtering. We note that the noise was assumed to have wide bandwidth, and thus from Section 6.5.2 we see that unmatchable improvements in estimation accuracy could have been achieved by the simple expedient of including a low-pass filter prior to sampling. The improvement in accuracy will be of the order of the ratio of the noise bandwidth to the system bandwidth, provided that the filter is chosen appropriately. (The variance of the discrete noise on the samples is roughly proportional to the noise bandwidth at the samples.) ∇

7.8. STOCHASTIC CONTROL

In this final section of the book we shall show how recursive parameter estimation algorithms may be combined with on-line control strategies to produce adaptive control algorithms. Our treatment of this topic will be brief as adaptive control is an area of interest in its own right and numerous books have been devoted to this topic [129]. Our motivation for including this section in the book is that adaptive control provides an interesting application of the ideas of recursive parameter estimation discussed earlier in this chapter.

It is usual to distinguish two classes of stochastic control strategy, viz., *feedback* and *closed loop*. The difference lies in the amount of information available to the controller. Controllers in the feedback class cannot make use of the fact that the loop will remain closed in the future. If the parameters are unknown, then both classes of controller can use past measurements to obtain parameter estimates. Only a controller in the closed loop class can, however, actively probe the system in an attempt to reduce parameter inaccuracies. Therefore, a closed loop controller will, in general, lead to better control on average than a controller of the feedback class. We shall refer to controllers of the feedback class as *passive* and to controllers of the closed loop class as *active* or *dual*. A more detailed exposition on the differences between feedback and closed loop strategies may be found in [130].

The controllers that we shall consider are based on the minimum variance controller [14] for which we now present a simple derivation:

Result 7.8.1 Consider the system

$$y_t = \frac{B(z)}{A(z)} u_t + \frac{D(z)}{C(z)} \varepsilon_t \qquad (7.8.1)$$

where

$$B(z) = b_k z^{-k} + \cdots + b_n z^{-n}, \qquad A(z) = 1 + a_1 z^{-1} + \cdots + a_n z^{-n}$$

For simplicity we assume that $B^*(z) = z^n B(z)$ has no roots outside the unit circle. Then the control strategy which minimizes the cost function

$$J = E\{y_{t+k}^2 | Y_t, U_t\} \qquad (7.8.2)$$

where

$$Y_t = \{y_t, y_{t-1}, \ldots\}, \qquad U_t = \{u_t, u_{t-1}, \ldots\}$$

is given by

$$u_t = -[A(z)G(z)/z^k B(z)C(z)F(z)]y_t \qquad (7.8.3)$$

where F and G satisfy the following identity

$$D(z)/C(z) = F(z) + z^{-k}G(z)/C(z) \qquad (7.8.4)$$

where

$$F(z) = 1 + f_1 z^{-1} + f_2 z^{-2} + \cdots + f_{k-1} z^{-(k-1)}$$

Proof   Substituting (7.8.4) into (7.8.1) yields

$$y_{t+k} = \frac{z^k B(z)}{A(z)} u_t + F(z)\varepsilon_{t+k} + \frac{G(z)}{C(z)} \varepsilon_t$$

$$= \frac{z^k B(z)}{A(z)} u_t + F(z)\varepsilon_{t+k} + \frac{G(z)}{C(z)} \left\{ \frac{C(z)}{D(z)} \left(y_t - \frac{B(z)}{A(z)} u_t \right) \right\}$$

Hence

$$E\{y_{t+k}^2 \,|\, Y_t, U_t\} = \left[\frac{B(z)}{A(z)} \left(z^k - \frac{G(z)}{D(z)} \right) u_t + \frac{G(z)}{D(z)} y_t \right]^2$$

$$+ (1 + f_1^2 + \cdots + f_{k-1}^2)\Sigma$$

Hence $E\{y_{t+k}^2\}$ is minimized when u_t is chosen to satisfy

$$\frac{B(z)}{A(z)} \left(z^k - \frac{G(z)}{D(z)} \right) u_t + \frac{G(z)}{D(z)} y_t = 0 \qquad (7.8.5)$$

Simplification of (7.8.5) yields (7.8.3).   ▽

Further discussion on the above control law including the case where $B^*(z)$ has roots outside the unit circle is given in [14].

We shall be particularly interested in models with the least squares structure:

$$A(z)y_t = B(z)u_t + \varepsilon_t \qquad (7.8.6)$$

where $\{\varepsilon_t\}$ is a sequence of independent identically distributed random variables and $\{u_t\}$, and $\{y_t\}$ are the input and output sequences, respectively. Also

$$A(z) = 1 + a_1 z^{-1} + \cdots + a_n z^{-n} \qquad (7.8.7)$$

$$B(z) = b_1 z^{-1} + \cdots + b_n z^{-n} \qquad (7.8.8)$$

We further define

$$\beta^T = (b_1, \theta^T) \qquad (7.8.9)$$

where

$$\theta^T = (b_2, b_3, \ldots, b_n, a_1, a_2, \ldots, a_n) \qquad (7.8.10)$$

Thus (7.8.6) may be written as

$$y_t = \xi_t^T \theta + b_1 u_{t-1} + \varepsilon_t \qquad (7.8.11)$$

where

$$\xi_t^T = (u_{t-2}, \ldots, u_{t-n}, -y_{t-1}, \ldots, -y_{t-n}) \qquad (7.8.12)$$

The minimum variance controller is simply obtained by noting that the u_t which minimizes $E\{y_{t+1}^2\}$ is

$$u_t = -\xi_{t+1}^T \theta / b_1 \qquad (7.8.13)$$

In order to implement the minimum variance controller, we need to know the parameter values. If these are not known, then we could try replacing the parameters by their current estimate based on past data. This leads to a passive controller from the feedback class.

7.8.1 *Passive adaptive control* A passive adaptive controller based on the minimum variance controller and recursive least squares parameter estimator is called a self-tuning regulator [60, 131]. This is obtained by simply replacing θ in (7.8.13) by $\hat{\theta}_t$, the recursive least squares estimate given by Eqs. (7.2.7)–(7.2.9). (The choice of b_1 is usually not critical [135] and this parameter need not be estimated.) Detailed analysis of the self-tuning regulator is given in [132]. One unexpected property is that the scheme can converge to the correct controller even if the noise $\{\varepsilon_t\}$ in (7.8.6) is not white (i.e., for systems not in the least squares structure).

It is evident that the performance of the self-tuning regulator will depend upon the parameter uncertainties. This can be taken into account to a certain extent by rederiving a controller similar to the minimum variance controller except for the inclusion of terms depending upon the parameter covariance [133, 134]. The resulting controller is still of the feedback class, however, and this imposes a restriction on the achievable performance. A first approach is simply to inject a perturbation signal into the loop to increase the "rate of learning" [136, 137]. The resulting controller is then of the closed loop class. We now make a more systematic study of controllers in the closed loop class.

7.8.2 *Active adaptive control*   We again consider the model of (7.8.6).
The least squares estimate for $\beta = (b_1, \theta^T)^T$ is given by Eqs. (7.2.7)–(7.2.9):

$$\hat{\beta}_{t+1} = \hat{\beta}_t + K_{t+1}(y_{t+1} - x_{t+1}^T \hat{\beta}_t) \tag{7.8.14}$$

$$K_{t+1} = P_t x_{t+1}/(1 + x_{t+1}^T P_t x_{t+1}) \tag{7.8.15}$$

$$P_{t+1} = \left(I - P_t \frac{x_{t+1} x_{t+1}^T}{1 + x_{t+1}^T P_t x_{t+1}}\right) P_t \tag{7.8.16}$$

where

$$x_t^T = (u_{t-1}, \xi_t^T) \tag{7.8.17}$$

and ξ_t is given by (7.8.12).

Consider the partition of P_t corresponding to the partition of β into b_1 and θ:

$$P_t = \left[\begin{array}{c|c} s_t & w_t^T \\ \hline w_t & Q_t \end{array}\right] \tag{7.8.18}$$

Then we have the following result.

Result 7.8.2   The control minimizing the cost function

$$J = E_{Y_{t+1}|\beta}[y_{t+1}^2] - \lambda[\det P_t/\det P_{t+1}] \tag{7.8.19}$$

is given by

$$u_t = -(b_{1t}\hat{\theta}_t^T\xi_{t+1} - \lambda w_t^T\xi_{t+1})/(\hat{b}_{1t}^2 - \lambda s_t) \tag{7.8.20}$$

Proof

$$E_{Y_{t+1}|\beta} = (\xi_{t+1}^T\hat{\theta}_t + \hat{b}_{1t}u_t)^2 + \sigma^2 \qquad \text{(from (7.8.11))}$$

$$-\lambda \det P_t/\det P_{t+1} = -\lambda(1 + x_{t+1}^T P_t x_{t+1}) \qquad \text{(from (7.8.16))}$$

Hence

$$J = \hat{\theta}_t^T\xi_{t+1}\xi_{t+1}^T\hat{\theta}_t + 2\hat{b}_{1t}\hat{\theta}_t^T\xi_{t+1}u_t + \hat{b}_{1t}^2 u_t^2$$
$$- \lambda(1 + \xi_{t+1}^T Q_t \xi_{t+1}) - 2\lambda w_t^T\xi_{t+1}u_t - \lambda s_t u_t^2$$

Differentiating with respect to u_t and equating to zero yields Eq. (7.8.20). $\triangledown$

The rational behind the cost function given in (7.8.19) is that the first term is aimed at providing good control while the second term is aimed at making P_{t+1} small relative to P_t. Thus the factor λ (the learning rate parameter) gives a compromise between good short term control and the rate of learning as measured by the decrease in size of the parameter covariance matrix. The advantage of the control strategy given in Eq. (7.8.20) is that it is only marginally more difficult to implement than the usual passive minimum variance control law and yet it provides an improved learning capability at the expense of short term deterioration in the control performance. Putting $\lambda = 0$ in (7.8.20) yields the normal self-tuning regulator, as expected. If desired, the criterion

$$J = E_{Y_{t+1}|Y_t, U_t}[y_{t+1}^2] + \lambda \log \det P_{t+1}$$

could be used in place of (7.8.19) but this leads to a much more complex control law in general [137]. Extension of the above ideas to more general models is relatively straightforward.

In a practical application, the exponential window form of Eqs. (7.8.14)–(7.8.16) would probably be used. The exponential weighting factor α and the learning rate parameter λ would then be chosen by the designer to give best tracking and control performance.

In closing, it is probably worth remarking that the optimal input and feedback for output power constraints (Result 6.4.9) can be interpreted as a type of dual or active controller. For example, if the parameters are known to drift at a certain rate, then the output power level can be increased (by increasing the probing input power) until the learning rate just exceeds the drift rate (i.e., until the rate of increase of P_t due to parameter drift is counterbalanced by the decrease in P_t due to feedback and probing signal). Further details may be found in [121].

7.9. CONCLUSIONS

In this chapter we have discussed recursive estimation and experiment design schemes, and shown how they can be used to obtain active or passive adaptive control schemes. Probably the main advantages of recursive schemes are their simplicity of implementation and their ability to operate in real time. A possible disadvantage is the fact that, in general, the recursive algorithms are not mathematically equivalent to the corresponding off-line versions. Thus the optimal properties discussed in earlier chapters do not necessarily hold.

PROBLEMS

7.1. Show that Ψ_N which satisfies Eq. (7.2.21) can be written as in Eqs. (7.2.12) and (7.2.13).

7.2. Show that substitution of Eq. (7.3.4) into Eq. (7.3.10) leads to Eqs. (7.3.2) and (7.3.3).

7.3.   Derive Eqs. (7.3.18)–(7.3.20), using Lemma E.1.1.
7.4.   Prove Eqs. (7.3.21)–(7.3.22).
7.5.   Relate the recursive estimator

$$\hat{\theta}_{N+1} = \hat{\theta}_N + K_{N+1}(y_{N+1} - x_{N+1}^{\mathrm{T}}\hat{\theta}_N)$$

where

$$K_{N+1} = x_{N+1}/N$$

to the stochastic approximation algorithm of Eq. (7.5.7). Are (7.5.3) and (7.5.4) satisfied?
7.6.   Derive an on-line input design algorithm for the model

$$A(z)y_t = B(z)u_t + C(z)\varepsilon_t$$

$(A, B, C, y_t, u_t, \varepsilon_t$ defined in Section 7.4.3) with cost function log det P_{N+1} and constraint $-1 \le u_t \le 1$ for all t. Hint: Use the appropriate maximum likelihood algorithm (Section 7.4.3) to obtain an expression for P_{N+1} and proceed as in Section 7.7.1.
7.7.   Rederive Result 7.7.1 for the cost function $J_2 = \text{trace } P_N^{-1}P_{N+1}$.
7.8.   Derive the minimum variance controller for the model of Problem 7.6. Show that the minimum variance control strategy can be expressed as

$$u_t = -\xi_{t+1}^{\mathrm{T}}\theta/b_1$$

(This latter result is particularly transparent if the model is written in the PEM form.)

Derive a suitable expression for ξ_{t+1}. (θ is the vector of coefficients in A, B, and C, excluding b_1.)
7.9.   Show that Eq. (7.8.4) can be solved for F and G by equating coefficients. What are the orders of F and G?
7.10.   Use results of Problems 7.6 and 7.8 to derive an active adaptive control strategy for the model $A(z)y_t = B(z)u_t + C(z)\varepsilon_t$.
7.11.   Recall that an optimal power constrained input for the model

$$y_t = B(z)u_t + \varepsilon_t, \qquad B(z) = b_0 + b_1 z^{-1} + \cdots + b_n z^{-n}$$

has autocorrelation Γ, with the property

$$\Gamma(0) = 1, \quad \Gamma(\tau) = 0, \quad \tau = 1, 2, \ldots, n$$

Derive the recursive algorithm for realizing the above autocorrelation with a binary signal (Section 7.7.1). Compare your result with the algorithm of Result 7.7.1.

APPENDIX

A

Summary of Results from Distribution Theory

A.1. CHARACTERISTIC FUNCTION†

Definition A.1.1   The *characteristic function* $\phi_X(t)$ for a scalar random variable X is defined to be

$$\phi_X(t) = E_X\{e^{itX}\} \qquad (A.1.1)$$

Theorem A.1.1   If X and Y are independent random variables, then the random variable $Z = X + Y$ has characteristic function given by

$$\phi_Z(t) = \phi_X(t)\phi_Y(t) \qquad (A.1.2)$$

Proof

$$\phi_Z(t) = E_Z\{e^{jtZ}\} = E_{X,Y}\{E^{jt(X+Y)}\}$$
$$= E_X\{e^{jtX}\}E_Y\{e^{jtY}\} \qquad \text{since } X \text{ and } Y \text{ independent}$$
$$= \phi_X(t)\phi_Y(t) \qquad \triangledown$$

† See [38].

Theorem A.1.2   (*Inversion*)   If $\int_{-\infty}^{\infty} |\phi_X(t)| \, dt < \infty$, then X has a unique density function given by

$$p_X(x) = \frac{1}{2\pi} \int_{-\infty}^{\infty} e^{-itx} \, \phi_X(t) \, dt \tag{A.1.3}$$

Proof   See Loève [37].

Definition A.1.2   Let X be a k-variate random variable, and let t be a k vector. Then the multivariable characteristic function for X is defined to be

$$\phi_X(t) = E_X[\exp(jt^{\mathsf{T}} X)] \tag{A.1.4}$$

Remark A.1.1   Analogous theorems to Theorems A.1.1 and A.1.2 exist for the multivariable case.

A.2.   THE NORMAL DISTRIBUTION

Definition A.2.1   A scalar random variable X is said to have an $N(\mu, \sigma^2)$ normal distribution if X has density function

$$p_X(x) = (\sigma\sqrt{2\pi})^{-1} e^{-(x-\mu)^2/2\sigma^2} \tag{A.2.1}$$

Result A.2.1   Let X_i, $i = 1, \ldots, N$ be random variables independent and identically distributed as $N(0, 1)$. Since they are independent, the joint density function of $X = (X_1, \ldots, X_N)^{\mathsf{T}}$ is

$$p_X(X) = (2\pi)^{-N/2} \exp\{-\tfrac{1}{2} X^{\mathsf{T}} X\} \tag{A.2.2}$$

Result A.2.2   Consider now the transformation

$$y = AX + \mu \tag{A.2.3}$$

where A is an $N \times N$ nonsingular matrix and μ is an $N \times 1$ vector. It follows from the rule for transformations of density functions [10] that

$$\begin{aligned} p_Y(y) &= (2\pi)^{-N/2} |\det A|^{-1} \exp[-\tfrac{1}{2}(y-\mu)^{\mathsf{T}} A^{-\mathsf{T}} A^{-1}(y-\mu)] \\ &= \{(2\pi)^N |\Sigma|\}^{-1/2} \exp\{-\tfrac{1}{2}(y-\mu)\Sigma^{-1}(y-\mu)]\} \end{aligned} \tag{A.2.4}$$

where $\Sigma = AA^{\mathsf{T}}$.

Definition A.2.2   A multivariable random variable Y is said to have an $N(\mu, \Sigma)$ normal distribution if Y has density function (A.2.4).

Result A.2.3 The characteristic function for $X \sim N(\mu, \Sigma)$ is given by

$$\phi_X(t) = \exp\{jt^T\mu - \tfrac{1}{2}t^T\Sigma t\} \tag{A.2.5}$$

Proof Left as an exercise (see Problem 2.6).

Result A.2.4 If Z is an N variate normal distribution $Z \sim N(\mu, \Sigma)$ and $y = BZ + v$, where B is an $M \times N$ matrix and v is an M vector, then $Y \sim N(B\mu + v, B\Sigma B^T)$.

Proof

$$\phi_Y(t) = E_Y[e^{jt^TY}] = E_Z[e^{jt^T(BZ+v)}] = e^{jt^Tv}E_Z[e^{it^TBZ}]$$

$$= e^{jt^Tv}E_Z[e^{is^TZ}] \quad \text{where } s = B^Tt$$

$$= e^{jt^Tv}\phi_Z(s) = e^{jt^Tv}e^{\{js^T\mu - \tfrac{1}{2}s^T\Sigma t\}} = e^{jt^T(B\mu+v) - \tfrac{1}{2}t^TB\Sigma B^T} \quad \text{(from A.2.5)}$$

Hence, from the inversion of the characteristic function and Result A.2.3

$$Y \sim N(B\mu + v, B\Sigma B^T) \tag{A.2.6}$$

Corollary A.2.4 Let $X \sim N(\mu, \Sigma)$ and partition X by

$$X = \begin{vmatrix} X_1 \\ X_2 \end{vmatrix}$$

with the corresponding partitions of μ and Σ given as

$$\mu = \begin{vmatrix} \mu_1 \\ \mu_2 \end{vmatrix}, \quad \Sigma = \begin{vmatrix} \Sigma_{11} & \Sigma_{12} \\ \Sigma_{21} & \Sigma_{22} \end{vmatrix}$$

then

$$X_1 \sim N(\mu_1, \Sigma_{11}), \quad X_2 \sim N(\mu_2, \Sigma_{22})$$

Proof Select a suitable B in Result A.2.4.

Result A.2.5 Using the partitions of X as in Corollary A.2.4, it follows that if $\Sigma_{12} = \Sigma_{21}^T = 0$, then x_1 and x_2 are statistically independent.

Proof

$$p_{X_1X_2}(x_1, x_2) = C \exp\left\{ -\frac{1}{2}\begin{pmatrix} x_1 - \mu_1 \\ x_2 - \mu_2 \end{pmatrix}^T \begin{vmatrix} \Sigma_{11} & 0 \\ 0 & \Sigma_{22} \end{vmatrix}^{-1} \begin{pmatrix} x_1 - \mu_1 \\ x_2 - \mu_2 \end{pmatrix} \right\}$$

$$= C_1 \exp\{-\tfrac{1}{2}(x_1 - \mu_1)^T\Sigma_{11}^{-1}(x_1 - \mu_1)\}$$

$$\times C_2 \exp\{-\tfrac{1}{2}(x_2 - \mu_2)^T\Sigma_{22}^{-1}(x_2 - \mu_2)\}$$

$$= p_{X_1}(x_1)p_{X_2}(x_2)$$

Hence x_1 and x_2 are independent as required. ∇

Result A.2.6   Let X be partitioned as in Corollary A.2.4, then the conditional distribution for x_1 given x_2 is normal with mean $\mu_1 + \Sigma_{12}\Sigma_{22}^{-1}$ $\times (x_2 - \mu_2)$ and covariance $\Sigma_{11} - \Sigma_{12}\Sigma_{22}^{-1}\Sigma_{21}$

Proof   We consider a multivariate process $\binom{z}{x_2}$ related to $\binom{x_1}{x_2}$ by the unimodular transformation

$$\begin{bmatrix} z \\ x_2 \end{bmatrix} = \begin{bmatrix} I & -\Sigma_{12}\Sigma_{22}^{-1} \\ 0 & I \end{bmatrix} \begin{bmatrix} x_1 \\ x_2 \end{bmatrix} \tag{A.2.7}$$

It now follows from Result A.2.4 that

$$\begin{bmatrix} z \\ x_2 \end{bmatrix} \sim N \begin{bmatrix} \mu_1 - \Sigma_{12}\Sigma_{22}^{-1}\mu_2, & \begin{vmatrix} \Sigma_{11} - \Sigma_{12}\Sigma_{22}^{-1}\Sigma_{21} & 0 \\ 0 & \Sigma_{22} \end{vmatrix} \end{bmatrix} \tag{A.2.8}$$

Thus from Result A.2.5,

$$p_{ZX_2}(z, x_2) = p_Z(z)p_{X_2}(x_2) \tag{A.2.9}$$

But since $\begin{bmatrix} I & -\Sigma_{12}\Sigma_{22}^{-1} \\ 0 & I \end{bmatrix}$ has unit determinant then the Jacobian of the transformation from $\binom{z}{x_2}$ to $\binom{x_1}{x_2}$ is unity. Hence, from the transformation rule,

$$p_{X_1X_2}(x_1, x_2) = p_{Z, x_2}(x_1 - \Sigma_{12}\Sigma_{22}^{-1}x_2, x_2)$$
$$= p_Z(x_1 - \Sigma_{12}\Sigma_{22}^{-1}x_2)p_{X_2}(x_2) \qquad \text{using (A.2.9)}$$

From Bayes' rule, it follows that

$$p_{X_1|X_2}(x_1|x_2) = \frac{p_{X_1X_2}(x_1, x_2)}{p_{X_2}(x_2)} = p_Z(x_1 - \Sigma_{12}\Sigma_{22}^{-1}x_2) \tag{A.2.10}$$

But from (A.2.8),

$$p_Z(z) = C \exp\{-\tfrac{1}{2}(z - \bar{z})^\mathsf{T}H^{-1}(z - \bar{z})\}$$

where C is the normalizing constant

$$\bar{z} = \mu_1 - \Sigma_{12}\Sigma_{22}^{-1}\mu_2, \qquad H = \Sigma_{11} - \Sigma_{12}\Sigma_{22}^{-1}\Sigma_{21}$$

Therefore,

$$p_{X_1|X_2}(x_1|x_2) = C \exp\{-\tfrac{1}{2}(x_1 - \Sigma_{12}\Sigma_{22}^{-1}x_2 - \bar{z})H^{-1}(x_1 - \Sigma_{12}\Sigma_{22}^{-1}x_2 - \bar{z})\}$$

We recognize this as a normal distribution with mean $\mu_1 + \Sigma_{12}\Sigma_{22}^{-1}$ $\times (x_2 - \mu_2)$ and convariance $H = \Sigma_{11} - \Sigma_{12}\Sigma_{22}^{-1}\Sigma_{21}$ as required. ∇

A.3.   THE χ^2 ("CHI SQUARED") DISTRIBUTION

A scalar random variable X is said to have a $\chi^2(k)$ distribution if X has density function

$$p_X(x) = \frac{1}{2^{k/2}\Gamma(k/2)} e^{-x/2} x^{(k/2)-1} \qquad (A.3.1)$$

Γ is the standard gamma function and k is known as the *degrees of freedom*. The characteristic function can be shown to be

$$\phi(t) = (1/(1-2it))^{k/2} \qquad (A.3.2)$$

Result A.3.1   Let X_i, $i = 1, \ldots, m$ be a set of independently distributed χ^2 random variables having degrees of freedom k_i, $i = 1, \ldots, m$, respectively. Then the sum of these variables $Y = \sum_{i=1}^{m} X_i$ has a $\chi^2(\sum_{i=1}^{m} k_i)$ distribution.

Proof   From independence and Theorem A.1.1,

$$\phi_Y(t) = \phi_{X_1}(t) \cdots \phi_{X_m}(t)$$
$$= (1/(1-2it))^{k_1/2} \cdots (1/(1-2it))^{k_m/2}$$
$$= (1/(1-2it))^{\Sigma k_i/2} \qquad (A.3.3)$$

The result follows from the uniqueness property of characteristic functions. ∇

Result A.3.2   If $X \sim N(0, 1)$ (random variable X has normal $N(0, 1)$ distribution), then the distribution of $Y = X^2$ is $\chi^2(1)$.

Proof   We use the rule for transformation of density functions (see Papoulis [10], p. 127), namely,

$$p_Y(y) = \frac{p_X(+\sqrt{y})}{2\sqrt{y}} + \frac{p_X(-\sqrt{y})}{2\sqrt{y}} \qquad (A.3.4)$$

$$= \frac{1}{2\sqrt{2\pi y}} e^{-\frac{1}{2}(\sqrt{y})^2} + \frac{1}{2\sqrt{2\pi y}} e^{-\frac{1}{2}(-\sqrt{y})^2}$$

$$= \frac{1}{\sqrt{2\pi}} e^{-y/2} y^{-\frac{1}{2}} \qquad (A.3.5)$$

Comparison with (A.3.1) establishes the result. ∇

A.4. THE "F" DISTRIBUTION

A scalar random variable X is said to have an $F(k_1, k_2)$ distribution if X has density function

$$p_X(x) = \frac{(k_1/k_2)^{k_1/2} x^{(k_1/2) - 1}}{\beta(k_1/2, k_2/2)(1 + (k_1 x/k_2))^{(k_1 + k_2)/2}} \tag{A.4.1}$$

The integers k_1 and k_2 are known as the *degrees of freedom*. β denotes beta function.

Result A.4.1 Let $X_1 \sim \chi^2(k_1)$ and $X_2 \sim \chi^2(k_2)$ be independent, then the distribution for $Y = (X_1/k_1)/(X_2/k_2)$ is $F(k_1, k_2)$.

Proof The joint density function for X_1 and X_2 is

$$p_{X_1 X_2}(x_1, x_2) = C_1 e^{-x_1/2} e^{-x_2/2} x_1^{(k_1/2) - 1} x_2^{(k_2/2) - 1} \tag{A.4.2}$$

where C_1 is a constant.

Now we introduce the following transformation

$$x_1 = r \sin^2 \theta \tag{A.4.3}$$

$$x_2 = r \cos^2 \theta \tag{A.4.4}$$

Using the rule for transformation of density functions [10]:

$$p_{R\Theta}(r, \theta) = C_2 e^{-r/2} r^{(k_1/2) + (k_2/2) - 1} \sin \theta^{k_1 - 1} \cos \theta^{k_2 - 1} \tag{A.4.5}$$

It can be seen from (A.4.5) that $p_{R\Theta}(r, \theta)$ can be written in the form

$$p_{R\Theta}(r, \theta) = p_R(r) p_\Theta(\theta) \tag{A.4.6}$$

Equation (A.4.6) shows that R and Θ are independent.

Comparison of (A.4.5) and (A.3.1) shows that the marginal distribution for R is clearly $\chi^2(k_1 + k_2)$. However, $R = X_1 + X_2$ and hence we have confirmation of Result A.3.1.

The marginal distribution for Θ is

$$p_\Theta(\theta) = C_3 \sin \theta^{k_1 - 1} \cos \theta^{k_2 - 1} \tag{A.4.7}$$

Now we have defined Y as

$$Y = (X_1/k_1)/(X_2/k_2) \tag{A.4.8}$$

Using (A.4.3) and (A.4.4),

$$Y = (k_2/k_1) \tan^2 \theta \tag{A.4.9}$$

The Jacobian of the transformation is

$$|\partial y/\partial\theta| = 2|(k_2/k_1)\tan\theta\sec^2\theta| = 2|k_2\sin\theta/k_1\cos^3\theta| \quad (A.4.10)$$

Hence from (A.4.7) and (A.4.9), the distribution for Y is

$$p_Y(y) = p_\Theta(\theta)/|\partial y/\partial\theta| \quad \text{for} \quad \theta = \tan^{-1}\sqrt{k_1 y/k_2}$$

$$= C_4\left(\frac{(k_1 y/k_2)^{1/2}}{1+(k_1 y/k_2)}\right)^{k_1-1}\left(\frac{1}{1+(k_1 y/k_2)}\right)^{k_2-1}\frac{(1+(k_1 y/k_2))^{-1}}{(k_1 y/k_2)^{1/2}/(1+(k_1 y/k_2))}$$

$$(A.4.11)$$

Therefore

$$p_Y(y) = C_5\frac{(y)^{(k_1/2)-1}}{(1+(k_1 y/k_2))^{k_1+k_2}} \quad (A.4.12)$$

Comparison of (A.4.12) with (A.4.1) establishes the result. ∇

A.5. THE STUDENT t DISTRIBUTION

A scalar random variable X is said to have student t distribution on k degrees of freedom $(t(k))$ if X has the density function

$$p_X(x) = \{\sqrt{k}\,\beta(\tfrac{1}{2}, k/2)\}^{-1}(1+(x^2/2))^{-(k+1)/2} \quad (A.5.1)$$

Result A.5.1 Let $Y \sim N(0, 1)$ and $X \sim \chi^2(k)$ be independent random variables. Then the ratio $Z = Y/(X/k)^{1/2}$ has a $t(k)$ distribution.

Proof The joint density for Y and X is

$$p_{XY}(x, y) = C_1 e^{-y^2/2}e^{-x/2}x^{(k/2)-1} \quad (A.5.2)$$

We now make the following change of variables:

$$y = r\sin\theta \quad (A.5.3)$$

$$x = r^2\cos^2\theta \quad (A.5.4)$$

The Jacobian of the transformation is

$$|\partial(x, y)/\partial(r, \theta)| = 2r^2\cos\theta \quad (A.5.5)$$

Hence, using the rule for transformation of density functions

$$p_{R\Theta}(r, \theta) = C_2 e^{-r^2/2}r^k(\cos\theta)^{k-1} \quad (A.5.6)$$

Hence the marginal distribution for Θ is clearly

$$p_{\Theta}(\theta) = C_3(\cos \theta)^{k-1} \tag{A.5.7}$$

Now (A.5.3) and (A.5.4) give

$$z = y/(x/k)^{1/2} = \sqrt{k} \tan \theta \tag{A.5.8}$$

Hence, again applying the rule for transformation of density functions,

$$p_Z(z) = C_4(1 + z^2/k)^{-(k+1)/2} \tag{A.5.9}$$

Comparison of (A.5.9) with (A.5.1) establishes the result. ∇

A.6.   THE FISHER–COCHRANE THEOREM

Theorem A.6.1   (Fisher–Cochrane)   Suppose X has a k dimensional multivariate normal distribution with zero mean and unit variance, i.e., $X \sim N_k(0, I)$ and $A_1, \ldots, A_q$ are symmetric nonnegative definite matrices of ranks $r_1, \ldots, r_q$, respectively, such that

$$A_1 + A_2 + \cdots + A_q = I \tag{A.6.1}$$

Then
   (i)   The quadratic forms $Z_j = X^T A_j X$ are independent $\chi^2(r_j)$ random variables if and only if
   (ii) $r_1 + \cdots + r_q = k$ $\tag{A.6.2}$

Proof   Suppose (i) is true. Then $Z_1 + \cdots + Z_q$ has a $\chi^2(r_1 + \cdots + r_q)$ distribution using Result A.3.1. However,

$$Z_1 + \cdots + Z_q = X^T(A_1 + \cdots + A_q)X = X^T X \quad \text{using (A.6.1)} \tag{A.6.3}$$

Therefore $X^T X$ has a $\chi^2(r_1 + \cdots + r_q)$ distribution. Now $X^T X = X_1^2 + \cdots + X_k^2$ and $X_i^2 \sim \chi^2(1)$ using Result A.3.2. Therefore $X^T X \sim \chi^2(k)$ using Result A.3.1. Hence

$$k = r_1 + \cdots + r_q \tag{A.6.4}$$

This completes the "only if" part of the theorem.
   Suppose (ii) is true. Then let B_j be a $k \times r_j$ matrix such that

$$A_j = B_j B_j^T \tag{A.6.5}$$

(A suitable matrix B_j such that (A.6.5) is true can always be found since A_j has rank r_j and A_j is symmetric.)
   Now define

$$B = [B_1 \vdots B_2 \vdots \ldots \vdots B_q] \tag{A.6.6}$$

Clearly B is a $k \times k$ matrix since

$$r_1 + \cdots + r_q = k \tag{A.6.7}$$

Then for any vector X,

$$\begin{aligned} X^T B B^T X &= X^T[B_1 B_1{}^T + \cdots + B_q B_q{}^T]X \\ &= X^T[A_1 + \cdots + A_q]X = X^T X \end{aligned} \tag{A.6.8}$$

Hence

$$BB^T = I \tag{A.6.9}$$

Now let

$$Y = B^T X$$

It follows from the standard rule for transformation of density functions that

$$Y \sim N_k(0, I) \tag{A.6.10}$$

Now we split Y as

$$Y = \begin{bmatrix} Y_1 \\ \hline Y_2 \\ \hline \vdots \\ \hline Y_q \end{bmatrix} \tag{A.6.11}$$

where Y_i is a vector of length r_i.

The Y_js are independent since the density function of Y (i.e., $N_k(0, I)$) can be split into the product of q terms each depending on one Y_j. It also follows that the marginal distributions of the Y_j are $N_{r_j}(0, I)$.

Now

$$Z_j = X^T A_j X = X^T B_j B_j{}^T X = Y_j{}^T Y_j \tag{A.6.12}$$

The Z_js are independent since the Y_js are. Also each Z_j is the sum of r_j squares of the elements of Y_j. Hence from Results A.3.1 and A.3.2, Z_j has a $\chi^2(r_j)$ distribution.

This establishes the "if" part of the theorem. $\triangledown$

Corollary A.6.2 If A is an idempotent matrix (i.e., $A^2 = A$) and $x \sim N(0, I)$, then $X^T A X$ has a $\chi^2(r)$ distribution, where $r = \text{trace } A = \text{rank } A$.

Proof We first show that rank $A = $ trace A for an idempotent matrix. Let

$$\text{rank } A = r$$

Then we can write $A = BB^T$, where B is $k \times r$. Now $A^2 = BB^T BB^T = A = BB^T$ and hence, because B has rank r,

$$B^T B = I$$

where I is $r \times r$. Hence

$$\text{trace } A = \text{trace } BB^T = \text{trace } B^T B = \text{trace } I_r = r \qquad \text{(A.6.13)}$$

We now consider the following quadratic form:

$$X^T X = X^T AX + X^T (I - A)X = X^T A_1 X + X^T A_2 X$$

where

$$A_1 + A_2 = I \qquad \text{(A.6.14)}$$

Now we have rank $A_1 = r$. A_2 can also be shown to be idempotent and hence

$$\text{rank } A_2 = \text{trace } A_2 = \text{trace}(I - A) = \text{trace } I - \text{trace } A = k - r$$

$$\text{rank } A_1 + \text{rank } A_2 = k \qquad \text{(A.6.15)}$$

(A.6.14) and (A.6.15) satisfy the conditions of Theorem A.6.1. Thus $X^T AX$ and $X^T (I - A)X$ are independent χ^2 random variables having r and $k - r$ degrees of freedom, respectively. $\triangledown$

A.7.   THE NONCENTRAL χ^2 DISTRIBUTION

By combining Results (A.3.1) and (A.3.2), we know that the sum of the squares of r independent normal random variables with zero mean and unit variance is distributed as χ^2 on r degrees of freedom. We now consider the distribution of the sum of squares when the means are not necessarily zero.

Result A.7.1   If x is an r-dimensional random variable having a normal distribution with mean μ and covariance I, then the random variable $z = x^T x$ has probability density function given by

$$p_z(z) = \frac{e^{-\frac{1}{2}(z+h)} z^{\frac{1}{2}(r-2)}}{2^{\frac{1}{2}r} \Gamma\{\frac{1}{2}(r-1)\} \Gamma\{\frac{1}{2}\}} \sum_{j=0}^{\infty} \frac{h^j x^j}{(2j)!} B\{\tfrac{1}{2}(r-1), \tfrac{1}{2} + j\} \qquad \text{(A.7.1)}$$

where

$$h = \mu^T \mu \qquad \text{(A.7.2)}$$

The above distribution is called the *noncentral χ^2 distribution with r degrees of freedom and noncentrality parameter h* (written $\chi'^2(r, h)$).

Proof Consider an orthogonal transformation B from x to y, i.e.,

$$y = Bx \tag{A.7.3}$$

where $B^T B = I$.

Then y has $N(\mu^1, I)$ distribution where $\mu^1 = B\mu$. We choose B so that the first $(r - 1)$ components of μ^1 are zero. Then since B preserves length, μ^1 must be given by

$$\mu^1 = (0, 0, \ldots, 0, \sqrt{h}) \tag{A.7.4}$$

where $h = (\mu^T \mu)$.

Hence $z = x^T x = y^T y$ is the sum of r independent normal random variables, the first $(r - 1)$ components of which have zero mean and the last component of which has mean $\sqrt{h}$. Write

$$u = \sum_{i=1}^{r-1} y_i^2, \qquad v = y_r^2 \tag{A.7.5}$$

The distribution of u is $\chi^2(r - 1)$, and the distribution of v can be obtained from the rule for transformation of density functions (see [10, p. 127]) as

$$
\begin{aligned}
p_V(v) &= (1/2\sqrt{v})p_{Y_r}(+\sqrt{y_r}) + (1/2\sqrt{v})p_{Y_r}(-\sqrt{y_r}) \\
&= (1/2\sqrt{2\pi v})e^{-\frac{1}{2}(\sqrt{v}-\sqrt{h})^2} + e^{-\frac{1}{2}(-\sqrt{v}-\sqrt{h})^2} \\
&\propto v^{-\frac{1}{2}}e^{-\frac{1}{2}(v+h)}\sum_{j=0}^{\infty}\frac{(vh)^j}{(2j)!}
\end{aligned}
\tag{A.7.6}
$$

Also from Eq. (A.3.1)

$$p_U(u) \propto e^{-\frac{1}{2}u}u^{\frac{1}{2}(r-3)} \tag{A.7.7}$$

Since u and v are independent, we have that the joint density is given by

$$p_{U,V}(u, v) = p_U(u)p_V(v) \tag{A.7.8}$$

Substituting (A.7.6) and (A.7.7) into (A.7.8) and using the following transformation:

$$z = u + v, \qquad w = u/(u + v) \tag{A.7.9}$$

leads to the following joint distribution for z and w

$$p_{Z,W}(z, w) \propto e^{-\frac{1}{2}(z+h)}z^{\frac{1}{2}(r-2)}w^{(\frac{1}{2}r-3)}(1 - w)^{-\frac{1}{2}}\sum_{j=0}^{\infty}\frac{h^j z^j}{(2j)!}(1 - w)^j \tag{A.7.10}$$

Hence integrating w over its range 0 to 1 gives the marginal distribution for z as in (A.7.1). ∇

APPENDIX

B

Limit Theorems

B.1. CONVERGENCE OF RANDOM VARIABLES

Definition B.1.1   Let $\{X_t; t = 1, 2, \ldots\}$ be a sequence of random variables having distribution functions $\{F_t; t = 1, 2, \ldots\}$. The sequence $\{X_t\}$ is said to *converge in law* (or distribution) to a random variable X with distribution function F if the sequence F_t converges to F (at the points of continuity of F).

We use the following notation for this type of convergence

$$X_t \xrightarrow{\text{law}} X \tag{B.1.1}$$

Definition B.1.2   Let $\{X_t; t = 1, 2, \ldots\}$ be a sequence of random variables, then $\{X_t\}$ is said to *converge in probability* to a random variable X; if given $\varepsilon, \delta > 0$, then there exists a $t_0(\varepsilon, \delta)$ such that $\forall t > t_0$

$$p(|X_t - X| > \varepsilon) < \delta \tag{B.1.2}$$

We use the following notation for this type of convergence

$$X_t \xrightarrow{\text{prob}} X \tag{B.1.3}$$

Definition B.1.3   Let $\{X_t; t = 1, 2, \ldots\}$ be a sequence of random variables possessing second moments. Then $\{X_t\}$ *converges in quadratic mean* to a randim variable X if

$$\lim_{t \to \infty} E(X_t - X)^{\mathrm{T}}(X_t - X) = 0 \qquad (\text{B.1.4})$$

We use the following notation for this type of convergence

$$X_t \xrightarrow{\text{q.m.}} X \qquad (\text{B.1.5})$$

Definition B.1.4   Let $\{X_t; t = 1, 2, \ldots\}$ be a sequence of random variables. Then $\{X_t\}$ is said to *converge almost surely* (or with probability 1) to a random variable X, if given $\varepsilon, \delta > 0$, then there exists a $t_0(\varepsilon, \delta)$ such that

$$p(|X_t - X| < \varepsilon \ \forall \ t \geq (t_0)) > 1 - \delta \qquad (\text{B.1.6})$$

We use the following notation for this type of convergence

$$X_t \xrightarrow{\text{a.s.}} X \qquad (\text{B.1.7})$$

B.2.   RELATIONSHIPS BETWEEN CONVERGENCE CONCEPTS

The interrelationships between the definitions of convergence is discussed in the following lemmas. The proofs of these results may be found in Loève [37].

Lemma B.2.1   Convergence in probability implies convergence in law, i.e.,

$$X_t \xrightarrow{\text{prob}} X \Longrightarrow X_t \xrightarrow{\text{law}} X \qquad (\text{B.2.1})$$

(Note the converse is false in general.)

Lemma B.2.2   Convergence in quadratic mean implies convergence in probability (and hence in law); i.e.,

$$X_t \xrightarrow{\text{q.m.}} X \Longrightarrow X_t \xrightarrow{\text{prob}} X \qquad (\text{B.2.2})$$

(Note the converse is false in general.)

Lemma B.2.3   Convergence almost surely implies convergence in probability (and hence in law); i.e.,

$$X_t \xrightarrow{\text{a.s.}} X \Longrightarrow X_t \xrightarrow{\text{prob}} X \qquad (\text{B.2.3})$$

(Again the converse is false in general.)

Lemma B.2.4   Convergence in quadratic mean does not imply and is not implied by convergence almost surely in general; i.e.,

$$X_t \xrightarrow{\text{a.s.}} X \rlap{\,\diagup\kern-0.3em\diagup} \rightleftarrows X_t \xrightarrow{\text{q.m.}} X \tag{B.2.4}$$

B.3.   SOME IMPORTANT CONVERGENCE THEOREMS

Theorem B.3.1   (*A weak law of large numbers*)   Let $\{X_t\,;\, t = 1, 2, \ldots\}$ be a sequence of uncorrelated random variables having common first and second central moments μ and Σ. Then

$$m_n \xrightarrow{\text{q.m.}} \mu \tag{B.3.1}$$

where

$$m_n = \frac{1}{n} \sum_{i=1}^{n} X_i \tag{B.3.2}$$

Proof $E[m_n] = \mu$

$$E[m_n - E(m_n)][m_n - E(m_n)]^{\mathrm{T}} = E\left[\frac{1}{n}\sum_{i=1}^{n}(X_i - \mu)\right]\left[\frac{1}{n}\sum_{j=1}^{n}(X_j - \mu)\right]^{\mathrm{T}}$$

$$= \frac{1}{n^2}\sum_{i=1}^{n} E(X_i - \mu)(X_i - \mu)^{\mathrm{T}} \quad \text{since the } X\text{s are} \atop \text{uncorrelated}$$

$$= (1/n)\Sigma$$

Hence

$$\lim_{n\to\infty} E[(m_n - E(m_n))^{\mathrm{T}}(m_n - E(m_n))] = \lim_{n\to\infty} \frac{1}{n}\,\text{trace}\,\Sigma$$

$$= 0 \qquad \text{since } \Sigma \text{ is finite} \qquad \nabla$$

Theorem B.3.2   (*A strong law of large numbers*)   If $\{X_t\}$ is a sequence of i.i.d. random variables with finite mean μ, then

$$m_n \xrightarrow{\text{a.s.}} \mu \tag{B.3.3}$$

where

$$m_n = \frac{1}{n} \sum_{i=1}^{n} X_i \tag{B.3.4}$$

Proof   See Feller [41, p. 244]

Theorem B.3.3   (*A central limit theorem*)   Let $\{X_t;\ t = 1, 2, \ldots\}$ be a sequence of i.i.d. random variables with finite mean μ and covariance Σ, then

$$z_t \xrightarrow{\text{law}} z \tag{B.3.5}$$

where

$$z_t = \frac{1}{\sqrt{t}} \sum_{i=1}^{t} (X_i - \mu) \tag{B.3.6}$$

and

$$z \sim N(0, \Sigma) \tag{B.3.7}$$

Proof   Let $Y_t = X_t - \mu;\ t = 1, 2, \ldots$. Then $E[Y_t] = 0$ and $E[Y_t Y_t^{\mathrm{T}}] = \Sigma$. We have

$$z_t = \frac{1}{\sqrt{t}} \sum_{i=1}^{t} Y_i$$

The characteristic function (Definition A.1.2) for z_t is

$$\phi_{z_t}(s) = E[\exp(js^{\mathrm{T}} z_t)] = E\left[\exp\left(\frac{j}{\sqrt{t}} s^{\mathrm{T}}(Y_1 + Y_2 + \cdots + Y_t)\right)\right]$$

$$= \{\phi_Y(s/\sqrt{t})\}^t \qquad \text{since } Y_1, \ldots, Y_t \text{ are i.i.d.}$$

Now define

$$\psi_Y(u) = \log \phi_Y(u) = \log E[e^{ju^{\mathrm{T}}Y}]$$

$$= \log E[1 + ju^{\mathrm{T}}Y - \tfrac{1}{2}(u^{\mathrm{T}}Y)^2 + 0(\|u\|^3)] \qquad \text{for small } \|u\|$$

$$= \log[1 - \tfrac{1}{2}u^{\mathrm{T}}\Sigma u + 0(\|u\|^3)] = -\tfrac{1}{2}u^{\mathrm{T}}\Sigma u + 0(\|u\|^3) \tag{B.3.8}$$

Then

$$\psi_{z_t}(s) = \log \phi_{z_t}(s) = t \log \phi_Y(1/\sqrt{t}\,s)$$

$$= t\psi_Y(s/\sqrt{t}) = -\tfrac{1}{2}s^{\mathrm{T}}\Sigma s + 0(1/\sqrt{t}\,\|s\|^3)$$

Hence

$$\lim_{t \to \infty} \psi_{z_t}(s) = -\tfrac{1}{2}s^{\mathrm{T}}\Sigma s$$

i.e.,

$$\lim_{t \to \infty} \phi_{z_t}(s) = \exp\{-\tfrac{1}{2}s^{\mathrm{T}}\Sigma s\}$$

The theorem follows from Result A.2.3, and the inversion property of the characteristic function.

Theorem B.3.4 (*Frechèt's theorem*)   Let $\{X_t\}$ be a sequence of random variables converging in probability to a constant ξ. Let $f(\cdot)$ be a rational function of the elements of x_t. Then

$$f(X_t) \xrightarrow{\text{prob}} f(\xi) \tag{B.3.9}$$

Proof   See Cramer [43, p. 254], [37]. $\triangledown$

APPENDIX

C

Stochastic Processes

C.1. BASIC RESULTS

Definition C.1.1   *A stochastic process* $\{X_t; t \in T\}$, $T \in R^1$, $X_t \in R^n$, is a family of random variables indexed by the parameter t and defined on a common probability space (Ω, A, P), where Ω is the sample space, A some σ-algebra, and P a probability measure defined on A.

For each t, $X_t(\omega)$, $\omega \in \Omega$ is a random variable.

For each ω, $X_t(\omega)$, $t \in T$ is called a sample function or realization of the process. The stochastic process is said to be *continuous* if T is a connected subset of R^1 and is said to be discrete if T is a finite set from R, i.e., $\{t_1, \ldots, t_n\}$. In most applications we shall think of $\{t_1, \ldots, t_n\}$ as being the points in time at which observations of the process are available. However, more general interpretations are possible, i.e., t_i may be a spatial coordinate.

Our main interest will be in discrete time stochastic processes because the data are normally required in sampled form for computational analysis. Occasionally the process is intrinsically discrete time but usually there is an underlying continuous process.

225

Definition C.1.2   A process $\{X_t, t \in T\}$ is said to be a *stationary process* if, for any $(t_1, \ldots, t_N)$, i.e., any finite subset of T, the joint distribution of $(X_{t_1+\tau}, X_{t_2+\tau}, \ldots, X_{t_N+\tau})$ does not depend upon τ. ∇

It follows that for a stationary stochastic process the distribution of $(X_{t_1+\tau}, X_{t_2+\tau}, \ldots, X_{t_N+\tau})$ depends only on the differences, $(t_2 - t_1, t_3 - t_1, t_4 - t_1, \ldots, t_N - t_1)$.

Definition C.1.3   A process $\{X_t, t \in T\}$, is said to be *wide sense stationary* if

    (a) $E\{X_t\} = \mu$ a finite constant.
    (b)   The covariance matrix $D(t, s) = E(X_t - \mu)(X_s - \mu)^T$ exists.
    (c) $D(t, s)$ depends only on $t - s$. ∇

A process satisfying properties (b) and (c) in Definition C.1.3 but not necessarily property (a) is called a *covariance stationary stochastic process*.

In the sequel we shall be primarily concerned with the first and second moments of the joint distribution function. Processes for which these quantities are finite are called *second-order stochastic processes*.

An important class of stochastic processes is the class of Gaussian or *normal stochastic processes* for which the joint distribution functions are normal.

We shall find two concepts, namely, those of an "independent sequence" and "white noise" play an important role in the modeling of stochastic processes.

Definition C.1.4   A discrete time stochastic process is said to be an *independent sequence* if for any set $(t_1, \ldots, t_N) \subset T$, the corresponding random variables $X_{t_1}, X_{t_2}, \ldots, X_{t_N}$ are independent; i.e., the joint distribution function F can be factored as follows:

$$F(X_{t_1}, X_{t_2}, \ldots, X_{t_N}) = F_1(X_{t_1})F_2(X_{t_2}) \cdots F_N(X_{t_N})$$

where $F_i(X_{t_i})$ is the marginal distribution function of X_{t_i}.

Further, if $F_1, \ldots, F_N$ are identical functions, then the sequence is said to be an *independent and identically distributed (i.i.d.) sequence*. ∇

Definition C.1.5   A discrete time stochastic process is said to be *white noise* if the covariance matrix $D(s, t)$ can be expressed in the form $\Sigma(t)\delta_{st}$ where δ_{st} is the Kronecker delta function and $\Sigma(t)$ is nonnegative definite. ∇

In the remainder of this section we restrict attention to discrete time stationary stochastic processes that admit a *spectral density matrix*. For our current purposes it suffices to define the spectral density matrix as the Fourier transform of the covariance function. We thus have the following Fourier transform pair:

$$\Phi(e^{j\omega}) = \sum_{-\infty}^{\infty} D(\tau)e^{-j\omega\tau} \tag{C.1.1}$$

and

$$D(\tau) = \frac{1}{2\pi} \int_{-\pi}^{\pi} \Phi(e^{j\omega})e^{j\omega\tau} \, d\omega \tag{C.1.2}$$

where $D(\tau) = D(t, t - \tau)$ is the covariance for shift τ, and $\Phi(e^{j\omega})$, $\omega \in [-\pi, \pi]$, is the spectral density matrix. Elementary properties of the spectral density matrix are

(i) $\Phi(e^{-j\omega}) = \overline{\Phi}(e^{j\omega})$ \hfill (C.1.3)
(ii) $\Phi^T(e^{j\omega}) = \overline{\Phi}(e^{j\omega})$ \hfill (C.1.4)
(iii) $\Phi(e^{-j\omega}) + \Phi(e^{j\omega})$ is positive semidefinite. \hfill (C.1.5)

Result C.1.1   The power density spectrum for wide sense stationary white noise with covariance $D(s, t) = \Sigma\delta_{st}$ is

$$\Phi_n(z) = \Sigma \tag{C.1.6}$$

Proof   It follows from the uniqueness of the Fourier transform that it suffices to show $\Phi_n(z) = \Sigma$ yields the required covariance. It is readily verified that

$$\frac{1}{2\pi} \int_{-\pi}^{\pi} e^{j\omega t}\Sigma \, d\omega = \Sigma\delta_{t0} \tag{C.1.7}$$

The result follows from wide sense stationarity of the process. ∇

Comment   Result C.1.1 explains the use of the term "white" noise to describe these processes since the power density spectrum is flat.

Result C.1.2   The output power density spectrum $\Phi(z)$ of an asymptotically stable linear system with transfer function $L(z)$ driven by a zero mean wide sense stationary process with power density spectrum $\Phi_n(z)$ is

$$\Phi_y(z) = L(z)\Phi_n(z)L^T(z^{-1}) \tag{C.1.8}$$

where $z = e^{j\omega}$.

Proof Let the system pulse response corresponding to $L(z)$ be Λ_0, Λ_1, ..., i.e.,

$$L(z) = \sum_{i=0}^{\infty} \Lambda_i z^{-i} \tag{C.1.9}$$

The output of the system is thus

$$y_t = \sum_{i=0}^{\infty} \Lambda_i \eta_{t-i} \tag{C.1.10}$$

where $\{\eta_t\}$ is the driving wide sense stationary process.

The covariance of y_k is then expressible as

$$D_y(s, t) = E\{y_s y_t^T\} = E\left\{ \sum_{i=0}^{\infty} \Lambda_i \eta_{s-i} \left(\sum_{k=0}^{\infty} \Lambda_k \eta_{t-k} \right)^T \right\}$$

$$= \sum_{i=0}^{\infty} \sum_{k=0}^{\infty} \Lambda_i E\{\eta_{s-i} \eta_{t-k}^T\} \Lambda_k^T = \sum_{i=0}^{\infty} \sum_{k=0}^{\infty} \Lambda_i D_n(s-i, t-k) \Lambda_k^T \tag{C.1.11}$$

Now due to stationarity we can rewrite this as

$$D_y(\tau, 0) = \sum_{i=0}^{\infty} \sum_{k=0}^{\infty} \Lambda_i D_n(\tau + k - i, 0) \Lambda_k^T \tag{C.1.12}$$

The appropriate inverse Fourier transform is

$$\Phi_y(e^{j\omega}) = \sum_{\tau=-\infty}^{\infty} D_y(\tau, 0) e^{-j\omega\tau}$$

$$= \sum_{\tau=-\infty}^{\infty} \sum_{i=0}^{\infty} \sum_{k=0}^{\infty} \Lambda_i D_n(\tau + k - i, 0) \Lambda_k^T e^{-j\omega\tau}$$

$$= \sum_{\tau=-\infty}^{\infty} \sum_{i=0}^{\infty} \sum_{k=0}^{\infty} \Lambda_i e^{-j\omega i} D_n(\tau + k - i, 0) \Lambda_k^T e^{+j\omega k} \tag{C.1.13}$$

And from asymptotic stability it follows that Φ_y can be written as

$$\Phi_y(e^{j\omega}) = \sum_{s=-\infty}^{\infty} \sum_{i=0}^{\infty} \sum_{k=0}^{\infty} \Lambda_i e^{-j\omega i} D_n(s, 0) e^{-j\omega s} \Lambda_k^T e^{+j\omega k} \tag{C.1.14}$$

where $s = \tau + k - i$. Therefore

$$\Phi_y(e^{j\omega}) = L(e^{j\omega}) \Phi_n(e^{j\omega}) L^T(e^{-j\omega}) \tag{C.1.15}$$

where $L(z)$ is given by Eq. (C.1.9) and $z = e^{j\omega}$. Thus the theorem is proved. ∇

Corollary If the driving stochastic process is white noise with power density spectrum Σ, then the output power density spectrum is

$$\Phi_y(z) = L(z)\Sigma L^T(z^{-1}) \tag{C.1.16}$$

for $z = e^{j\omega}$. $\quad \nabla$

C.2. CONTINUOUS TIME STOCHASTIC PROCESSES

In the last section, we have defined the concept of white noise for a discrete time stochastic process. This concept is extremely useful in the modeling of general stochastic processes. However, the corresponding concept for continuous processes leads to mathematical difficulties. These difficulties may be overcome by considering stochastic processes having uncorrelated increments.

Definition C.2.1 A stochastic process $\{X_t, t \in T\}$ is said to be a *process with uncorrelated increments*, if for any $(t_1, t_2, \ldots, t_N) \subset T$ such that $t_1 < t_2 < t_3 < \cdots < t_N$, the random variables $X_{t_1}, X_{t_2} - X_{t_1}, X_{t_3} - X_{t_2}, \ldots,$ $X_{t_N} - X_{t_{N-1}}$ are uncorrelated. $\quad \nabla$

Property C.2.1 Given a zero mean stochastic process $\{X_t, t \in T\}$ having *uncorrelated increments*, then for any $t_i, t_j \in T$ such that $t_i > t_j$, it follows that

$$D(t_i, t_i) - D(t_j, t_j)$$

is nonnegative definite where $D(t_i, t_j)$ is the covariance of X_{t_i} and X_{t_j}.

Proof

$$
\begin{aligned}
D(t_i, t_i) &= E\{X_{t_i} X_{t_i}^T\} \\
&= E\{(X_{t_i} - X_{t_j} + X_{t_j} - X_{t_1} + X_{t_1})(X_{t_i} - X_{t_j} + X_{t_j} - X_{t_1} + X_{t_1})^T\} \\
&= E\{(X_{t_i} - X_{t_j})(X_{t_i} - X_{t_j})^T\} + E\{(X_{t_j} - X_{t_1})(X_{t_j} - X_{t_1})^T\} \\
&\quad + E\{X_{t_1} X_{t_1}^T\} \\
&= E\{(X_{t_i} - X_{t_j})(X_{t_i} - X_{t_j})^T\} + E\{X_{t_j} X_{t_j}^T\} \tag{C.2.1}
\end{aligned}
$$

Hence

$$D(t_i, t_i) - D(t_j, t_j) = E\{(X_{t_i} - X_{t_j})(X_{t_i} - X_{t_j})^T\} \quad \nabla \tag{C.2.2}$$

Property C.2.2   Given a zero mean stochastic process $\{X_t, t \in T\}$ having a *wide sense stationary uncorrelated increments*, then the covariance matrix $D(t, t)$ is linear in t.

Proof   Using the wide sense stationary property we can write

$$E\{(X_{t_i} - X_{t_j})(X_{t_i} - X_{t_j})^T\} = C_{t_i - t_j} \tag{C.2.3}$$

Hence using Eq. (C.2.2), we have the following two results:

$$D(t_i, t_i) = D(t_1, t_1) + C_{t_i - t_1} \tag{C.2.4}$$

$$D(t_j, t_j) = D(t_1, t_1) + C_{t_j - t_1} \tag{C.2.5}$$

Substituting these into (C.2.2) gives

$$C_{t_i - t_1} - C_{t_j - t_1} = C_{t_i - t_j} = C_{(t_i - t_1) - (t_j - t_1)} \tag{C.2.6}$$

Therefore C_t can be expressed as

$$C_{t - t_1} = \Sigma(t - t_1) \tag{C.2.7}$$

and hence $D(t, t)$ is linear in t as required. ∇

Equation (C.2.3) can now be expressed formally as

$$E\{dx(t)\, dx(t)^T\} = \Sigma\, dt \tag{C.2.8}$$

We are now in a position to define an extremely important class of stochastic process.

Definition C.2.6   A continuous time normal stochastic process having stationary uncorrelated (and hence independent) increments is known as a *Wiener process*. ∇

We define spectral densities for the continuous time case analogously to the discrete time case by the following transform pair

$$\Phi(j\omega) = \int_{-\infty}^{\infty} D(\tau) e^{-j\omega\tau}\, d\tau \tag{C.2.9}$$

and

$$D(\tau) = \frac{1}{2\pi} \int_{-\infty}^{\infty} \Phi(j\omega) e^{j\omega\tau}\, d\omega \tag{C.2.10}$$

where $D(\tau) = D(t, t - \tau)$ is the covariance for shift τ and $\Phi(j\omega)$, $\omega \in (-\infty, \infty)$ is the spectral density matrix.

Definition C.2.7   We formally define a linear continuous stochastic model by the following equations

$$dx = Ax \, dt + K \, d\eta \qquad \text{(C.2.11)}$$

$$y = Cx \qquad \text{(C.2.12)}$$

where η is a process with wide sense stationary uncorrelated increments having incremental covariance $E\{d\eta \, d\eta^T\} = \Sigma \, dt$.

By Eqs. (C.2.11) and (C.2.12), we mean that y can be expressed in terms of the following stochastic integral (see Itô [79]),

$$y(t) = \int_{-\infty}^{\infty} H(\tau) \, d\omega(t - \tau) \qquad \text{(C.2.13)}$$

where $H(\tau)$ is the *system impulse response* and is given by

$$H(\tau) = Ce^{A\tau}K, \qquad \tau \geq 0$$
$$= 0 \qquad \nabla \qquad \text{(C.2.14)}$$

The *system transfer function* is the Laplace transform of the impulse response

$$G(s) = \int_{-\infty}^{\infty} H(\tau)e^{-s\tau} \, d\tau \qquad \text{(C.2.15)}$$

$$= C(sI - A)^{-1}K \qquad \text{(C.2.16)}$$

We can now find the power density spectrum of the process $y(t)$ modeled by Eqs. (C.2.11) and (C.2.12).

Result C.2.1   The output power density spectrum of the system modeled by Eqs. (C.2.11) and (C.2.12) is given by

$$\Phi_y(s) = G(s)\Sigma G^*(s), \qquad s = j\omega \qquad \text{(C.2.17)}$$

where * denotes complex conjugate transpose and

$$G(s) = C(sI - A)^{-1}K$$

Proof   From (C.2.13),

$$y(t) = \int_{-\infty}^{\infty} H(\tau) \, d\omega(t - \tau)$$

where

$$H(\tau) = Ce^{A\tau}K, \qquad \tau \geq 0$$
$$= 0, \qquad \tau < 0$$

Hence the output covariance is given by

$$D_y(\gamma) = E\{y(t + \gamma)y(t)^T\} = E \int_{-\infty}^{\infty} \int_{-\infty}^{\infty} H(\tau) \, d\omega(t + \gamma - \tau) \, d\omega^T(t - \tau')H^T(\tau')$$

but

$$E\{d\omega(t + \gamma - \tau) \, d\omega^T(t - \tau')\} = \Sigma \, dt' \qquad \text{if} \quad \tau = \tau' + \gamma$$
$$= 0 \qquad \text{otherwise}$$

Therefore

$$D_y(\gamma) = \int_{-\infty}^{\infty} H(\tau' + \gamma) \, \Sigma H^T(\tau') \, d\tau'$$

Thus

$$\Phi_y(j\omega) = \int_{-\infty}^{\infty} D_y(\gamma)e^{-j\omega\gamma} \, d\gamma = \int_{-\infty}^{\infty} \int_{-\infty}^{\infty} H(\tau' + \gamma)\Sigma H^T(\tau') \, d\tau' e^{-j\omega\gamma} \, d\gamma$$

$$= \int_{-\infty}^{\infty} \int_{-\infty}^{\infty} H(\tau' + \gamma)e^{-j\omega(\tau' + \gamma)}\Sigma H(\tau')e^{j\omega\tau'} \, d\gamma \, d\tau'$$

$$= \int_{-\infty}^{\infty} H(\tau)e^{-j\omega\tau} \, d\tau\Sigma \int_{-\infty}^{\infty} H(\tau')e^{j\omega\tau'} \, d\tau' = G(j\omega)\Sigma G^T(-j\omega)$$

where

$$G(s) = \int_{-\infty}^{\infty} H(\tau)e^{-s\tau} \, d\tau \qquad \nabla$$

C.3.   SPECTRAL REPRESENTATION OF STOCHASTIC PROCESSES

For a general stationary discrete stochastic process the covariance function will be a nonnegative definite matrix. It then follows from Bochner's theorem [37] that the covariance $D(\tau)$ can be uniquely represented in the following form:

$$D(\tau) = \frac{1}{2\pi} \int_{-\pi}^{\pi} e^{j\omega\tau} \, dF(\omega) \tag{C.3.1}$$

The integral in Eq. (C.3.1) is of the Stieltjes type. The function $F(\omega)$ is called the *spectral distribution function* and has the following properties:

(i) $F(\omega_1) - F(\omega_2)$ is Hermitian positive semidefinite for all $\omega_1 \geq \omega_2$.
(ii) $F(\omega)$ is of bounded variation on $(-\pi, \pi)$.
(iii) $F(\omega)$ is right continuous.
(iv) $F(-\pi)$ is zero.

Result C.3.1   (*Lebesgue decomposition lemma* [13])   A spectral distribution function can always be factored into the sum of three components

$$F(\omega) = F^{\mathrm{I}}(\omega) + F^{\mathrm{II}}(\omega) + F^{\mathrm{III}}(\omega) \qquad (C.3.2)$$

where $F^{\mathrm{I}}(\omega)$ is absolutely continuous and can be expressed in terms of a *spectral density matrix* $\Phi(j\omega)$ as follows:

$$F^{\mathrm{I}}(\omega) = \int_{-\pi}^{\omega} \Phi(j\omega')\, d\omega' \qquad (C.3.3)$$

$F^{\mathrm{II}}(\omega)$ is piecewise constant with a countable number of discontinuities, its value at ω being equal to the sum of all the jumps of $F(\lambda)$ for λ lying to the left of ω.

$F^{\mathrm{III}}(\omega)$ is a continuous function with a derivative which vanishes almost everywhere. ∇

The corresponding continuous time results are obtained by simply replacing the interval $[-\pi, \pi]$ by $(-\infty, \infty)$.

APPENDIX
D
Martingale Convergence Results

D.1. TOEPLITZ AND KRONECKER LEMMAS

Lemma D.1.1 (*Toeplitz lemma [37, p. 238], [102]*) Let ϕ_{nk}, $k = 1, 2, \ldots, \infty$ be a sequence of real ($p \times p$) matrices such that for every fixed k

$$\lim_{n \to \infty} \phi_{nk} = 0 \qquad (D.1.1)$$

and for all n,

$$\sum_{k=1}^{\infty} \|\phi_{nk}\| \leq c_1 < \infty \qquad (D.1.2)$$

Let

$$y_n = \sum_{k=1}^{\infty} \phi_{nk} x_k \qquad (D.1.3)$$

where $\{x_k\}$ is any sequence of real (p) vectors. Then

(i) $\lim_{k \to \infty} x_k = 0$ $\qquad (D.1.4)$

implies

$$\lim_{n \to \infty} y_n = 0 \qquad (D.1.5)$$

(ii)   If

$$\lim_{n \to \infty} \sum_{k=1}^{\infty} \phi_{nk} = I \qquad \text{(D.1.6)}$$

then

$$\lim_{k \to \infty} x_k = x < \infty \qquad \text{(D.1.7)}$$

implies

$$\lim_{n \to \infty} y_n = x \qquad \text{(D.1.8)}$$

Proof (i)   From (D.1.4) we have

$$\lim_{k \to \infty} x_k = 0$$

Hence for given $\varepsilon > 0$, there exists a $k(\varepsilon)$ such that for all $k > k(\varepsilon)$, $\|x_k\| < \varepsilon/c_1$ so that

$$y_n = \sum_{k=1}^{\infty} \phi_{nk} x_k \qquad \text{(from (D.1.3))}$$

$$= \sum_{k=1}^{k(\varepsilon)} \phi_{nk} x_k + \sum_{k=k(\varepsilon)+1}^{\infty} \phi_{nk} x_k$$

$$\|y_n\| \le \left\| \sum_{k=1}^{k(\varepsilon)} \phi_{nk} x_k \right\| + \sum_{k=k(\varepsilon)+1}^{\infty} \|\phi_{nk}\| \, \|x_k\|$$

$$\le \left\| \sum_{k=1}^{k(\varepsilon)} \phi_{nk} x_k \right\| + \varepsilon \qquad \text{(from (D.1.2))}$$

Thus letting $n \to \infty$, and using (D.1.1),

$$\lim_{n \to \infty} \|y_n\| \le \varepsilon$$

Now, letting $\varepsilon \to 0$ establishes (i).

(ii)

$$y_n = \sum_{k=1}^{\infty} \phi_{nk} x + \sum_{k=1}^{\infty} \phi_{nk}(x_k - x) \qquad \text{(from (D.1.3))}$$

Thus, taking limit and using part (i) and (D.1.6) yields

$$\lim_{n \to \infty} y_n = x + 0 = x \qquad \nabla$$

Remark D.1.1 In the above lemma $\| \ \|$ denotes matrix norm, e.g., $\|A\| = [\lambda_{\max}[A^T A]]^{1/2}$. For the scalar case $\| \ \|$ becomes absolute value.

Lemma D.1.2 (*Kronecker lemma* [37, p. 238], [102]) If

$$S_n = \sum_{k=1}^{n} A_k, \qquad A_k \text{ is a } p \times p \text{ positive semidefinite matrix} \qquad \text{(D.1.9)}$$

$$v_n = \sum_{k=1}^{n} S_k^{-1} x_k, \qquad \text{for some sequence of } p \text{ vectors } \{x_k\} \qquad \text{(D.1.10)}$$

Then

$$\lim_{n \to \infty} v_n = v, \qquad \|v\| < \infty \qquad \text{(D.1.11)}$$

$$\lim_{n \to \infty} S_n^{-1} A_k = 0, \qquad \text{for all} \quad k \qquad \text{(D.1.12)}$$

$$\sum_{k=1}^{n} \|S_n^{-1} A_k\| < \infty, \qquad \text{for all} \quad n \qquad \text{(D.1.13)}$$

imply

$$\lim_{n \to \infty} S_n^{-1} \sum_{k=1}^{n} x_k = 0 \qquad \text{(D.1.14)}$$

Proof Set $v_0 = 0$. Then

$$S_n^{-1} \sum_{k=1}^{n} x_k = S_n^{-1} \sum_{k=1}^{n} S_k (v_k - v_{k-1}) = v_n - S_n^{-1} \sum_{k=1}^{n} (S_k - S_{k-1}) v_{k-1}$$

$$= v_n - S_n^{-1} \sum_{k=1}^{n} A_k v_{k-1}$$

Now define

$$\phi_{nk} = S_n^{-1} A_k, \qquad 1 < k \le n$$

$$= 0, \qquad k > n$$

We then see that

$$\lim_{n \to \infty} \phi_{nk} = 0 \qquad \text{(from (D.1.12))}$$

$$\sum_{k=1}^{\infty} \phi_{nk} = \sum_{k=1}^{n} S_n^{-1} A_k = I \qquad \text{(from (D.1.9))}$$

$$\sum_{k=1}^{\infty} \|\phi_{nk}\| = \sum_{k=1}^{n} \|S_n^{-1} A_k\| < \infty \qquad \text{(from (D.1.13))}$$

Hence, applying the Toeplitz lemma,

$$\lim_{n\to\infty} S_n^{-1} \sum_{k=1}^{n} A_k v_{k-1} = \lim_{n\to\infty} \sum_{k=1}^{\infty} \phi_{nk} v_{k-1} = v$$

$$\lim_{n\to\infty} S_n^{-1} \sum_{k=1}^{n} x_k = \lim_{n\to\infty} v_n - \lim_{n\to\infty} S_n^{-1} \sum_{k=1}^{n} A_k v_{k-1} = v - v = 0 \qquad \triangledown$$

D.2. MARTINGALES

Consider a probability space $(\Omega, \mathscr{A}, P)$, where Ω is some nonempty set (the *sample space*), $\mathscr{A}$ a *σ-algebra* of subsets of Ω, and P a *probability measure* defined on $\mathscr{A}$.

Let $\{A_t, t \in T\}$ be a sequence of σ-algebra with the property that

$$A_t \subset \mathscr{A} \qquad \text{for all} \quad t \in T$$

$$A_t \subset A_s \qquad \text{whenever} \quad t \le s$$

A sequence $\{A_t\}$ with the above properties is said to be *increasing*. (If we think of A_t as being the information at time t, then the increasing property corresponds to increasing information at time elapses.)

Definition (D.2.1) A stochastic process (see Appendix C) $\{X_t, t \in T\}$ is said to be *adapted* to the sequence of σ-algebras $\{A_t\}$ if x_t is A_t *measurable* for all t. In other words, the information contained in A_t is sufficient to completely specify x_t, i.e.,

$$E(x_t|A_t) = x_t \qquad \triangledown \tag{D.2.1}$$

Definition (D.2.2) A stochastic process $\{x_t, t \in T\}$ is said to be a *Martingale* with respect to a sequence of σ-algebras $\{A_t, t \in T\}$ if

   (i) $\{A_t\}$ is increasing.
  (ii) $\{x_t\}$ is adapted to A_t.
 (iii) $E(|x_t|) < \infty$ (D.2.2)
 (iv) $x_s = E(x_t|A_s, s < t)$ $\triangledown$ (D.2.3)

If property (iv) is replaced by

(iva) $x_s \le E(x_t|A_s, s < t)$ (D.2.4)

then $\{x_t\}$ is said to be a submartingale with respect to $\{A_t\}$ and if property (iv) is replaced by

(ivb) $x_s \ge E(x_t|A_s, s < t)$ (D.2.5)

then $\{x_t\}$ is said to be a *supermartingale* with respect to $\{A_t\}$.

Theorem D.2.1   (*Doob decomposition*) $\{x_t\}$ is a submartingale with respect to $\{A_t\}$ if and only if there exists a Martingale $\{x_t'\}$ with respect to $\{A_t\}$ and a positive increasing sequence $\{z_t\}$ such that

$$x_t = z_t + x_t' \tag{D.2.6}$$

Proof   (See also [103], [104])   Let $\{x_t\}$ be a submartingale and

$$y_{k+1} = x_{k+1} - E(x_{k+1}|A_k), \qquad y_1 = x_1$$

and

$$v_{k+1} = -x_k + E(x_{k+1}|A_k), \qquad v_1 = 0$$

then $E(y_{k+1}|A_k) = 0$ and $v_{k+1} \geq 0$ since $\{x_t\}$ is a submartingale. Then

$$x_n = x_1 + \sum_{k=1}^{n-1}(x_{k+1} - x_k) = y_1 + \sum_{k=2}^{n} y_k + \sum_{k=1}^{n} v_k = x_n' + z_n$$

where $x_n' = \sum_{k=1}^{n} y_k$ and $z_n = \sum_{k=1}^{n} v_k$. Now,

$$E(x_{n+1}'|A_n) = E\left(\sum_{k=1}^{n} y_k + y_{n+1}\Big|A_n\right) = \sum_{k=1}^{n} y_k = x_n'$$

Hence $\{x_n'\}$ is a Martingale and z_n is obviously increasing. To show the converse, we simply note that

$$E(x_{n+1}|A_n) = E(z_{n+1} + x_{n+1}'|A_n) = z_{n+1} + x_n' \geq z_n + x_n' = x_n$$

since $\{z_k\}$ is increasing. ∇

Theorem D.2.2   Let $\{x_t\}$ be a submartingale with respect to $\{A_t\}$ and let f be an increasing convex function defined on R. Then $\{f(x_t)\}$ is a submartingale with respect to $\{A_t\}$.

Proof

$$x_t \leq E(x_{t+1}|A_t)$$

Since $f(\cdot)$ is increasing, we have

$$f(x_t) \leq f(E(x_{t+1}|A_t))$$

Now, convexity of f implies that

$$f(E(x_{t+1}|A_t)) \leq E(f(x_{t+1})|A_t)$$

(Jensen's inequality [11]) and the result follows. ∇

Theorem D.2.3 Let $\{x_t\}$ be a scalar Martingale with respect to $\{A_t\}$ such that $E([x_t]^2) < \infty$ for all t. Then, there exists a random variable x having bounded variance such that

(a) $x_t \xrightarrow{\text{a.s.}} x$ $\qquad\qquad$ (D.2.7)

(b) $x_t \xrightarrow{\text{q.m.}} x$ $\qquad\qquad$ (D.2.8)

(c) $x_t = E[x | A_t]$ $\quad$ for every t $\qquad\qquad$ (D.2.9)

Proof See [105], [106]. ∇

Corollary D.2.3 Let $\{x_t\}$ be a vector Martingale with respect to $\{A_t\}$ such that $\|E\{x_t x_t^T\}\| < \infty$ for all t. Then there exists a vector random variable x having bounded covariance such that

(a) $x_t \xrightarrow{\text{a.s.}} x$ $\qquad\qquad$ (D.2.10)

(b) $x_t \xrightarrow{\text{q.m.}} x$ $\qquad\qquad$ (D.2.11)

(c) $x_t = [Ex | A_t]$ $\quad$ for every t $\qquad\qquad$ (D.2.12)

Proof Apply Theorem D.2.3 element by element and use properties of covariance matrix. ∇

Definition (*D.2.3*) Given an increasing sequence of σ-algebras $\{A_t\}$ in the probability space $(\Omega, \mathscr{A}, P)$, a sequence $\{\varepsilon_t\}$ of vector random variables adapted to $\{A_t\}$ is said to be an *innovations sequence* if

$$E[\varepsilon_t | A_{t-1}] = 0 \qquad \text{for all} \quad t \qquad \nabla \qquad (\text{D.2.13})$$

Theorem D.2.4 Let $\{\varepsilon_t\}$ be an innovations sequence adapted to $\{A_t\}$. Define $\{z_n\}$ as

$$z_n = \sum_{t=1}^{n} S_t^{-1} \varepsilon_t, \qquad n = 1, 2, \dots \qquad (\text{D.2.14})$$

$$z_0 = 0$$

where

$$S_t = \sum_{k=1}^{t} \Psi_k, \qquad \Psi_k \text{ is a } p \times p \text{ positive semidefinite symmetric matrix}$$

$$(\text{D.2.15})$$

such that $\{S_t\}$ is adapted to $\{A_{t-1}\}$,

$$\left\| E \left| \sum_{t=1}^{n} S_t^{-1} \varepsilon_t \varepsilon_t^{\mathsf{T}} S_t^{-1} \right| \right\| < \infty \tag{D.2.16}$$

$$\lim_{n \to \infty} S_n^{-1} \Psi_k = 0 \qquad \text{for all} \quad k \qquad (\text{a.e. } \omega \in \Omega) \tag{D.2.17}$$

and

$$\sum_{k=1}^{n} \left\| S_n^{-1} \Psi_k \right\| < \infty \qquad \text{for all} \quad n \qquad (\text{a.e. } \omega \in \Omega) \tag{D.2.18}$$

Then

(1) z_n is a Martingale.
(2) z_n converges q.m. and a.s.

(3) $S_n^{-1} \sum_{t=1}^{n} \varepsilon_t \xrightarrow{\text{a.s.}} 0$

Proof

(i) $E[z_n | A_{n-1}] = E[S_n^{-1} \varepsilon_n + z_{n-1} | A_{n-1}]$

$\qquad\qquad = z_{n-1} \qquad$ since S_n is adapted to A_{n-1} and ε_n is an
$\qquad\qquad\qquad\qquad$ innovations sequence adapted to A_n

Hence z_n is a Martingale.

(ii) $E[z_n z_n^{\mathsf{T}}] = E \left[\left(\sum_{k=1}^{n} S_k^{-1} \varepsilon_k \right) \left(\sum_{j=1}^{n} \varepsilon_j^{\mathsf{T}} S_j^{-1} \right) \right]$

$\qquad = E \left[\sum_{t=1}^{n} S_t^{-1} \varepsilon_t \varepsilon_t^{\mathsf{T}} S_t^{-1} + \sum_{k=1}^{n} \sum_{j=k+1}^{n} S_k^{-1} \varepsilon_k \varepsilon_j^{\mathsf{T}} S_j^{-1} \right.$

$\qquad\qquad \left. + \sum_{k=1}^{n} \sum_{j=1}^{k-1} S_k^{-1} \varepsilon_k \varepsilon_j^{\mathsf{T}} S_j^{-1} \right]$

$\qquad = E \left[\sum_{t=1}^{n} S_t^{-1} \varepsilon_t \varepsilon_t^{\mathsf{T}} S_t^{-1} + \sum_{k=1}^{n} \sum_{j=k+1}^{n} E\{S_k^{-1} \varepsilon_k \varepsilon_j^{\mathsf{T}} S_j^{-1} | A_{j-1}\} \right.$

$\qquad\qquad \left. + \sum_{k=1}^{n} \sum_{j=1}^{k-1} E\{S_k^{-1} \varepsilon_k \varepsilon_j^{\mathsf{T}} S_j^{-1} | A_{k-1}\} \right]$

$\qquad = E \left[\sum_{t=1}^{n} S_t^{-1} \varepsilon_t \varepsilon_t^{\mathsf{T}} S_t^{-1} \right] \qquad$ using the fact that S_n is
$\qquad\qquad\qquad\qquad\qquad\qquad$ adapted to A_{n-1} and ε_n
$\qquad\qquad\qquad\qquad\qquad\qquad$ is adapted to A_n

Therefore, $\| E[z_n z_n^{\mathsf{T}}] \| < \infty$ by (D.2.16). Hence, z_n converges q.m. and a.s.
via Corollary D.2.3.

(iii)   Follows from Lemma D.1.2. $\triangledown$

APPENDIX
E
Mathematical Results

E.1. MATRIX RESULTS

Lemma E.1.1   *(Matrix inversion lemma)*

$$(A + BC)^{-1} = A^{-1} - A^{-1}B(I + CA^{-1}B)^{-1}CA^{-1} \qquad \text{(E.1.1)}$$

Proof   See Problem 5.2. ∇

Remark E.1.1   We note that, if B is a column $(= g$, say) and C is a row $(= h^{\mathrm{T}}$, say), then Lemma E.1.1 reduces to

$$(A + gh^{\mathrm{T}})^{-1} = \left| I - A^{-1} \frac{gh^{\mathrm{T}}}{(1 + h^{\mathrm{T}}A^{-1}g)} \right| A^{-1} \qquad \text{(E.1.2)}$$

The implication of Eq. (E.1.2) is that the inversion of a matrix plus a diad reduces to a simple scalar division provided the inverse of the original matrix is known. (See also the solution to Problem 5.2). ∇

Result E.1.1   If A is partitioned as

$$A = \begin{vmatrix} A_{11} & A_{12} \\ A_{21} & A_{22} \end{vmatrix} \qquad \text{(E.1.3)}$$

where A, A_{11}, and A_{22} are all square and nonsingular, then

$$A^{-1} =$$

$$\left[\begin{array}{c|c} (A_{11} - A_{12}A_{22}^{-1}A_{21})^{-1} & -(A_{11} - A_{12}A_{22}^{-1}A_{21})^{-1}A_{12}A_{22}^{-1} \\ \hline -(A_{22} - A_{21}A_{11}^{-1}A_{12})^{-1}A_{21}A_{11}^{-1} & (A_{22} - A_{21}A_{11}^{-1}A_{12})^{-1} \end{array}\right]$$

$$\text{(E.1.4)}$$

Proof Simply show $AA^{-1} = I$ (exercise). ∇

Result E.1.2   If A is as defined in (E.1.3), then

$$\det A = \det A_{22} \cdot \det(A_{11} - A_{12}A_{22}^{-1}A_{21}) \qquad \text{(E.1.5)}$$

$$= \det A_{11} \cdot \det(A_{22} - A_{21}A_{11}^{-1}A_{12}) \qquad \text{(E.1.6)}$$

Proof

$$\left|\begin{array}{cc} A_{11} & A_{12} \\ A_{21} & A_{22} \end{array}\right| \left|\begin{array}{cc} I & 0 \\ -A_{22}^{-1}A_{21} & A_{22}^{-1} \end{array}\right| = \left|\begin{array}{cc} (A_{11} - A_{12}A_{22}^{-1}A_{21}) & A_{12}A_{22}^{-1} \\ 0 & I \end{array}\right|$$

Therefore $\det A \cdot \det A_{22}^{-1} = \det(A_{11} - A_{12}A_{22}^{-1}A_{21})$.   Equation   (E.1.5) follows since $\det A_{22}^{-1} = 1/(\det A_{22})$. Similarly, Eq. (E.1.6) may be established. ∇

Result E.1.3

$$\det(I + BC) = \det(I + CB) \qquad \text{(E.1.7)}$$

Proof Consider the matrix

$$A = \left|\begin{array}{cc} I & B \\ C & I \end{array}\right|$$

The result follows from Result E.1.2. ∇

Result E.1.4

$$(I + BC)^{-1}B = B(I + CB)^{-1} \qquad \text{(E.1.8)}$$

Proof From Lemma E.1.1,

$$(I + BC)^{-1}B = B - B(I + CB)^{-1}CB = B(I - (I + CB)^{-1}CB)$$

$$= B(I + CB)^{-1}(I + CB - CB) = B(I + CB)^{-1} \qquad \nabla$$

E.2. VECTOR AND MATRIX DIFFERENTIATION RESULTS

The following conventions are adopted throughout the book:

(1) $\partial\phi/\partial\theta$, where ϕ is a scalar and θ a vector, denotes a *row vector* with ith element $\partial\phi/\partial\theta_i$, θ_i being the ith component of θ.

(2) $\partial\beta/\partial\theta$, where θ and β are vectors, denotes a matrix with ijth element $\partial\beta_i/\partial\theta_j$ where β_i, θ_j are the ith and jth elements of β and θ, respectively.

(3) $\partial\phi/\partial M$, where ϕ is a scalar and M is a matrix, denotes a matrix with ijth element $\partial\phi/\partial M_{ij}$ where M_{ij} is the ijth element of M.

We now have the following results. (See also [70].)

Result E.2.1

$$\frac{\partial}{\partial M} \log \det M = M^{-\mathsf{T}} \tag{E.2.1}$$

where M is any square nonsingular matrix.

Proof

$$M^{-1} = \operatorname{adj} M/\det M$$

Therefore $M(\operatorname{adj} M) = (\det M)I$. Taking log of iith element gives

$$\log \det M = \log\left(\sum_k M_{ik}(\operatorname{adj} M)_{ki}\right)$$

or

$$\frac{\partial \log \det M}{\partial M_{ij}} = \frac{(\operatorname{adj} M)_{ji}}{\sum_k M_{ik}(\operatorname{adj} M)_{ki}} \qquad \text{(since } (\operatorname{adj} M)_{ki} \text{ does not depend upon } M_{ij})$$

$$= \frac{(\operatorname{adj} M)_{ji}}{\det M} = M^{-\mathsf{T}} \qquad \triangledown$$

Result E.2.2

$$\frac{\partial}{\partial M} (\operatorname{trace} WM^{-1}) = -(M^{-1}WM^{-1})^{\mathsf{T}} \tag{E.2.2}$$

Proof

$$MM^{-1} = I$$

Hence

$$\frac{\partial M}{\partial M_{ij}} M^{-1} + M \frac{\partial M^{-1}}{\partial M_{ij}} = 0$$

Hence

$$\frac{\partial M^{-1}}{\partial M_{ij}} = -M^{-1} \frac{\partial M}{\partial M_{ij}} M^{-1} = -M^{-1} e_i e_j^{\mathsf{T}} M^{-1}$$ (where e_i is a null vector save for 1 in the ith position)

Hence

$$\frac{\partial}{\partial M_{ij}} \text{trace } WM^{-1} = \text{trace} \left| W \frac{\partial M^{-1}}{\partial M_{ij}} \right| = -\text{trace}[WM^{-1} e_i e_j^{\mathsf{T}} M^{-1}]$$

$$= -e_j^{\mathsf{T}} M^{-1} WM^{-1} e_i = -e_i^{\mathsf{T}} [M^{-1} WM^{-1}]^{\mathsf{T}} e_j$$

Hence

$$\frac{\partial}{\partial M} \text{trace } WM^{-1} = -[M^{-1} WM^{-1}]^{\mathsf{T}} \qquad \triangledown$$

Result E.2.3 If M is a positive definite matrix with unit diagonal ($M_{ii} = 1$, $i = 1, \ldots, n$), then det M achieves its maximum value for $M = I$.

Proof For all positive x we have the following inequality:

$$\log x < x - 1 \qquad \text{with equality iff} \quad x = 1$$

Now $\log \det M = \sum_{i=1}^{n} \lambda_i$, where $\lambda_1 \cdots \lambda_n$ are the eigenvalues of M. Hence using the above inequality,

$$\log \det M \leq \text{trace } M - n$$

with equality iff $\lambda_1 = \lambda_2 = \cdots = \lambda_n$. Hence $\log \det M$ achieves its maximum value for $M = I$. $\triangledown$

E.3. CARATHEODORY'S THEOREM

Throughout this section S denotes a subset of R^n.

Definition E.3.1 The *convex hull* of the set S is the set $\mathscr{C}$ of all convex combinations of elements of S, i.e.,

$$\mathscr{C} = \left\{ c \,\middle|\, c = \sum_{k=1}^{\infty} \lambda_i s_i, \, 0 \leq \lambda_i \leq 1, \, \sum_{k=1}^{\infty} \lambda_i = 1, \, s_i \in S \right\}$$

or, more generally,

$$\mathscr{C} = \left\{ c \,\middle|\, c = \int_S s \, d\lambda(s), \; d\lambda(s) \geq 0, \; \int_S d\lambda(s) = 1, \; s \in S \right\}$$

An alternative definition is that $\mathscr{C}$ is the smallest convex set containing S. ∇

We now have the following theorem due to Caratheodory [210]:

Theorem E.3.1   Let S be any set in R^n and $\mathscr{C}$ its convex hull. Then $c \in \mathscr{C}$ if and only if c can be expressed as a convex combination of $n + 1$ elements of S, i.e., if and only if

$$c = \sum_{i=1}^{n+1} \lambda_i s_i, \qquad s_i \in S, \qquad 0 \leq \lambda_i \leq 1, \qquad \sum_{i=1}^{n+1} \lambda_i = 1$$

Proof   See [210]. ∇

Theorem E.3.2   If S is closed and bounded, then its convex hull $\mathscr{C}$ is closed and bounded.

Proof   See [210]. ∇

Definition E.3.2   A *face* of a convex set $\mathscr{C}$ is a convex subset $\mathscr{C}'$, such that every line segment in $\mathscr{C}$ with interior point $\in \mathscr{C}'$ also has both endpoints in $\mathscr{C}'$. ∇

Definition E.3.3   An *extreme point* of $\mathscr{C}$ is a zero-dimensional face of $\mathscr{C}$. ∇

Theorem E.3.3   A closed bounded convex set is the convex hull of its extreme points.

Proof   See [210]. ∇

Theorem E.3.4   Any boundary point of a closed bounded convex set $\mathscr{C}$ can be expressed as a convex combination of n or less extreme points of $\mathscr{C}$.

Proof   See [210]. ∇

E.4.   INEQUALITIES

Result E.4.1   If $M = \alpha M_1 + (1 - \alpha)M_2$, $0 < \alpha < 1$, then

$$\det M > \det M_1{}^\alpha \det M_2^{(1-\alpha)}$$

Proof   See [156]. ∇

Problem Solutions

CHAPTER 1

1.1

$$E_{x,y}(X + Y) = \int_{-\infty}^{\infty} \int_{-\infty}^{\infty} (x + y)\, dF_{x,y}(x, y)$$

Prove for $F_{x,y}$ absolutely continuous only:

$$E_{x,y}(X + Y) = \int_{-\infty}^{\infty} \int_{-\infty}^{\infty} x p_{x,y}(x, y)\, dx\, dy + \int_{-\infty}^{\infty} \int_{-\infty}^{\infty} y p_{x,y}(x, y)\, dx\, dy$$

$$= \int_{-\infty}^{\infty} x p_x(x)\, dx + \int_{-\infty}^{\infty} y p_y(y)\, dy = E_x(x) + E_y(y)$$

1.2

$$E_\theta\{g(\theta)E_{y\mid\theta}\{f(Y)\}\} = \int_{-\infty}^{\infty} g(\theta) p_\theta(\theta) \left\{ \int_{-\infty}^{\infty} f(y) p_{y\mid\theta}(y/\theta)\, dy \right\} d\theta$$

$$= \int_{-\infty}^{\infty} \int_{-\infty}^{\infty} f(y) g(\theta) p_{Y\mid\theta}(y/\theta) p_\theta(\theta)\, dy\, d\theta$$

$$= \int_{-\infty}^{\infty} \int_{-\infty}^{\infty} f(y) g(\theta) p_Y(y) p_{\theta\mid Y}(\theta/y)\, dy\, d\theta \qquad \text{(Bayes)}$$

$$= \int_{-\infty}^{\infty} f(y) p_Y(y) \left\{ \int_{-\infty}^{\infty} g(\theta) p_{\theta\mid Y}(\theta/y)\, d\theta \right\} dy$$

$$= E_Y\{f(Y) E_{\theta\mid Y}\{g(\theta)\}\}$$

1.3

$$\int_{-\infty}^{\infty} \frac{1}{\sigma\sqrt{2\pi}} \exp\left\{-\frac{1}{2\sigma^2}(y-\mu)^2\right\} dy = 1 \qquad (*)$$

differentiate (*) with respect to μ:

$$\int_{-\infty}^{\infty} 2\sigma^2(y-\mu) \exp\left\{-\frac{1}{2\sigma^2}(y-\mu)^2\right\} dy = 0$$

hence $E_Y(y) = \mu$.

Differentiate (*) with respect to σ:

$$\int_{-\infty}^{\infty} \frac{(y-\mu)^2}{\sigma^3} \exp\left\{-\frac{1}{2\sigma^2}(y-\mu)^2\right\} dt = \sqrt{2\pi}$$

hence $E_y(y-\mu)^2 = \sigma^2$.

1.4

$$E\{\bar{X}\} = \frac{1}{N} \sum_{i=1}^{N} E\{X_i\} = \mu \qquad \text{from Problem 1.1}$$

$$E\tfrac{1}{2}\{X_1 + X_N\} = \tfrac{1}{2}(E\{X_1\} + E\{X_N\}) = \mu$$

Comment: The average of any subset of the N variables is an unbiased estimator or μ.

1.5

$$p(y_1, \ldots, y_N|\mu, \sigma^2) = \prod_{i=1}^{N} p(y_i|\mu, \sigma^2)$$

$$= \left(\frac{1}{\sigma\sqrt{2\pi}}\right)^N \exp\left\{-\frac{1}{2\sigma^2} \sum_{i=1}^{N}(y_i-\mu)^2\right\}$$

$$= \left(\frac{1}{\sigma\sqrt{2\pi}}\right)^N \exp\left\{-\frac{1}{2\sigma^2}\left\{\sum_{i=1}^{N} y_i^2 - 2\mu \sum_{i=1}^{N} y_i + N\mu^2\right\}\right\}$$

The required result follows from Theorem 1.4.1.

1.6

$$\log p(y_1, \ldots, y_N|\mu, \sigma^2) = -\frac{N}{2}\log(2\pi\sigma^2) - \frac{1}{2\sigma^2}\sum_{i=1}^{N}(y_i-\mu)^2 = l$$

$$\frac{\partial l}{\partial \mu} = \frac{1}{\sigma^2}\sum_{i=1}^{N}(y_i-\mu), \qquad \frac{\partial l}{\partial \sigma^2} = -\frac{N}{2\sigma^2} + \frac{1}{2\sigma^4}\sum_{i=1}^{N}(y_i-\mu)^2$$

$$\frac{\partial^2 l}{\partial \mu^2} = -\frac{N}{\sigma^2}, \qquad \frac{\partial^2 l}{\partial \mu \partial \sigma^2} = -\frac{1}{\sigma^4}\sum_{i=1}^{N}(y_i-\mu),$$

$$\frac{\partial^2 l}{\partial(\sigma^2)^2} = \frac{N}{2\sigma^4} - \frac{1}{\sigma^6}\sum_{i=1}^{N}(y_i-\mu)^2$$

Hence

$$
M_\theta = - E_{Y|\theta} \begin{bmatrix} \dfrac{\partial^2 l}{\partial \mu^2} & \dfrac{\partial^2 l}{\partial \mu \, \partial \sigma^2} \\[2ex] \dfrac{\partial^2 l}{\partial \mu \, \partial \sigma^2} & \dfrac{\partial^2 l}{\partial (\sigma^2)^2} \end{bmatrix} = \begin{bmatrix} \dfrac{N}{\sigma^2} & 0 \\[2ex] 0 & \dfrac{N}{2\sigma^4} \end{bmatrix}
$$

1.7

$$
\mathrm{cov}(g) = E\left\{ \left(\frac{1}{N} \sum_{i=1}^{N} y_i - \mu \right)^2 \right\} = E\left\{ \frac{1}{N^2} \sum_{i=1}^{N} (y_i - \mu)^2 \right\} = \frac{\sigma^2}{N}
$$

cf. (1,1) element of M_θ^{-1} as given in Problem 1.6, i.e., g is efficient.

1.8 (Derivation of Kalman filter equations)

(a) $\begin{vmatrix} x_{k+1} \\ y_k \end{vmatrix} = \begin{vmatrix} A_k & I & 0 \\ C_k & 0 & I \end{vmatrix} \begin{vmatrix} x_k \\ \omega_k \\ v_k \end{vmatrix}$

Hence from Result A.2.4:

$$
\begin{vmatrix} x_{k+1} \\ y_k \end{vmatrix} \sim N \left| \begin{vmatrix} A_k \hat{x}_k \\ C_k \hat{x}_k \end{vmatrix}, \begin{vmatrix} A_k P_k A_k^{\mathsf{T}} + Q_k & A_k P_k C_k^{\mathsf{T}} \\ C_k P_k A_k^{\mathsf{T}} & C_k P_k C_k^{\mathsf{T}} + R_k \end{vmatrix} \right|
$$

(b) From Result A.2.6, $p(x_{k+1}|y_k)$ is normal with mean $\hat{x}_{k+1}$ and covariance P_{k+1} where

$$
\hat{x}_{k+1} = A_k \hat{x}_k + A_k K_k (y_k - C_k \hat{x}_k)
$$
$$
P_{k+1} = A_k (P_k - K_k C_k P_k) A_k^{\mathsf{T}} + Q_k
$$
$$
K_k = P_k C_k^{\mathsf{T}} (C_k P_k C_k^{\mathsf{T}} + R_k)^{-1}
$$

(c) Follows immediately.

1.9 (a) Self-evident.

(b) $p(\theta = W | \varepsilon = \varepsilon_A, y = W)$

$$
= \frac{p(y = W | \varepsilon = \varepsilon_A, \theta = W) p(\theta = W | \varepsilon = \varepsilon_A)}{p(y = W | \varepsilon = \varepsilon_A)}
$$

$p(y = W | \varepsilon = \varepsilon_A)$

$$
= p(y = W | \varepsilon = \varepsilon_A, \theta = W) p(\theta = W | \varepsilon = \varepsilon_A)
$$
$$
+ p(y = W | \varepsilon = \varepsilon_A, \theta = H) p(\theta = H | \varepsilon = \varepsilon_A)
$$

Hence

$$
p(\theta = W | \varepsilon = \varepsilon_A, y = W) = \frac{0.9 \times 0.6}{0.9 \times 0.6 + 0.8 \times 0.4} \quad \frac{0.54}{0.86} \quad \text{and so on}
$$

(c) $\bar{J}(d = d_W, y = W, \varepsilon = \varepsilon_A) = \$ \left| 100 \times \dfrac{0.54}{0.86} - 100 \times \dfrac{0.32}{0.86} \right|$

$$= \$ \dfrac{22}{0.86} \quad \text{and so on}$$

(d)   Optimal experiment $= \varepsilon_A$.
Optimal decision rule is d_W if $y = W$, d_H if $y = H$. Hence expected return is

$$\$ \left| \dfrac{22}{.86} p(y = W | \varepsilon = \varepsilon_A) + \dfrac{10}{0.14} P(y = H | \varepsilon = \varepsilon_A) \right| = \$[22 + 10] = \$32$$

1.10   First apply Result 1.7.2. Then for Gaussian case, we wish to maximize log det Σ subject to trace Σ = constant. Using Lagrange multiplier, we seek stationary point of

$$J_C = \log \det \Sigma + \lambda \, (\text{trace } \Sigma - C)$$

Differentiating w.r.t. Σ using Result E.2.1 yields

$$\Sigma^{-1} + \lambda[I] = 0, \quad \text{i.e.,} \quad \Sigma = (1/\lambda), [I] = (C/m) [I]$$

CHAPTER 2

2.1   (i)

$$\begin{vmatrix} \hat{\theta}_0 \\ \hat{\theta}_1 \end{vmatrix} = \begin{vmatrix} 12 & \sum_{i=1}^{12} u_i \\ \sum_{i=1}^{12} u_i & \sum_{i=1}^{12} u_i^2 \end{vmatrix}^{-1} \begin{vmatrix} \sum_{i=1}^{12} y_i \\ \sum_{i=1}^{12} u_i y_i \end{vmatrix}$$

$$= \begin{vmatrix} 12 & 78 \\ 78 & 650 \end{vmatrix}^{-1} \begin{vmatrix} 17.1 \\ 123.0 \end{vmatrix} = \begin{vmatrix} 0.8864 \\ 0.0829 \end{vmatrix}$$

(ii)

$$\hat{\varepsilon} = \begin{bmatrix} 0.1308 \\ -0.1521 \\ -0.0349 \\ 0.1822 \\ -0.0007 \\ -0.2835 \\ -0.2664 \\ 0.0508 \\ 0.4679 \\ 0.2850 \\ -0.1978 \\ -0.1807 \end{bmatrix},$$

$$S(\hat{\theta}) = \hat{\varepsilon}^{\mathrm{T}}\hat{\varepsilon} = 0.6005$$

$$t_{0.025} \left(\dfrac{C_{11} S(\hat{\theta})}{10} \right)^{1/2} = 2.228 \left(\dfrac{650}{1716} \times \dfrac{0.6005}{10} \right)^{1/2}$$

$$= 0.3360$$

$$t_{0.025} \left(\dfrac{C_{22} S(\hat{\theta})}{10} \right)^{1/2} = 2.228 \left(\dfrac{12}{1716} \times \dfrac{0.6005}{10} \right)^{1/2}$$

$$= 0.0457$$

Hence 95 % confidence intervals are given by $0.8864 - 0.3360 < \theta_0 < 0.8864 + 0.3360$, i.e., $0.5504 < \theta_0 < 1.2024$; $0.0829 - 0.0457 < \theta_1 < 0.0829 + 0.0457$, i.e., $0.0372 < \theta_1 < 0.1286$.

(iii)   Null hypothesis: $\theta_i = 0$

$$\tilde{\varepsilon} = \begin{bmatrix} -0.325 \\ -0.525 \\ -0.325 \\ -0.025 \\ -0.125 \\ -0.325 \\ -0.225 \\ -0.175 \\ 0.675 \\ 0.575 \\ 0.175 \\ 0.275 \end{bmatrix}, \quad \begin{array}{l} S(\tilde{\theta}) = \tilde{\varepsilon}^{\mathrm{T}}\tilde{\varepsilon} = 1.5825 \\[2mm] S_{\mathrm{H}} = 1.5825 - 0.6005 = 0.9820 \end{array}$$

Hence

$$\frac{S_{\mathrm{H}}/s}{S(\hat{\theta})/(N - p)} = \frac{0.9820 \times 10}{0.6005} = 16.35$$

But from tables $F_{0.95}(1,10) = 4.96 < 16.35$. Hence reject the null hypothesis at 5 % risk level.

2.2

$$X = \begin{bmatrix} 1 & 1 & 1 \\ 4 & 2 & 1 \\ 9 & 3 & 1 \\ \vdots & \vdots & \vdots \\ 100 & 10 & 1 \end{bmatrix}, \quad X^{\mathrm{T}}X = \begin{bmatrix} 25333 & 3025 & 385 \\ 3025 & 385 & 55 \\ 385 & 55 & 10 \end{bmatrix}, \quad X^{\mathrm{T}}Y = \begin{bmatrix} 1406.25 \\ 167.65 \\ 30.85 \end{bmatrix}$$

$$\hat{\theta} = \frac{1}{4.356 \times 10^5} \begin{bmatrix} 825 & -9075 & 18150 \\ -9075 & 105105 & -228690 \\ 18150 & -228690 & 602580 \end{bmatrix} \begin{bmatrix} 1406.25 \\ 167.65 \\ 30.85 \end{bmatrix} = \begin{bmatrix} 0.4561 \\ -5.0412 \\ 13.2533 \end{bmatrix}$$

2.3   (a)

$$e_i = A(j\omega_i)\hat{H}(j\omega_i) - B(j\omega_i)$$
$$= [1 + a_1 j\omega_i + \cdots + a_n(j\omega_i)^n]\hat{H}(j\omega_i) - [b_0 + b_1 j\omega_i + \cdots + b_{n-1}(j\omega_i)^{n-1}]$$
$$= \hat{H}(j\omega_i) - [-j\omega_i\hat{H}(j\omega_i), \ldots, -(j\omega_i)^n\hat{H}(j\omega_i), 1, j\omega_i, \ldots, (j\omega_i)^{n-1}]\theta$$

The result follows from $J = \Sigma e_i{}^*e_i$.

(b)

$$J = (Y - X\theta)*(Y - X\theta)$$
$$\partial J/\partial\theta = -X*(Y - X\theta) - X^\mathsf{T}(\overline{Y - X\theta})$$
$$= -(X*Y + X^\mathsf{T}\overline{Y}) + (X*X + X^\mathsf{T}\overline{X})\theta$$

Hence the value of θ minimizing J is given by

$$\hat\theta = (X*X + X^\mathsf{T}\overline{X})^{-1}(X*Y + X^\mathsf{T}\overline{X}) = (X*X + \overline{X*X})^{-1}(X*Y + \overline{X*Y})$$
$$= \operatorname{Re}(X*X))^{-1}(\operatorname{Re}(X*Y))$$

2.4 $X*X$ is quadratic in the data and hence $\hat\theta$ is nonlinear in the data. The theory of Section 2.3 is not applicable since $\hat\theta$ is neither linear nor, in general, unbiased.

2.5

$$\operatorname{cov}\hat\theta = E\{(\hat\theta - \theta)(\hat\theta - \theta)^\mathsf{T}\} = E\{(X^\mathsf{T}\Sigma^{-1}X)^{-1}X^\mathsf{T}\Sigma^{-1}\varepsilon\varepsilon^\mathsf{T}\Sigma^{-1}X(X^\mathsf{T}\Sigma^{-1}X)^{-1}\}$$
$$= (X^\mathsf{T}\Sigma^{-1}X)^{-1}X^\mathsf{T}\Sigma^{-1}\Sigma\Sigma^{-1}X(X^\mathsf{T}\Sigma^{-1}X)^{-1} = (X^\mathsf{T}\Sigma^{-1}X)^{-1}$$

2.6

$$\phi_X(t) = E_X\{\exp\{jt^\mathsf{T}X\}\}$$

Consider only the case where Σ is nonsingular. Then

$$\phi_X(t) = \int \exp\{jt^\mathsf{T}x\}\{(2\pi)^k|\Sigma|\}^{-1/2}\exp\{-\tfrac{1}{2}(x - \mu)^\mathsf{T}\Sigma^{-1}(x - \mu)\}\,dx$$

$$= \exp\{jt^\mathsf{T}\mu - \tfrac{1}{2}t^\mathsf{T}\Sigma t\}\int \{(2\pi)^k|\Sigma|\}^{-1/2}$$

$$\times \exp\{-\tfrac{1}{2}(x - \mu - j\Sigma t)^\mathsf{T}\Sigma^{-1}(x - \mu - j\Sigma t)\}\,dx$$

$$= \exp\{jt^\mathsf{T}\mu - \tfrac{1}{2}t^\mathsf{T}\Sigma t\}$$

2.7

$$\phi_X(t) = E_X\{\exp\{jt^\mathsf{T}X\}\}$$
$$\partial\phi_X(t)/\partial t_{i_1} = E_X\{jX_{i_1}\exp\{jt^\mathsf{T}X\}\}$$

$$\frac{\partial^n\phi_X(t)}{\partial t_{i_1}\cdots\partial t_{i_n}} = E_X\{(j)^nX_{i_1}\cdots X_{i_n}\exp\{jt^\mathsf{T}X\}\}$$

Hence

$$\frac{1}{(j)^n} \frac{\partial^n \phi_X(t)}{\partial t_{i_1} \cdots \partial t_{i_n}}\bigg|_{t=0} = E_X\{X_{i_1} \cdots X_{i_n}\}$$

2.8 Define $Y = X - \mu$. Then

$$E_Y(Y_l Y_m Y_n Y_p) = \frac{1}{j^4} \frac{\partial^4 \phi_Y(t)}{\partial t_l \partial t_m \partial t_n \partial t_p}\bigg|_{t=0} = \frac{\partial^4}{\partial t_l \partial t_m \partial t_n \partial t_p} \exp\left\{-\frac{1}{2t} {}^T\Sigma_t\right\}\bigg|_{t=0}$$

$$= -\frac{\partial^3}{\partial t_l \partial t_m \partial t_n} \sum_s \Sigma_{ps} t_s \exp\left\{-\frac{1}{2} t^T\Sigma t\right\}\bigg|_{t=0}$$

$$= \frac{\partial^2}{\partial t_l \partial t_m} \left[-\Sigma_{pn} + \sum_s \Sigma_{ps} t_s \sum_q \Sigma_{nq} t_q\right] \exp\left\{-\frac{1}{2} t^T\Sigma t\right\}\bigg|_{t=0}$$

and so on until putting $t = 0$ yields the required result.

2.9 $\mathrm{Cov}(\hat{\theta}) \geq (X^T X)^{-1}\sigma^2$ follows directly from Eq. (2.5.11). Also from Eqs. (2.5.11) and (2.5.6)

$$\mathrm{cov}(\hat{\sigma}^2) \geq \left|E_{Y|\beta}\left|-\frac{N}{2\sigma^2} + \frac{1}{2\sigma^4} (Y - X\theta)^T(Y - X\theta)\right|^2\right|^{-1}$$

$$= \left|\frac{N^2}{4\sigma^4} - \frac{N^2}{2\sigma^4} + \frac{1}{4\sigma^8} E_{Y|\theta}\{(Y - X\theta)^T(Y - X\theta)\}^2\right|^{-1}$$

Denote $Y - X\theta$ by ε with ith component ε_i. Then

$$\mathrm{cov}(\hat{\sigma}^2) \geq \left|-\frac{N^2}{4\sigma^4} + \frac{1}{4\sigma^8} E_{Y|\theta}\left|\sum_{i=1}^{N} \sum_{j=1}^{N} \varepsilon_i^2 \varepsilon_j^2\right|\right|^{-1}$$

$$= \left|-\frac{N^2}{4\sigma^4} + \frac{1}{4\sigma^8} E_{Y|\theta}\left[\sum_{\substack{i-1 \\ }}^{N} \sum_{\substack{j=1 \\ j\neq i}}^{N} \varepsilon_i^2 \varepsilon_j^2 + \sum_{i=1}^{N} \varepsilon_i^4\right]\right|^{-1}$$

$$= \left|-\frac{N^2}{4\sigma^4} + \frac{1}{4\sigma^8} [N(N-1)\sigma^4 + 3N\sigma^4]\right|^{-1} \qquad \text{(from Problem 2.8)}$$

$$= \frac{2\sigma^4}{N}$$

2.10

$$\hat{V} = \frac{1}{N-p} (Y - X\hat{\theta})^T(Y - X\hat{\theta}) = \frac{1}{N-p} \varepsilon^T A\varepsilon$$

where $A = I_N - X(X^TX)^{-1}X^T$ is given by Eq. (2.5.16) and $\varepsilon = Y - X\theta$ which is $N(0, \sigma^2 I)$ from Eq. (2.5.14).

$$\text{var}(\hat{V}) = E\{(\hat{V} - \sigma^2)^2\} = E\left\{\left(\frac{1}{N-p}\right)\varepsilon^T A\varepsilon\right\}^2 - \sigma^4$$

$$= \left(\frac{1}{N-p}\right)^2 E\left\{\sum_{i=1}^N \sum_{j=1}^N \sum_{k=1}^N \sum_{l=1}^N \varepsilon_i \varepsilon_j \varepsilon_k \varepsilon_l a_{ij} a_{kl}\right\} - \sigma^4$$

$$= \left(\frac{1}{N-p}\right)^2 \sum_{i=1}^N \sum_{j=1}^N \sum_{k=1}^N \sum_{l=1}^N a_{ij} a_{kl}(\Sigma_{ij}\Sigma_{kl} + \Sigma_{il}\Sigma_{jk} + \Sigma_{ik}\Sigma_{jl}) - \sigma^4,$$

(Problem 2.8)

where $\Sigma = \sigma^2 I$, i.e.,

$$\text{var}(\hat{V}) = \left(\frac{1}{N-p}\right)^2 \sum_{i=1}^N \sum_{k=1}^N \sigma^4[a_{ii} a_{kk} + a_{ik} a_{ki} + a_{ik} a_{ik}] - \sigma^4$$

$$= \left(\frac{1}{N-1}\right)^2 \sigma^4[(\text{trace } A)^2 + 2 \text{ trace } A^2] - \sigma^4$$

$$= \sigma^4 + \frac{2\sigma^4}{N-p} - \sigma^4 = \frac{2\sigma^4}{N-p} \qquad \text{(from Eq. (2.5.17))}$$

$(A^2 = A$ since A is idempotent).

2.11   From the proof of Theorem 2.5.7, $S(\hat{\theta})/\sigma^2$ and S_H/σ^2 are independent $\chi^2(N - P)$ and $\chi^2(s)$ random variables, respectively. Hence from Result A.3.1,

$$\frac{S(\tilde{\theta})}{\sigma^2} \triangleq \frac{S_H}{\sigma^2} + \frac{S(\hat{\theta})}{\sigma^2} \quad \text{is a } \chi^2(N - p + s) \text{ random variable}$$

2.12

$$p(Y|\theta) = (2\pi\sigma^2)^{-N/2} \exp\left\{-\frac{1}{2\sigma^2}(Y - X\theta)^T(Y - X\theta)\right\} \qquad \text{(Eq. (2.5.4))}$$

$$= (2\pi\sigma^2)^{-N/2} \exp\left\{-\frac{1}{2\sigma^2}[(Y - X\hat{\theta})^T(Y - X\hat{\theta}) + (\hat{\theta} - \theta)^T X^T X(\hat{\theta} - \theta)]\right\}$$

$$= (2\pi\sigma^2)^{-N/2} \exp\left\{-\frac{1}{2\sigma^2}[\hat{V}(N - p) + (\hat{\theta} - \theta)^T X^T X(\hat{\theta} - \theta)]\right\}$$

The result follows from Theorem 1.4.1.

CHAPTER 3

3.1

$$\int \theta^{2N} \exp\left\{-\theta \sum_{k=1}^{N} x_k\right\} \prod_{k=1}^{N} x_k \, dx = 1$$

Integrating with respect to θ yields

$$\int -\frac{1}{\sum_k x_k} \exp\left\{-\theta \sum_{k=1}^{N} x_k\right\} \prod_{k=1}^{N} x_k \, dx = -\frac{1}{(2N-1)\theta^{2n-1}}$$

i.e.,

$$\int \theta \theta^{2N} \exp\{-\theta \sum_k x_k\} \prod_{k}^{N} x_k \, dx = \frac{2N\theta}{2N-1}$$

i.e.,

$$E\{\hat{\theta}\} - \theta = \frac{2N}{2N-1}\theta - \theta = \frac{\theta}{2N-1} = \text{bias}$$

3.2

$$p(y|\theta) = \{(2\pi)^N |\Sigma|\}^{-1/2} \exp\{-\tfrac{1}{2}(Y - X\theta)^T \Sigma^{-1}(Y - X\theta)\}$$

where $Y^T = (Y_1, \ldots, Y_N)$, $X^T = (u_1, \ldots, u_N)$ and

$$\Sigma_{ij} = E\{(e_i + ce_{i-1})(e_j + ce_{j-1})\}$$
$$= 1 + c^2 \quad \text{if} \quad |i - j| = 0$$
$$= c \quad \text{if} \quad |i - j| = 1$$
$$= 0 \quad \text{otherwise}$$

From Result 3.3.1 the MLE of θ is $\hat{\theta}$ given by

$$\hat{\theta} = (X^T\Sigma^{-1}X)^{-1}(X^T\Sigma^{-1}Y)$$

3.3

$$p(Y|\theta, \sigma^2) = \{2\pi\sigma^2\}^{-m/2} \exp\{-(1/2\sigma^2)(Y - X\theta)^T(Y - X\theta)\}$$
$$(\partial/\partial\theta) \log p(Y|\theta, \sigma^2) = -(1/2\sigma^2)[-2X^T(Y - X\theta)]$$

equating to zero yields $\hat{\theta} = (X^TX)^{-1}X^TY$ as required.

$$(\partial/\partial\sigma^2) \log p(Y|\hat{\theta}, \sigma^2) = -(m/2\sigma^2) + (1/2\sigma^4)(Y - X\hat{\theta})^T(Y - X\hat{\theta})$$

equating to zero yields $\hat{\sigma}^2 = (1/m)(Y - X\hat{\theta})^T(Y - X\hat{\theta})$ as required.

3.4 (a)

$$X_t = \log Y_t \sim N(\mu, \sigma^2)$$

Hence

$$E\{Y_t\} = \int \exp\{x_t\}\{2\pi\sigma^2\}^{-1/2} \exp\{-(1/2\sigma^2)(x_t - \mu)^2\} \, dx_t$$

$$= \exp\left\{\mu + \frac{\sigma^2}{2}\right\} \int (2\pi\sigma^2)^{-1/2} \exp\{-(1/2\sigma^2)(x_t - \mu - \sigma^2)^2\} \, dx_t$$

$$= \exp\{\mu + (\sigma^2/2)\} = \xi$$

$$E\left\{\left(\exp\{X_t\} - \exp\left\{\mu + \frac{\sigma^2}{2}\right\}\right)^2\right\} = E\{\exp\{2X_t\}\} - \exp\{2\mu + \sigma^2\}$$

$$= \exp\{2\mu + 2\sigma^2\} - \exp\{2\mu + \sigma^2\}$$

$$= \exp\{2\mu + \sigma^2\}[\exp\{\sigma^2\} - 1]$$

$$= \xi^2[\exp\{\sigma^2\} - 1]$$

(b) MLE of μ and σ^2 are given, respectively, by

$$\hat{\mu} = \frac{1}{N}\sum_{t=1}^{N} x_t \quad \text{and} \quad \hat{\sigma}^2 = \frac{1}{N}\sum_{t=1}^{N}(x_t - \hat{\mu})^2 \quad \text{(from Problem 3.3)}$$

The result follows from Theorem 3.4.1.

3.5

$$\log p(y|\hat{\theta}, \hat{\Sigma}) = \log p(y|\hat{\theta}, R(\hat{\theta}))$$

$$= -\frac{mN}{2}\log(2\pi) - \frac{N}{2}\log \det R(\hat{\theta})$$

$$-\tfrac{1}{2}\sum_{t=1}^{N}(Y_t - X_t\hat{\theta})^T R(\hat{\theta})^{-1}(Y_t - X_t\hat{\theta})$$

$$= -\frac{mN}{2}\log(2\pi) - \frac{N}{2}\log \det R(\hat{\theta})$$

$$-\tfrac{1}{2}\sum_{t=1}^{N}\text{trace}\{R(\hat{\theta})^{-1}(Y_t - X_t\hat{\theta})(Y_t - X_t\hat{\theta})^T\}$$

$$= -(mN/2)\log(2\pi) - (N/2)\log \det R(\hat{\theta})$$

$$- (N/2)\text{trace}\{R(\hat{\theta})^{-1}R(\hat{\theta})\}$$

$$= -(Nm/2)\{\log(2\pi) + 1\} - (N/2)\log \det R(\hat{\theta})$$

3.6

$$J = \log \det\left|\frac{1}{N}\sum_{t=1}^{N}(Y_t - X_t\theta)(Y_t - X_t\theta)^{\mathrm{T}}\right|$$

(a)   Hence

$$\frac{\partial J}{\partial\theta_i} = \sum_{k=1}^{M}\sum_{j=1}^{m}(R(\theta)^{-1})_{kj}\frac{\partial R_{kj}(\theta)}{\partial\theta_i} = \mathrm{trace}\left\{R(\theta)^{-1}\frac{\partial R(\theta)}{\partial\theta_i}\right\}$$

$$\frac{\partial R(\theta)}{\partial\theta_i} = -\frac{1}{N}\sum_{t=1}^{N}(X_t e_i(Y_t - X_t\theta)^{\mathrm{T}} + (Y_t - X_t\theta)e_i^{\mathrm{T}}X_t^{\mathrm{T}})$$

where e_i is a zero vector except for element i which $= 1$. Hence

$$\frac{\partial J}{\partial\theta_i} = -\frac{1}{N}\sum_{t=1}^{N}(\mathrm{trace}\{R^{-1}X_t e_i(Y_t - X_t\theta)^{\mathrm{T}}\} + \mathrm{trace}\{R^{-1}(Y_t - X_t\theta)e_i^{\mathrm{T}}X_t^{\mathrm{T}}\}$$

$$= -\frac{1}{N}\sum_{t=1}^{N}\{(Y_t - X_t\theta)^{\mathrm{T}}R^{-1}X_t e_i + e_i^{\mathrm{T}}X_t^{\mathrm{T}}R^{-1}(Y_t - X_t\theta)\}$$

$$= -\frac{2}{N}\sum_{t=1}^{N}e_i^{\mathrm{T}}X_t^{\mathrm{T}}R^{-1}(Y_t - X_t\theta)$$

Hence

$$\frac{\partial J}{\partial\theta} = -\frac{2}{N}\sum_{t=1}^{N}(Y_t - X_t\theta)^{\mathrm{T}}R(\theta)^{-1}X_t$$

3.7   (i)   Let $x_i = 1$ denote success at the ith trial and $x_i = 0$ denote failure. Then $p(x_i|\mu) = \mu^{x_i}(1 - \mu)^{1 - x_i}$. Hence

$$\text{likelihood} = \prod_{i=1}^{N}\mu^{x_i}(1 - \mu)^{1 - x_i}$$

(ii) $l = \log \text{likelihood} = \sum_{i=1}^{N}\{x_i \log\mu + (1 - x_i)\log(1 - \mu)\}$
Hence

$$\frac{\partial l}{\partial\mu} = \sum_{i=1}^{N}\left[\frac{x_i}{\mu} - \frac{(1 - x_i)}{1 - \mu}\right] = \sum_{i=1}^{N}\frac{x_i - \mu}{\mu(1 - \mu)}$$

Hence

$$\hat{\mu} = \frac{1}{N}\sum_{i=1}^{N}x_i$$

(iii) $E\{\hat{\mu}\} = \dfrac{1}{N}\sum_{i=1}^{N}(0 \cdot \mu^0(1 - \mu)^1 + 1 \cdot \mu^1(1 - \mu)^0) = \mu$

hence unbiased.

(iv) $\dfrac{\partial^2 l}{\partial \mu^2} = \displaystyle\sum_{i=1}^{N} \left| \dfrac{x_i - \mu}{\mu^2(1-\mu)^2}(1-2\mu) - \dfrac{1}{\mu(1-\mu)} \right|$

Hence

$E \dfrac{\partial^2 l}{\partial \mu^2} = \displaystyle\sum_{i=1}^{N} \left| \dfrac{\mu - \mu}{\mu^2(1-\mu)^2}(1-2\mu) - \dfrac{1}{\mu(1-\mu)} \right|$

$\qquad = \dfrac{-N}{\mu(1-\mu)}$

Hence Cramer–Rao lower bound $= \mu(1-\mu)/N$.

3.8

$$\hat{\theta} = \sum_{t=1}^{N}[X_t^T \Sigma^{-1} X_t]^{-1} \sum_{t=1}^{N} X_t^T \Sigma^{-1} Y_t \qquad \text{for arbitrary } \Sigma$$

$$E\{\hat{\theta}\} = \sum_{t=1}^{N}[X_t^T \Sigma^{-1} X_t]^{-1} \sum_{t=1}^{N} X_t^T \Sigma^{-1} E\{Y_t\}$$

$$= \sum_{t=1}^{N}[X_t^T \Sigma^{-1} X_t]^{-1} \sum_{t=1}^{N} X_t^T \Sigma^{-1} X_t \theta = \theta$$

CHAPTER 4

4.1 (a)

$$y_k = \frac{1}{1+\alpha z^{-1}}[1, \; 1+\beta z^{-1}]\begin{vmatrix} v_k \\ w_k \end{vmatrix}$$

Hence

$$\phi_y(z) = \left(\frac{1}{1+\alpha z^{-1}}\right)\left(\frac{1}{1+\alpha z}\right)[1, \; 1+\beta z^{-1}]\begin{vmatrix} Q & C \\ C & R \end{vmatrix}\begin{vmatrix} 1 \\ 1+\beta z \end{vmatrix}$$

$$\text{from (C.1.6) and (C.1.8)}$$

Hence

$$\phi_y(e^{j\omega}) = \frac{1}{1+\alpha^2 + 2\alpha \cos \omega}[Q + R(1+\beta^2 + 2\beta \cos \omega) + 2C(1+\beta \cos \omega)]$$

(b) $\; y_k = \dfrac{1}{1-az^{-1}}[1+dz^{-1}]\eta_k$

Hence

$$\phi_y(z) = \left(\frac{1}{1 - az^{-1}}\right)\left(\frac{1}{1 - az}\right)(1 + dz^{-1})\Sigma(1 + dz)$$

$$\phi_y(e^{j\omega}) = \frac{1}{1 + a^2 - 2a\cos\omega}[\Sigma(1 + d^2 + 2d\cos\omega)]$$

which can be made equal to the $\phi_y(e^{j\omega})$ of part (a) by setting $a = -\alpha$ and

$$\Sigma(1 + d^2) = Q + R(1 + \beta^2) + 2C \tag{1}$$

$$2d\Sigma = 2\beta R + 2C\beta \tag{2}$$

(c)   Solving for d and Σ:

$$\text{from (2) } d = \frac{\beta}{\Sigma}(R + C) \tag{3}$$

$$\text{substituting into (1) } \Sigma\left(1 + \frac{\beta^2(R + C)^2}{\Sigma^2}\right) = Q + R(1 + \beta^2) + 2C$$

i.e.,

$$\Sigma^2 - (Q + R(1 + \beta^2) + 2C)\Sigma + \beta^2(R + C)^2 = 0$$

whence

$$\Sigma = \tfrac{1}{2}\{(Q + R(1 + \beta^2) + 2C) \pm [(Q + R(1 + \beta^2) + 2C)^2 - 4\beta^2(R + C)^2]^{1/2}\}$$

and d follows from (3).

4.2

$$\phi(j\omega) = \int_{-\infty}^{\infty} e^{-|\tau|}e^{-j\omega\tau}\,d\tau = \int_{-\infty}^{0} e^{\tau - j\omega\tau}\,d\tau + \int_{0}^{\infty} e^{-\tau - j\omega\tau}\,d\tau$$

$$= \frac{e^{(1 - j\omega)\tau}}{1 - j\omega}\bigg|_{-\infty}^{0} + \frac{e^{-(1 + j\omega)\tau}}{-(1 + j\omega)}\bigg|_{0}^{\infty} = \frac{1}{1 - j\omega} + \frac{1}{1 + j\omega} = \frac{2}{1 + \omega^2}$$

4.3   From (C.2.16) transfer function $G(s) = 1/(s + 1)$. Hence from Result C.2.1,

$$\phi_Y(j\omega) = \frac{1}{j\omega + 1}\cdot\frac{1}{-j\omega + 1} = \frac{1}{1 + \omega^2}$$

4.4   (a)

$$E\{y_k\} = 0, \; E\{y_k\,y_l\} = E\{(\varepsilon_k + C\varepsilon_{k-1})(\varepsilon_l + C\varepsilon_{l-1})\}$$

$$= E\{\varepsilon_k\,\varepsilon_l\} + C^2E\{\varepsilon_{k-1}\varepsilon_{l-1}\} + CE\{\varepsilon_{k-1}\varepsilon_l\} + CE\{\varepsilon_k\,\varepsilon_{l-1}\}$$

$$= (1 + C^2)\delta_{kl} + C(\delta_{k\overline{l-1}} + \delta_{\overline{k-1}l})$$

i.e.

$$D_y(\tau) = 1 + C^2 \qquad \text{for } \tau = 0$$
$$= C \qquad \text{for } \tau = \pm 1$$
$$= 0 \qquad \text{otherwise}$$

(b)

$$\phi(e^{j\omega}) = \sum_{-\infty}^{\infty} D_y(\tau)e^{-j\omega\tau}$$

$$= 1 + C^2 + C(e^{j\omega\tau} + e^{-j\omega\tau}) = 1 + C^2 + 2C \cos \omega$$

(Also from $\phi(z) = (1 + Cz^{-1})(1 - Cz^{-1})$.)

4.5 System obviously first-order, say $L(z) = (c + dz)/(a + bz)$. Then

$$\phi_Y(e^{j\omega}) = \frac{c^2 + d^2 + 2cd \cos \omega}{a^2 + b^2 + 2ab \cos \omega}$$

Hence

$$c^2 + d^2 = 10, \qquad 2cd = 6$$
$$a^2 + b^2 = 5, \qquad 2ab = 4$$

Now stability implies $|a| < |b|$ and min phase implies $|c| < |d|$. Hence choose $c = 1$, $d = 3$, and $a = 1$, $b = 2$, i.e., $L(z) = (1 + 3z)/(1 + 2z)$ or $L(z) = (3 + z^{-1})/(2 + z^{-1})$.

4.6

$$y_k = \left| \frac{1}{1 - 0.2z^{-1}}, \frac{1}{1 - 0.5z^{-1}} \right| \left| \begin{matrix} \varepsilon_k^{(1)} \\ \varepsilon_k^{(2)} \end{matrix} \right|$$

Hence analogously to Problem 4.1, we have

$$\phi_y(z) = \frac{v^{(1)}}{(1 - 0.2z^{-1})(1 - 0.2z)} + \frac{v^{(2)}}{(1 - 0.5z^{-1})(1 - 0.5z)} \qquad (1)$$

The second process has the spectral density

$$\phi_y(z) = \frac{(z + C)(z^{-1} + C)\lambda}{(z - 0.2)(z^{-1} - 0.2)(z - 0.5)(z^{-1} - 0.5)} \qquad (2)$$

equating (1) and (2) yields

$$v^{(1)}(1.25 - 0.5(z + z^{-1})) + v^{(2)}(1.04 - 0.2(z + z^{-1}))$$
$$= \lambda(1 + C) + \lambda C(z + z^{-1})$$

whence $\lambda C = 0.5v^{(1)} + 0.2v^{(2)}$

$$\lambda(1 + C) = 1.25v^{(1)} + 1.04v^{(2)}$$

which may be solved for C and λ (cf Problem 4.1).

4.7 See [5, p. 26].

4.8 See [5, pp. 268–270].

4.9 Substituting the given expressions for $H(z)$ and Σ yields

$$H(z)\Sigma H^\mathsf{T}(z^{-1}) = [C(zI - A)^{-1}APC^\mathsf{T}(CPC^\mathsf{T} + R)^{-1} + I](CPC^\mathsf{T} + R)[\,\cdots\,]^\mathsf{T}$$
$$= C(zI - A)^{-1}APC^\mathsf{T}(CPC^\mathsf{T} + R)^{-1}CPA^\mathsf{T}(z^{-1}I - A^\mathsf{T})^{-1}C^\mathsf{T}$$
$$+ C(zI - A)^{-1}APC^\mathsf{T} + CPA^\mathsf{T}(zI - A)^{-1}C^\mathsf{T} + CPC^\mathsf{T} + R$$

But from the matrix Riccati equation (4.5.16),

$$APC^\mathsf{T}(CPC^\mathsf{T} + R)CPA^\mathsf{T} = APA^\mathsf{T} - P + Q$$

Hence

$$H(z)\Sigma H^\mathsf{T}(z^{-1}) = C(zI - A)^{-1}(APA^\mathsf{T} - P + Q)(z^{-1}I - A^\mathsf{T})^{-1}C^\mathsf{T} + \cdots$$
$$= C(zI - A)^{-1}\{-P + Q + AP(zI - A^\mathsf{T}) + (zI - A)PA^\mathsf{T}$$
$$+ APA^\mathsf{T}\}(zI - A^\mathsf{T})^{-1}C^\mathsf{T} + CPC^\mathsf{T} + R$$
$$= C(zI - A)^{-1}[Q - (zI - A)P(z^{-1}I - A^\mathsf{T})](z^{-1}I - A^\mathsf{T})^{-1}C^\mathsf{T}$$
$$+ CPC^\mathsf{T} + R$$
$$= C(zI - A)^{-1}Q(z^{-1}I - A^\mathsf{T})^{-1}C^\mathsf{T} - CPC^\mathsf{T} + CPC^\mathsf{T} + R$$
$$= \phi_n(z)$$

CHAPTER 5

5.1

$$\hat{a} = (y_0{}^2 + y_1{}^2)^{-1}(y_0 y_1 + y_1 y_2)$$
$$= [y_0{}^2 + (ay_0 + \xi_1)^2]^{-1}(y_0(ay_0 + \xi_1) + (ay_0 + \xi_1)(a(ay_0 + \xi_1) + \xi_2)$$
$$= a + \frac{y_0 \xi_1 + ay_0 \xi_2 + \xi_1 \xi_2}{y_0{}^2 + a^2 y_0{}^2 + 2ay_0 \xi_1 + \xi_1{}^2}$$

Hence

$$E\hat{a} = a + \frac{1}{4}\left(\underbrace{\frac{-y_0 - ay_0 + 1}{y_0{}^2 + a^2y_0{}^2 - ay_0 + 1}}_{(-1,\,-1)} + \underbrace{\frac{-y_0 + ay_0 - 1}{y_0{}^2 + a^2y_0{}^2 - 2ay_0 + 1}}_{(-1,\,+1)}\right.$$

$$\left. + \underbrace{\frac{y_0 - ay_0 - 1}{y_0{}^2 + a^2y_0{}^2 + 2ay_0 + 1}}_{(+1,\,-1)} + \underbrace{\frac{y_0 + ay_0 + 1}{y_0{}^2 + a^2y_0{}^2 + 2ay_0 + 1}}_{(+1,\,+1)}\right)$$

Hence

$$\text{bias} = \frac{-y_0}{2(y_0{}^2 + a^2y_0{}^2 - 2ay_0 + 1)} + \frac{y_0}{2(y_0{}^2 + a^2y_0{}^2 + 2ay_0 + 1)}$$

$$= \frac{-2ay_0{}^2}{(y_0{}^2 + a^2y_0{}^2 + 1)^2 - 4a^2y_0{}^2} \neq 0 \text{ in general}$$

5.2

$$\left|I + D^{-1}\frac{gh^\mathsf{T}}{1 - h^\mathsf{T}D^{-1}g}\right|D^{-1}(D - gh^\mathsf{T}) = \left(I + \frac{D^{-1}gh^\mathsf{T}}{1 - h^\mathsf{T}D^{-1}g}\right)(I - D^{-1}gh^\mathsf{T})$$

$$= I + \frac{D^{-1}gh^\mathsf{T}}{1 - h^\mathsf{T}D^{-1}g} - D^{-1}gh^\mathsf{T}$$

$$- \frac{D^{-1}gh^\mathsf{T}D^{-1}gh^\mathsf{T}}{1 - h^\mathsf{T}D^{-1}g}$$

$$= I + D^{-1}g\left(\frac{1}{1 - h^\mathsf{T}D^{-1}g} - 1\right.$$

$$\left. - \frac{h^\mathsf{T}D^{-1}g}{1 - h^\mathsf{T}D^{-1}g}\right)h^\mathsf{T} = I$$

thus establishing (*) (the matrix inversion lemma). Hence

$$B^{-1} = (A - xx^\mathsf{T})^{-1} = A^{-1} + A^{-1}\frac{xx^\mathsf{T}}{1 - x^\mathsf{T}A^{-1}x}A^{-1}$$

whence $C = A^{-1}xx^\mathsf{T}A^{-1}xx^\mathsf{T}A^{-1}/(1 - x^\mathsf{T}A^{-1}x)$. Also,

$$A^{-1} = (B + xx^\mathsf{T}) = B^{-1} - \frac{B^{-1}xx^\mathsf{T}B^{-1}}{1 + x^\mathsf{T}B^{-1}x}$$

so

$$x^T A^{-1} x = x^T B^{-1} x - \frac{x^T B^{-1} x x^T B^{-1} x}{1 + x^T B^{-1} x}$$

$$= \frac{x^T B^{-1} x}{1 + x^T B^{-1} x} \left(1 + x^T B^{-1} x - x^T B^{-1} x\right)$$

$$< 1$$

It follows that $C = (\alpha/\beta) A^{-1} x x^T A^{-1}$, where $0 \le \alpha < 1$, $0 < \beta \le 1$, and hence C is positive semidefinite.

5.3 It follows from the arguments of Example 4.2.1 that exponentially bounded fourth moment of ε_t implies that the least squares estimator is weakly consistent, i.e., $\hat{\theta}_N \xrightarrow{\text{prob}} \theta$. Hence from Theorem B.3.4,

$$E\{S_N(\hat{\theta}_N)\} \to E\{S_N(\theta)\} = \sigma^2$$

and $E\{(S_N(\hat{\theta}_N) - \sigma^2)\}^2 \to 0$ because of exponential bound on $E\{\varepsilon_i \varepsilon_j \varepsilon_k \varepsilon_l\}$. Hence

$$S_N \xrightarrow{\text{q.m.}} \sigma^2 \quad \text{and hence} \quad S_N \xrightarrow{\text{prob}} \sigma^2$$

5.4

$$\bar{\theta}_N = [Z_N^T X_N]^{-1} Z_N^T Y_N = \theta + (Z_N^T X_N)^{-1} Z_N^T \rho_N$$

$$= \theta + \left(\frac{1}{N} Z_N^T X_N\right)^{-1} \left(\frac{1}{N} Z_N^T \rho_N\right)$$

$$\xrightarrow{\text{prob}} \theta + \Psi_{ZX} \cdot 0 = \theta \quad \text{from Theorem B3.4}$$

5.5

$$\frac{1}{N} Z_N^T X_N = \begin{bmatrix} \dfrac{1}{N} \sum_{t=1}^{N} u_{t-1} y_{t-1} & \dfrac{1}{N} \sum_{t=1}^{N} u_t u_{t-1} \\ \dfrac{1}{N} \sum_{t=1}^{N} u_t y_{t-1} & \dfrac{1}{N} \sum_{t=1}^{N} u_t^2 \end{bmatrix} \xrightarrow{\text{prob}} \begin{bmatrix} R_{uy}(0) & R_{uu}(1) \\ R_{UY}(1) & R_{UU}(0) \end{bmatrix}$$

which will be nonsingular provided $R_{UY}(0) R_{UU}(0) - R_{UU}(1) R_{UY}(1) \neq 0$. Furthermore, $Z_N^T \rho_N / N \to 0$ provided $\{u_t\}$ is generated independently of $\{\varepsilon_t\}$.

$$\frac{1}{N} Z_N^T Y_N = \begin{bmatrix} \dfrac{1}{N} \sum_{t=1}^{N} u_{t-1} y_t \\ \dfrac{1}{N} \sum_{t=1}^{N} u_t y_t \end{bmatrix} \xrightarrow{\text{prob}} \begin{bmatrix} R_{UY}(-1) \\ R_{UY}(0) \end{bmatrix}$$

Hence from Theorem B.3.4,

$$\begin{vmatrix} \bar{a} \\ \bar{b} \end{vmatrix} \xrightarrow{\text{prob}} \frac{1}{\Delta} \begin{vmatrix} R_{UU}(0) & -R_{UU}(1) \\ -R_{UY}(1) & R_{UY}(0) \end{vmatrix} \begin{vmatrix} R_{UY}(-1) \\ R_{UY}(0) \end{vmatrix}$$

$$= \frac{1}{\Delta} \begin{vmatrix} R_{UU}(0)R_{UY}(-1) - R_{UU}(1)R_{UY}(0) \\ -R_{UY}(1)R_{UY}(-1) + R_{UY}(0)^2 \end{vmatrix}$$

But $R_{UY}(-1) = aR_{UY}(0) + bR_{UU}(1)$ and $R_{UY}(0) = aR_{UY}(1) + bR_{UU}(0)$; hence

$$\begin{vmatrix} \bar{a} \\ \bar{b} \end{vmatrix} \xrightarrow{\text{prob}} \frac{1}{\Delta} \begin{vmatrix} R_{UU}(0)[aR_{UY}(0) + bR_{UU}(1)] - R_{UU}(1)[aR_{UY}(1) + bR_{UU}(0)] \\ -R_{UY}(1)[aR_{UY}(0) + bR_{UU}(1)] + R_{UY}(0)[aR_{UY}(1) + bR_{UU}(0)] \end{vmatrix}$$

$$= \begin{vmatrix} a \\ b \end{vmatrix}$$

5.6   From Section 6: $\hat{\Sigma} = D(\hat{\theta})$ and $\hat{\theta}$ minimizes $\log \det D(\theta)\ = D(\theta) = (1/N) \sum_{t=1}^{N} w_t^2$, where, from (5.7.15),

$$w_t = A(z)y_t - B(z)u_t$$
$$= y_t + a_1 y_{t-1} + \cdots + a_n y_{t-n} - b_1 u_{t-1} - \cdots - b_m u_{t-m} = y_t - x_t^{\mathsf{T}}\theta$$

Hence $\hat{\theta}$ is the value of θ which minimizes

$$D(\theta) = (1/N)\ (Y - X\theta)^{\mathsf{T}}(Y - X\theta)$$

which, by differentiation yields (cf. Section 5.2 and Chapter 2)

$$\theta = (X^{\mathsf{T}}X)^{-1}X^{\mathsf{T}}Y$$

Further, $\hat{\Sigma} = D(\hat{\theta}) = (1/N)(Y - X\hat{\theta})^{\mathsf{T}}(Y - X\hat{\theta})$.

5.7   From Eq. (5.6.3), $\hat{\theta}$ minimizes $J_1(\theta) = (1/N) \sum_{t=1}^{N} w_t^{N}\Sigma^{-1}w_t$, where

$$w_t = A(z)y_t - B(z)u_t \qquad \text{(from 5.7.15)} \qquad (1)$$

(1) is obviously linear in the parameters (coefficients of elements of A and B) and hence it follows that $J(\theta)$ is quadratic in θ, i.e.,

$$J_1(\theta) = \tfrac{1}{2}\theta^{\mathsf{T}}C\theta + b^{\mathsf{T}}\theta + a \qquad \text{(C symmetric w.l.o.g.) and}$$
$$\theta = -C^{-1}b \qquad \text{(C assumed full rank)}$$

Now C can always be expressed in the form $C = X^{\mathsf{T}}X$, where X is some matrix of full rank. Furthermore, given any b it is always possible to find a Y such that $X^{\mathsf{T}}Y = -b$ since X has full rank. The result follows.

5.8   The term $\tfrac{1}{2} \sum_{t=1}^{N} w_t^{N} D(\theta)^{-1}w_t$ in (5.6.9) may be rewritten as

$$\frac{1}{2} \text{trace}\left\{\left(\sum_{t=1}^{N} w_t w_t^{\mathsf{T}}\right)D(\theta)^{-1}\right\} = \frac{N}{2} \text{trace } I_m = \frac{Nm}{2}$$

the result follows.

5.9   The result we shall need is (cf. Eq. (3.3.11)

$$\frac{\partial}{\partial M} \log \det M = M^{-\mathrm{T}} \tag{1}$$

Therefore

$$\frac{\partial J_2}{\partial \theta_i} = \sum_k \sum_j \left(\frac{\partial}{\partial D(\theta)} \log \det D(\theta) \right)_{kj} \frac{\partial}{\partial \theta_i} D_{kj}(\theta) \tag{2}$$

$$= \operatorname{trace} D(\theta)^{-1} \cdot \frac{2}{N} \sum^N \frac{\partial w_t}{\partial \theta_i} w_t^{\mathrm{T}}$$

$$= \frac{2}{N} \sum_{t=1}^N w_t^{\mathrm{T}} D(\theta)^{-1} \frac{\partial w_t}{\partial \theta_i} \qquad \text{since trace } AB = \operatorname{trace} BA$$

$\partial J_1 / \partial \theta_i$ follows without use of matrix differentiation results.

5.10

$$e_k = y_k - C x_k - D u_k$$

Substituting into $x_{k+1} = A x_k + B u_k + K e_k$ yields the following PEM model:

$$x_{k+1} = (A - KC)x_k + (B - KD)u_k + K y_k$$
$$y_k = C x_k + D u_k + e_k$$

Hence, from Section 5.15,

$$p(y_N, u_N | \theta) = k((2\pi)^m \det \Sigma)^{-N/2} \exp\left| -\frac{1}{2} \sum_{k=1}^N w_k^{\mathrm{T}} \Sigma^{-1} w_k \right|$$

where $w_k(\theta) = y_k - C\bar{x}_k - D u_k$ and $\bar{x}_{k+1} = (A - KC)\bar{x}_k + (B - KD)u_k$ $+ K y_k$.

5.11   From Eq. (5.6.9), using Result E.1.4

$$G_1(I - G_3 G_1)^{-1} G_4 = H_1(I - H_3 H_1)^{-1} H_4$$

Hence using (5.6.11), $G_1 = H_1$. Similarly from Eq. (5.6.10), using Result E.1.4,

$$G_3(I - G_1 G_3)^{-1} G_2 = H_3(I - H_1 H_3)^{-1} H_2$$

Hence using (5.6.8), $G_3 = H_3$. Thus $(I - G_1 G_3) = (I - H_1 H_3)$. Therefore,

$$G_2 = H_2 \qquad \text{from (5.6.8)}$$
$$G_4 = H_4 \qquad \text{from (5.6.11)}$$

5.12 The feedback law is $u_k = g y_k$. Hence $\lambda(u_{k-1} - g y_{k-1}) = 0$ for arbitrary λ. Substituting into the system model gives

$$y_k = -a y_{k-1} + b u_{k-1} + \lambda(u_{k-1} - g y_{k-1}) + \varepsilon_k$$
$$= -(a + \lambda g) y_{k-1} + (b + \lambda) u_{k-1} + \varepsilon_k$$

This model is the same form as the original model but with a and b replaced by $(a + \lambda g)$ and $(b + \lambda)$, respectively.

5.13 The minimum variance control law for the given system is (see Eq. (7.8.3)) $u_k = \hat{G}_3 y_k$, where

$$\hat{G}_3 = \frac{z(1 - C)}{zB} = \frac{c_1 + \cdots + c_n z^{-n+1}}{z(b_1 z^{-1} + \cdots + b_n z^{-n})} = \frac{c_1 + \cdots + c_n z^{1-n}}{b_1 + \cdots + b_n z^{1-n}}$$

Clearly Eq. (5.6.79) is not satisfied by the above control law.

5.14 The closed loop transfer functions are

$$y = H_1 \varepsilon, \qquad u = H_2 \varepsilon$$

where

$$H_1 = (1 - G_1 G_3)^{-1} G_2, \qquad H_2 = (1 - G_1 G_3)^{-1} G_2 G_3$$

For system (a) and (b),

$$H_1 = \frac{1 + 0.5 z^{-1}}{1 + 0.4 z^{-1}}, \qquad H_2 = \frac{1 + 0.5 z^{-1}}{1 + 0.4 z^{-1}}$$

CHAPTER 6

6.1 Let $\theta = f(\beta)$, where f is one-to-one and differentiable. From Eq. (6.2.2)

$$M_\beta = E_{Y|\beta} \left\{ \left(\frac{\partial \log p(Y|\beta)}{\partial \beta} \right)^{\mathrm{T}} \left(\frac{\partial \log p(Y|\beta)}{\partial \beta} \right) \right\}$$

$$= E_{Y|\beta} \left\{ \left(\frac{\log p(Y|\theta)}{\partial \theta} \frac{\partial \theta}{\partial \beta} \right)^{\mathrm{T}} \left(\frac{\partial \log p(Y|\theta)}{\partial \theta} \frac{\partial \theta}{\partial \beta} \right) \right\}$$

$$= \left[\frac{\partial \theta}{\partial \beta} \right]^{\mathrm{T}} E_{Y|\theta} \left\{ \left(\frac{\log p(Y|\theta)}{\partial \theta} \right)^{\mathrm{T}} \left(\frac{\partial \log p(Y|\theta)}{\partial \theta} \right) \right\} \left[\frac{\partial \theta}{\partial \beta} \right]$$

$$= S^{\mathrm{T}} M_\theta S$$

where $S = [\partial \theta / \partial \beta]$ and is nonsingular by assumption. Hence from (6.2.7),

$$J_\beta = -2 \log \det S - \log \det M_\theta$$

Hence, optimizing J_β is equivalent to optimizing $J_\theta = -\log \det M_\theta$.

6.2

$$\text{trace } \overline{M} = \sum_{i=1}^{n} \left[\frac{1}{N} \sum_{t=1}^{N} u_{t-i}^2 \right]$$

$$= \text{constant for all inputs satisfying 6.3.12}$$

Hence $u_t = 1$ is "optimal" in the sense that trace M is maximized. However, $u_t = 1$ when substituted into (6.3.9) and (6.3.10) yields $\overline{M}$ singular unless $n = 1$. This means that the parameters are unidentifiable in general.

6.3 (a) This model is discussed in Section 6.3.1 where it is shown that the ijth element of M_θ is

$$[M_\theta]_{ij} = \sum_{t=0}^{N} \mu_{t-1} \mu_{t-j}$$

(b) For m-sequence

$$M = \begin{bmatrix} N & -1 & \cdots & -1 \\ -1 & & \ddots & \vdots \\ \vdots & & \ddots & -1 \\ -1 & \cdots & -1 & N \end{bmatrix} = (N+1)I - \mathbf{1}\,\mathbf{1}^{\mathsf T}$$

where $\mathbf{1}$ is an m-vector of ones. Using Lemma E.1.1,

$$\text{cov }\hat{\theta} = M^{-1} = \left[I + \frac{1}{N+1} I \frac{\mathbf{1}\,\mathbf{1}^{\mathsf T}}{\left(1 - \mathbf{1}^{\mathsf T} \frac{1}{N+1} I\, \mathbf{1}\right)} \right] \frac{1}{N+1} I$$

which has diagonal elements $(N - m + 2)/(N^2 - Nm + 2N - m + 1)$.

(c)

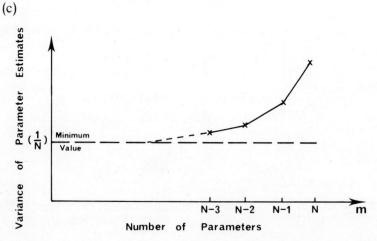

6.4 For input power α, the optimum information matrix is proportional to α^2. Hence we should attempt to maximize the input power subject to the given constraint. For $|u_t| \leq 1$, the maximum input power is achieved when u_t is always either $+1$ or -1.

6.5

$$w_t = G_2^{-1}(z)[y_t - G_1(z)]u_t$$

$$\frac{\partial w_t}{\partial \beta_i} = -G_2^{-2}(z)\frac{\partial G_2(z)}{\partial \beta_i}[y_t - G_1(z)]u_t - G_2^{-1}(z)\frac{\partial G_1(z)}{\partial \beta_i}u_t$$

$$= -G_2^{-1}(z)\left[\frac{\partial G_2(z)}{\partial \beta_i}w_t + \frac{\partial G_1(z)}{\partial \beta_i}u_t\right]$$

6.6 Since $G_2(\infty) = I$ (independent of θ). Hence

$$\left.\frac{\partial G_2(z)}{\partial \beta_i}\right|_{z=\infty} = 0$$

Hence from (6.3.22) there is no direct connection between $\{w_t\}$ and $\{\partial w_t/\partial \beta_i\}$.

6.7

$$M = E_{Y|\beta}\left\{\left[-\frac{1}{\Sigma}\sum_{t=1}^{N}w_t\frac{\partial w_t}{\partial \beta} - \frac{1}{2\Sigma}\frac{\partial \Sigma}{\partial \beta}\left(N - \frac{1}{\Sigma}\sum_{t=1}^{N}w_t^2\right)\right]^{\mathrm{T}}\right.$$

$$\times \left.\left[-\frac{1}{\Sigma}\sum_{s=1}^{N}w_s\frac{\partial w_s}{\partial \beta} - \frac{1}{2\Sigma}\frac{\partial \Sigma}{\partial \beta}\left(N - \frac{1}{\Sigma}\sum_{s=1}^{N}w_s^2\right)\right]\right\}$$

$$= E_{Y|\beta}\left\{\frac{1}{\Sigma^2}\sum_{t=1}^{N}\sum_{s=1}^{N}\frac{\partial w_t^{\mathrm{T}}}{\partial \beta}w_t w_s\frac{\partial w_s}{\partial \beta} + \frac{1}{4\Sigma^2}\left(N - \frac{1}{\Sigma}\sum_{t=1}^{N}w_t^2\right)\frac{\partial \Sigma^{\mathrm{T}}}{\partial \beta}\frac{\partial \Sigma}{\partial \beta}\right.$$

$$\times \left.\left(N - \frac{1}{\Sigma}\sum_{s=1}^{N}w_s^2\right)\right\}$$

Using independence of $\{w_t\}$ sequence and Problem 6.6:

$$M = E_{Y|\beta}\left\{\frac{1}{\Sigma}\sum_{t=1}^{N}\frac{\partial w_t^{\mathrm{T}}}{\partial \beta}\frac{\partial w_s}{\partial \beta} + \frac{1}{4\Sigma^2}\left[N^2 - \frac{N}{\Sigma}\sum_{t=1}^{N}w_t^2 - \frac{N}{\Sigma}\sum_{s=1}^{N}w_s^2\right.\right.$$

$$\left.\left. + \frac{1}{\Sigma^2}\sum_{t=1}^{N}\sum_{s=1}^{N}w_t^2 w_s^2\right]\frac{\partial \Sigma^{\mathrm{T}}}{\partial \beta}\frac{\partial \Sigma}{\partial \beta}\right\}$$

$$= E_{Y|\beta}\left\{\frac{1}{\Sigma}\sum_{t=1}^{N}\frac{\partial w_t^{\mathrm{T}}}{\partial \beta}\frac{\partial w_s}{\partial \beta}\right.$$

$$\left. + \frac{1}{4\Sigma^2}\left[N^2 - N^2 - N^2 + (N^2 - N) + 3N\right]\frac{\partial \Sigma^{\mathrm{T}}}{\partial \beta}\frac{\partial \Sigma}{\partial \beta}\right\}$$

$$= E_{Y|\beta}\left\{\frac{1}{\Sigma}\sum_{t=1}^{N}\frac{\partial w_t^{\mathrm{T}}}{\partial \beta}\frac{\partial w_s}{\partial \beta} + \frac{N}{2\Sigma^2}\frac{\partial \Sigma^{\mathrm{T}}}{\partial \beta}\frac{\partial \Sigma}{\partial \beta}\right\}$$

6.8

$$\tilde{M}(w) = \text{real part} \left\{ \frac{1}{\Sigma} \left| \frac{\partial G_1(e^{jw})}{\partial \beta} \right|^{\mathrm{T}} G_2^{-1}(e^{jw}) G_2^{-1}(e^{-j\omega}) \left| \frac{\partial G_1(e^{-jw})}{\partial \beta} \right| \right\}$$

$$\frac{\partial G_1}{\partial a_i} = -\frac{B}{A^2} z^{-i} = -\frac{1}{A^2} \sum_{j=1}^{n} b_j z^{-i-j}$$

$$\frac{\partial G_1}{\partial b_i} = \frac{1}{A} z^{-i} = \frac{A}{A^2} z^{-i} = \frac{1}{A^2} \sum_{j=0}^{n} a_j z^{-i-j}$$

Let $\theta^{\mathrm{T}} = (b_1, \ldots, b_n, a_1, \ldots, a_n)$

$$\frac{\partial G_1(e^{jw})^{\mathrm{T}}}{\partial \theta} = \frac{1}{A^2(e^{jw})}
\begin{bmatrix}
1 & a_1 \text{------------} a_n & 0 & & 0 \\
0 & 1 \quad a_1 & & a_n & \\
& & & & 0 \\
0 \text{------------} 1 & a_1 \text{------------} a_n \\
0 & -b \text{------------} b_n & 0 \text{------} 0 \\
& & & & 0 \\
0 & 0 \quad -b_1 \text{------------} b_n
\end{bmatrix}
\begin{bmatrix}
e^{-jw} \\
e^{-2jw} \\
\\
\vdots \\
\\
\\
e^{-2njw}
\end{bmatrix}$$

$$= \frac{1}{A^2(e^{jw})} S
\begin{bmatrix}
e^{-jw} \\
e^{-2jw} \\
\vdots \\
e^{-2njw}
\end{bmatrix}$$

Hence

$$\tilde{M}_\theta(w) = \frac{1}{\Sigma} S \Gamma_X S^{\mathrm{T}}$$

where

$$\Gamma_X = H(w) \text{ real part} \left\{
\begin{bmatrix}
e^{-jw} \\
e^{-2jw} \\
\vdots \\
e^{-2njw}
\end{bmatrix}
\begin{bmatrix} e^{jw} & e^{2jw} & \ldots & e^{2njw} \end{bmatrix} \right.$$

$$= H(w)
\begin{bmatrix}
1 & & \cos w & \ldots & \cos(\overline{2n-1}w) \\
\cos w & & & & \vdots \\
\vdots & & & & \cos w \\
\cos(\overline{2n-1}w) & \ldots & & \cos w & 1
\end{bmatrix}$$

with

$$H(w) = C(e^{jw})C(e^{-jw})/D(e^{jw})A^2(e^{jw})A^2(e^{-jw})D(e^{-jw})$$

6.9   From Eq. (6.5.5)

$$L = \log p(Y\,|\,\beta) = \frac{-Nm}{2}\log 2\pi - \frac{1}{2}\sum_{k=1}^{N}\log \det S_k - \frac{1}{2}\sum_{k=1}^{N} w_k^{\mathsf{T}} S_k^{-1} w_k$$

where $w_k = y_k - \bar{y}_k$ with $E\{w_k w_k^{\mathsf{T}}\} = S_k$. Hence,

$$-\frac{\partial L}{\partial \beta_i} = \frac{1}{2}\sum_{k=1}^{N}\frac{\partial}{\partial \beta_i}[\log \det S_k] + \frac{1}{2}\sum_{k=1}^{N} w_k^{\mathsf{T}}\frac{\partial S_k^{-1}}{\partial \beta_i} w_k + \sum_{k=1}^{N} w_k^{\mathsf{T}} S_k^{-1}\frac{\partial w_k}{\partial \beta_i}$$

$$= \frac{1}{2}\sum_{k=1}^{N}\text{trace}\left|S_k^{-1}\frac{\partial S_k}{\partial \beta_i}\right| - \frac{1}{2}\sum_{k=1}^{N} w_k^{\mathsf{T}} S_k^{-1}\frac{\partial S_k}{\partial \beta_i} S_k^{-1} w_k + \sum_{k=1}^{N} w_k^{\mathsf{T}} S_k^{-1}\frac{\partial \omega_k}{\partial \beta_i}$$

Hence

$$M_{ij} = E\left|\frac{\partial L}{\partial \beta_i}\frac{\partial L}{\partial \beta_j}\right| = E\Big|\frac{1}{4}\sum_{k=1}^{N}\sum_{l=1}^{N}\left(\text{trace}\, S_k^{-1}\frac{\partial S_k}{\partial \beta_i}\right)\left(\text{trace}\, S_l^{-1}\frac{\partial S_l}{\partial \beta_j}\right)$$

$$+ \frac{1}{4}\sum_{k=1}^{N}\sum_{l=1}^{N} w_k^{\mathsf{T}} S_k^{-1}\frac{\partial S_k}{\partial \beta_i} S_k^{-1} w_k\, w_l^{\mathsf{T}} S_l^{-1}\frac{\partial S_l}{\partial \beta_j} S_l^{-1} w_l$$

$$+ \sum_{k=1}^{N}\sum_{l=1}^{N} w_k^{\mathsf{T}} S_k^{-1}\frac{\partial w_k}{\partial \beta_i}\, w_l^{\mathsf{T}} S_l^{-1}\frac{\partial w_l}{\partial \beta_j}$$

$$- \frac{1}{4}\sum_{k=1}^{N}\sum_{l=1}^{N}\left(\text{trace}\, S_k^{-1}\frac{\partial S_k}{\partial \beta_i}\right)\left(\text{trace}\, S_l^{-1}\frac{\partial S_l}{\partial \beta_j} S_l^{-1} w_l w_l^{\mathsf{T}}\right)$$

$$- \frac{1}{4}\sum_{k=1}^{N}\sum_{l=1}^{N}\left(\text{trace}\, S_k^{-1}\frac{\partial S_k}{\partial \beta_i} S_k^{-1} w_k w_k^{\mathsf{T}}\right)\left(\text{trace}\, S_l^{-1}\frac{\partial S_l}{\partial \beta_j}\right)\Big|$$

$$= E\Big|\frac{1}{4}\sum_{k=1}^{N}\sum_{l=1}^{N} w_k^{\mathsf{T}} S_k^{-1}\frac{\partial S_k}{\partial \beta_i} S_k^{-1} w_k\, w_l^{\mathsf{T}} S_l^{-1}\frac{\partial S_l}{\partial \beta_j} S_l^{-1} w_l$$

$$+ \sum_{k=1}^{N}\sum_{l=1}^{N} w_k^{\mathsf{T}} S_k^{-1}\frac{\partial w_k}{\partial \beta_i}\, w_l^{\mathsf{T}} S_l^{-1}\frac{\partial w_l}{\partial \beta_j}$$

$$- \frac{1}{4}\sum_{k=1}^{N}\sum_{l=1}^{N}\text{trace}\, S_k^{-1}\frac{\partial S_k}{\partial \beta_i}\,\text{trace}\, S_l^{-1}\frac{\partial S_l}{\partial \beta_j}\Big| \tag{1}$$

The second term in (1) reduces to

$$\text{second term} = E\Big|\sum_{k=1}^{N}\left[\frac{\partial w_k^{\mathsf{T}}}{\partial \beta_i} S_k^{-1}\frac{\partial w_k}{\partial \beta_j}\right]\Big| \tag{2}$$

We now look at the first and third terms in (1) for $k \neq l$

$$E\left\{\frac{1}{4} \sum_{k=1}^{N} \sum_{\substack{l=1 \\ k \neq l}}^{N} w_k{}^{\mathrm{T}} S_k^{-1} \frac{\partial S_k}{\partial \beta_i} S_k^{-1} w_k w_l{}^{\mathrm{T}} S_l^{-1} \frac{\partial S_l}{\partial \beta_j} S_l^{-1} w_l \right.$$

$$\left. - \frac{1}{4} \sum_{k=1}^{N} \sum_{\substack{l=1 \\ k \neq l}}^{N} \left(\mathrm{trace}\ S_k^{-1} \frac{\partial S_k}{\partial \beta_i} \right)\left(\mathrm{trace}\ S_l^{-1} \frac{\partial S_l}{\partial \beta_j} \right) \right\} = 0 \qquad (3)$$

Substituting (2) and (3) into (1) gives

$$M_{ij} = E\left\{ \sum_{k=1}^{N} \left| \frac{\partial w_k}{\partial \beta_i}^{\mathrm{T}} S_k^{-1} \frac{\partial w_k}{\partial \beta_j} \right| + \frac{1}{4} \sum_{k=1}^{N} w_k{}^{\mathrm{T}} S_k^{-1} \frac{\partial S_k}{\partial \beta_i} S_k^{-1} w_k w_k{}^{\mathrm{T}} S_k^{-1} \frac{\partial S_k}{\partial \beta_j} S_k^{-1} w_k \right.$$

$$\left. - \frac{1}{4} \sum_{k=1}^{N} \left(\mathrm{trace}\ S_k^{-1} \frac{\partial S_k}{\partial \beta_i} \right)\left(\mathrm{trace}\ S_k^{-1} \frac{\partial S_k}{\partial \beta_j} \right) \right) \qquad (4)$$

We next look at the second term in (4) with k fixed. We use the Einstein dummy suffix notation (repeated subscript implies summation) and we define

$$P = S^{-1}, \qquad Q = \partial S / \partial \beta_i, \qquad R = \partial S / \partial \beta_j$$

$$\mathrm{term} = E\{w_i P_{ij} Q_{jl} P_{lm} w_m w_n P_{nr} P_{rs} P_{st} w_t\}$$

From Problem 2.8,

$$E\{w_i w_m w_n w_t\} = S_{im} S_{nt} + S_{in} S_{mt} + S_{it} S_{mn}$$

Hence

$$\mathrm{term} = (P_{ij} Q_{jl} P_{lm} P_{nr} R_{rs} P_{st})(S_{im} S_{nt} + S_{in} S_{mt} + S_{it} S_{mn})$$

Now using the fact that $P_{ij} S_{im} = S_{mi} P_{ij} = \delta_{mj}$, etc., we have

$$\mathrm{term} = 3 Q_{jl} P_{jl} R_{rn} P_{rn} = 3 \left(\mathrm{trace}\ S_k^{-1} \frac{\partial S_k}{\partial \beta_i} \right)\left(\mathrm{trace}\ S_k^{-1} \frac{\partial S_k}{\partial \beta_j} \right) \qquad (5)$$

Substituting (5) into (4) yields

$$M_{ij} = E\left\{ \sum_{k=1}^{N} \left| \frac{\partial w_k}{\partial \beta_i}^{\mathrm{T}} S_k^{-1} \frac{\partial w_k}{\partial \beta_j} \right| \right\} + \frac{3}{4} \sum_{k=1}^{N} \left(\mathrm{trace}\ S_k^{-1} \frac{\partial S_k}{\partial \beta_i} \right)\left(\mathrm{trace}\ S_k^{-1} \frac{\partial S_k}{\partial \beta_j} \right)$$

$$- \frac{1}{4} \sum_{k=1}^{N} \left(\mathrm{trace}\ S_k^{-1} \frac{\partial S_k}{\partial \beta_i} \right)\left(\mathrm{trace}\ S_k^{-1} \frac{\partial S_k}{\partial \beta_j} \right)$$

$$= E\left\{ \sum_{k=1}^{N} \left| \frac{\partial w_k}{\partial \beta_i}^{\mathrm{T}} S_k^{-1} \frac{\partial w_k}{\partial \beta_j} \right| \right\} + \frac{1}{2} \sum_{k=1}^{N} \left(\mathrm{trace}\ S_k^{-1} \frac{\partial S_k}{\partial \beta_i} \right)\left(\mathrm{trace}\ S_k^{-1} \frac{\partial S_k}{\partial \beta_j} \right)$$

This establishes Eq. (6.5.6).

6.10 (a) Rewrite model as

$$y_k = G_1(z)u_k + G_2(z)\varepsilon_k$$

with

$$G_1(z) = 1/(1 - az^{-1}), \qquad G_2(z) = 1/(1 - az^{-1})$$

Hence

$$\overline{M} = \frac{1}{\pi\sigma^2} \int_0^\pi \frac{\partial G_1(e^{jw})}{\partial a} G_2^{-1}(e^{jw})G_2^{-1}(e^{-jw}) \frac{\partial G_1(e^{-jw})}{\partial a} \Phi_u(jw)\, dw$$

$$+ \frac{1}{\pi} \int_0^\pi \frac{G_2(e^{jw})}{\partial a} G_2^{-1}(e^{jw})G_2^{-1}(e^{-jw}) \frac{\partial G_2(e^{-jw})}{\partial a}\, dw$$

$$= \frac{1}{\pi\sigma^2} \int_0^\pi \frac{\Phi_u(w)\, dw}{(1 + a^2) - 2a \cos w}$$

$$+ \frac{1}{\pi} \int_0^\pi \frac{dw}{(1 + a^2) - 2a \cos w}$$

$$= \frac{1}{\pi\sigma^2} \int_0^\pi \frac{\Phi_u(w)\, dw}{(1 + a^2) - 2a \cos w} + \frac{1}{1 - a^2}$$

(b) Sinusoidal input gives

$$\overline{M} = \frac{1}{\pi\sigma^2\{(1 + a^2) - 2a \cos w\}} + \frac{1}{1 - a^2}$$

If $a > 0$, then $\overline{M}$ maximal for $w \Rightarrow 0$; if $a < 0$, then $\overline{M}$ maximal for $w \Rightarrow \pi$. In both cases,

$$\overline{M}^* = [1/\pi\sigma^2\{1 - |a|\}^2] + [1/(1 - a^2)]$$

(c) Straightforward substitution into expression for u_k.
(d) Optimal information matrix is achievable with a single sinewave (cf. Theorem 6.4.3).

6.11 Rewrite model as

$$y_k = G_2(z)\varepsilon_k$$

where

$$G_2(z) = \frac{1}{1 + a_1 z^{-1} + a_2 z^{-2}}, \qquad \beta = (a_1, a_2, v)^\mathsf{T}$$

We can ignore the parameter v since this term is always decoupled from the other parameters in the information matrix. Hence look at

$$\overline{M}_\theta = \frac{1}{2\pi} \int_{-\pi}^{\pi} \frac{\partial G_2(e^{jw})^{\mathrm{T}}}{\partial \beta} G_2^{-1}(e^{jw}) G_2^{-1}(e^{-jw}) \frac{\partial G_2(e^{jw})}{\partial \beta} dw$$

$$\theta = (a_1, a_2)^{\mathrm{T}}$$

$$\frac{\partial G_2^{\mathrm{T}}(z)}{\partial \theta} = \frac{-1}{[1 + a_1 z^{-1} + a_2 z^{-2}]^2} \begin{bmatrix} z^{-1} \\ z^{-2} \end{bmatrix}$$

Hence

$$M_\theta = \frac{1}{2\pi} \int_{-\pi}^{\pi} \frac{1}{(1 + a_1 e^{-jw} + a_2 e^{-2jw})(1 + a_1 e^{jw} + a_2 e^{2jw})} \begin{bmatrix} 1 & e^{jw} \\ e^{-jw} & 1 \end{bmatrix} dw$$

Evaluating the integral using Cauchy's theorem for the case $a_1 = 0$ yields $\overline{M}_{12} = \overline{M}_{21} = 0$.

CHAPTER 7

7.1 We apply the general transformation described in Lemmas 2.6.3 and 2.6.4 to

$$\begin{bmatrix} R_N \\ x_{N+1}^{\mathrm{T}} \end{bmatrix}$$

where R_N is upper triangular. The appropriate orthonormal transformation is a product of p transformations: $Q_p Q_{p-1} Q_{p-2} \cdots Q_1$, where $Q_i = I - 2\mu(i)\mu(i)^{\mathrm{T}}$ and $\mu(i)$ is given by Eqs. (2.6.10) and (2.6.11). The result follows.

7.2 Substituting for P_{N+1} in (7.3.10) yields

$$\hat{\theta}_{N+1} = \frac{1}{\alpha} \left(I - P_N \frac{x_{N+1} x_{N+1}^{\mathrm{T}}}{(\alpha + x_{N+1}^{\mathrm{T}} P_N x_{N+1})} \right) P_N (\alpha P_N^{-1} \hat{\theta}_N + x_{N+1} y_{N+1})$$

$$= \hat{\theta}_N + \frac{1}{\alpha} P_N x_{N+1} y_{N+1}$$

$$- \frac{1}{\alpha} P_N \frac{x_{N+1} x_{N+1}^{\mathrm{T}}}{(\alpha + x_{N+1}^{\mathrm{T}} P_N x_{N+1})} P_N x_{N+1} y_{N+1} - K_{N+1} x_{N+1} \hat{\theta}_N$$

where K_{N+1} is given by (7.3.3). Hence

$$\hat{\theta}_{N+1} = \hat{\theta}_N + \frac{1}{\alpha} \frac{P_N x_{N+1}(\alpha + x_{N+1}^T P_N x_{N+1} - x_{N+1}^T P_N x_{N+1})y_{N+1}}{(\alpha + x_{N+1}^T P_N x_{N+1})}$$

$$- K_{N+1}x_{N+1}\hat{\theta}_N$$

$$= \hat{\theta}_N + K_{N+1}[y_{N+1} - x_{N+1}\hat{\theta}_N]$$

as required.

7.3

$$\hat{\theta}_{i+N, i+1} = P_{i+N, i+1}(X_{i+1}^{i+N})^T Y_{i+1}^{i+N}$$

where

$$X_{i+1}^{i+N} = \begin{bmatrix} x_{i+1}^T \\ \vdots \\ x_{i+N}^T \end{bmatrix}, \qquad Y_{i+1}^{i+N} = \begin{bmatrix} y_{i+1} \\ \vdots \\ y_{i+N} \end{bmatrix}$$

i.e.,

$$\hat{\theta}_{i+N, i+1} = (P_{i+N, i}^{-1} - x_i x_i^T)^{-1}[(X_i^{i+N})^T Y_i^{i+N} - x_i y_i]$$

Application of Lemma E.1.1 yields

$$P_{i+N, i+1} = \left(I + P_{i+N, i}\frac{x_i x_i^T}{(1 - x_i^T P_{i+N, i}x_i)}\right)P_{i+N, i}$$

Now following exactly the steps of Problem 7.2 or the proof of Result 7.2.1 leads to the required result.

7.4 In the notation of Problem 7.3,

$$\hat{\theta}_{i+2N, i+N+1} = P_{i+2N, i+N+1}(X_{i+N+1}^{i+2N})^T Y_{i+N+1}^{i+2N}$$

where

$$P_{i+2N, i+N+1}^{-1} = [(X_{i+N+1}^{i+2N})^T(X_{i+N+1}^{i+2N})]$$

$$= [(X_{i+1}^{i+2N})^T(X_{i+1}^{i+2N}) - (X_{i+1}^{i+N})^T(X_{i+1}^{i+N})]$$

$$= P_{i+2N, i+1}^{-1} - P_{i+N, i+1}^{-1}$$

thus establishing Eq. (7.3.22). Further,

$$\hat{\theta}_{i+2N, i+N+1} = P_{i+2N, i+N+1}^{-1}[X_{i+1}^{i+2N} Y_{i+1}^{i+2N} - X_{i+1}^{i+N} Y_{i+1}^{i+N}]$$

$$= P_{i+2N, i+N+1}^{-1}[P_{i+2N, i+1}^{-1}\hat{\theta}_{i+2N, i+1} - P_{i+N, i+1}^{-1}\hat{\theta}_{i+N, i+1}]$$

thus completing the derivation.

7.5 The estimator may be rewritten as

$$\hat{\theta}_{N+1} = \hat{\theta}_N + \frac{1}{N}x_{N+1}(y_{N+1} - x_{N+1}^T\hat{\theta}_N)$$

Comparison with (7.5.7) indicates that the estimator is a stochastic approximation scheme with $\gamma_N = 1/N$. (7.5.3) and (7.5.4) are satisfied since Σ/N diverges and Σ/N^2 is finite.

7.6

$$\log \det P_{N+1} = \log \det\left(I + P_N \frac{\bar{x}_{N+1}\bar{x}_{N+1}^{\mathrm{T}}}{(1 + \bar{x}_{N+1}^{\mathrm{T}}P_N \bar{x}_{N+1})}\right) + \log \det P_N$$

where $\bar{x}_N$ is given by Eqs. (7.4.39)–(7.4.43). The remainder of the derivation of Section 7.7.1 is unaltered since none of the $\bar{w}_t, \bar{w}_{t-1}, \ldots$, depends upon u_t.

7.7

$$J_2 = \operatorname{trace} P_N^{-1}P_{N+1} = \operatorname{trace} I_p - \operatorname{trace} \frac{x_{N+1}x_{N+1}^{\mathrm{T}}P_N}{(1 + x_{N+1}^{\mathrm{T}}P_N x_{N+1})}$$

$$= p - \frac{x_{N+1}^{\mathrm{T}}P_N x_{N+1}}{1 + x_{N+1}^{\mathrm{T}}P_N x_{N+1}} = p - 1 + \frac{1}{1 + x_{N+1}^{\mathrm{T}}P_N x_{N+1}}$$

which is minimized when $x_{N+1}^{\mathrm{T}}P_N x_{N+1}$ is maximized, i.e., the same result is obtained as for $J_1 = \log \det P_{N+1}$.

7.8 Converting to PEM form

$$y_t = B(z)u_t + [1 - A(z)]y_t + [C(z) - 1]C(z)^{-1}[A(z)y_t - B(z)u_t] + \varepsilon_t$$
$$= b_1 u_{t-1} + \xi_t^{\mathrm{T}}\theta + \varepsilon_t$$

where $\theta^{\mathrm{T}} = (b_2, \ldots, b_n, a_1, \ldots, a_n, c_1, \ldots, c_n)$ and $\xi_t^{\mathrm{T}} = (u_{t-2}, \ldots, u_{t-n}, y_{t-1}, \ldots, y_{t-n}, \bar{w}_{t-1}, \ldots, \bar{w}_{t-n})$ where $\bar{w}_t = C(z)^{-1}[A(z)y_t - B(z)u_t]$ the minimum variance strategy follows immediately analogously to (7.8.13).

7.9

$$D(z) = C(z)F(z) + z^{-k}G(z)$$

equating coefficients gives

$$1 = f_0$$

$$d_1 = c_1 f_0 + f_1$$

$$d_2 = c_2 f_0 + c_1 f_1 + f_2$$

$$\vdots$$

$$d_{k-1} = c_{k-1}f_0 + c_{k-2}f_1 + \cdots + f_{k-1}$$

which can obviously be solved for $f_0, f_1, \ldots, f_{k-1}$. Also

$$g_0 = d_k - c_k f_0 - c_{k-1}f_1 - \cdots - c_1 f_{k-1}$$

$$g_1 = d_{k+1} - c_{k+1}f_0 - c_k f_1 - \cdots - c_2 f_{k-1}$$

$$\vdots$$

$$g_{m-k} = d_m - c_m f_0 - c_{m-1}f_1 - \cdots - c_{m-k+1}f_{k-1}$$

$$\vdots$$

$$g_{m-1} = c_m f_{k-1}$$

7.10   The strategy of Result 7.8.2 is directly applicable provided ξ_t and $\hat{\theta}_t$ are given the interpretations discussed in Problems 7.6 and 7.8.

7.11.   If we work directly with the sample correlation, then at each time N we choose the value of u_N to minimize

$$V_N = \sum_{\tau=1}^{n} [\Gamma_N(\tau) - \Gamma^*(\tau)]^2$$

where

$$\Gamma^*(\tau) = 0, \qquad \tau = 1, 2, \ldots, n$$

i.e.

$$V_N = \sum_{\tau=1}^{n} \left| \frac{1}{N} [(N-1)\Gamma_{N-1}(\tau) + u_{N-\tau}\, u_N] \right|^2$$

$$= \sum_{\tau=1}^{n} \frac{1}{N^2} [(N-1)^2 \Gamma_{N-1}(\tau)^2 + 2(N-1)\Gamma_{N-1}(\tau)u_{N-\tau}\, u_N + u_{N-\tau}^2 u_N^2]$$

Clearly this is minimized if u_N is chosen as

$$u_N = -\text{sign} \sum_{\tau=1}^{n} \Gamma_{N-1}(\tau)u_{N-\tau}$$

On the other hand Result 7.7.1 yields

$$u_N = \text{sign} \sum_{\tau=1}^{n} P_{12}(\tau)u_{N-\tau}$$

where

$$\begin{bmatrix} P_{11} & P_{12}(1) & \cdots & P_{12}(n) \\ \hline P_{12}(1) & & & \\ \vdots & & P_{22} & \\ P_{12}(n) & & & \end{bmatrix} = \begin{bmatrix} 1 & \Gamma_{N-1}(1) & \cdots & \Gamma_{N-1}(n) \\ \hline \Gamma_{N-1}(1) & & & \\ \vdots & & \Gamma^1 & \\ \Gamma_{N-1}(n) & & & \end{bmatrix}^{-1}$$

References

1. Youla, D. C., On the factorization of rational matrices, *IEEE Trans. Inform. Theory,* **IT-7** (1961), 172–189.
2. Caines, P. E., Ph.D. Thesis, Imperial College, 1970.
3. McLane, S., and Birkhoff, G., "Algebra." MacMillan, New York, 1967.
4. Gohberg, I. C., and Krein, M. G., On the factorization of operators in Hilbert spaces, *Amer. Math. Soc. Trans.* **51** (1966).
5. Chen, C. T., "Introduction to Linear System Theory." Holt, Reinhart, and Winston, New York, 1970.
6. Kailath, T., An innovations' approach to least-squares estimation, Part I, *IEEE Trans. Auto. Control,* **AC-13** (1968), 646–660.
7. Glover, K., and Willems, J. C., Parametrizations of linear dynamical systems: Canonical forms and identifiability, *IEEE Trans. Auto. Control,* **AC-19**, No. 5 (1974), 640–645.
8. Denham, M. J., Canonical forms for the identification of multivariable linear systems, *IEEE Trans. Auto. Control,* **AC-19**, No. 5 (1974), 646–655.
9. Dickinson, B. W., Kailath, T., and Morf, M., Canonical matrix fraction and state space descriptions for deterministic and stochastic linear systems, *IEEE Trans. Auto. Control* **AC-19**, No. 5 (1974), 656–666.
10. Papoulis, A., "Probability Random Variables and Stochastic Processes." McGraw-Hill, 1965.
11. Silvey, S. D., "Statistical Inference." Penguin, 1970.
12. Eykhoff, P., "System Parameter and State Estimation." Wiley, London, 1974.
13. Hannan, E. J., "Multiple Time Series." Wiley, New York, 1970.
14. Åström, K. J., "Introduction to Stochastic Control Theory." Academic Press, New York, 1970.
15. Kendall, M. G., and Stuart, A., "The Advanced Theory of Statistics," Vols. 1–3. Griffin, London, 1961.

16. Cox, D. R., and Miller, H. D., "The Theory of Stochastic Processes." Methuen, London, 1965.
17. Drymes, P. J., "Econometrics." Harper and Row, New York, 1970.
18. Aoki, M., "Optimization of Stochastic Systems." Academic Press, New York, 1967.
19. Zellner, A., "An Introduction to Bayesian Inference in Econometrics." Wiley, New York, 1971.
20. Anderson, T. W., "The Statistical Analysis of Time Series." Wiley, New York, 1971.
21. Sage, A. P., and Melsa, J. L., "System Identification." Academic Press, New York, 1971.
22. Parzen, E. (ed.), "Time Series Analysis Papers." Holden–Day, San Francisco, 1967.
23. Graupe, D., "Identification of Systems." Van Nostrand, New York, 1972.
24. Nahi, N. E., "Estimation Theory and Applications." Wiley, New York, 1969.
25. Deutsch, R., "Estimation Theory." Prentice-Hall, Englewood Cliffs, New Jersey, 1965.
26. Goldberger, A. S., "Econometric Theory." Wiley, New York, 1964.
27. Zacks, S., "The Theory of Statistical Inference." Wiley, New York, 1971.
28. Bendat, J. S., and Piersol, A. G., "Measurement and Analysis of Random Data." Wiley, New York, 1966.
29. Box, G. E. P., and Jenkins, G. M., "Time Series Analysis, Forecasting, and Control." Holden–Day, San Francisco, 1970.
30. Rao, C. R., "Linear Statistical Inference and its Applications." Wiley, New York, 1965.
31. Hannan, E. J., "Time Series Analysis." Metheun, London, 1960.
32. Wasan, M. T., "Parametric Estimation." McGraw-Hill, New York, 1970.
33. Lewis, T. O., and Odell, P. L., "Estimation in Linear Models." Prentice-Hall, Englewood Cliffs, New Jersey, 1971.
34. Ferguson, T. S., "Mathematical Statistics—A Decision Theoretic Approach." Academic Press, New York, 1967.
35. Lindley, D. V., "Introduction to Probability and Statistics, from a Bayesian Viewpoint." Cambridge Univ. Press, London, 1965.
36. Raiffa, R., and Schlaifer, "Applied Statistical Decision Theory." M.I.T. Press, Cambridge, 1961.
37. Lòeve, M. M., "Probability Theory," 3rd ed. Van Nostrand, New York, 1963.
38. Lukacs. E., "Characteristic Functions." 2nd ed. Griffin, London, 1970.
39. Zadeh, L. A., and Desoer, C. A., "Linear System Theory." McGraw-Hill, New York, 1963.
40. Feller, W., "An Introduction to Probability Theory and its Applications," Vol. I. Wiley, New York, 1957.
41. Feller, W., "An Introduction to Probability Theory and its Applications," Vol. II. Wiley, New York, 1966.
42. Zehna, P. W., Invariance of maximum likelihood estimation, Ann. Math. Stat. 37 (1966), 755.
43. Cramer, H., "Mathematical Methods of Statistics." Princeton Univ. Press, Princeton, New Jersey, 1946.
44. Commins, W., Asymptotic variance as an approximation to expected loss for maximum likelihood estimates, Tech. Rep. No. 46, Contract N60, nr-2S140 (NR-342022), Dept. of Statistics, Stanford Univ.
45. Caines, P. E., and Rissanen, J., Maximum likelihood estimation of parameters in multivariate Gaussian stochastic processes, IEEE Trans. Inform. Theory IT-20, No. 1 (1974), 102–104.
46. Mayne, D. Q., Calculation of derivatives for parameter estimation, Publication 71/9, Dept. of Computing and Control, Imperial College, London, 1971.
47. Denery, D. G., Simplification in the computation of sensitivity functions for constant coefficient linear systems, IEEE Trans. Auto. Control AC-19 (1971).
48. Aoki, M., "Introduction to Optimization Techniques: Fundamentals and Applications of Nonlinear Programming." MacMillan, New York, 1971.

49. Gupta, N. K., and Mehra, R. K., Computational aspects of maximum likelihood estimation and reduction in sensitivity function calculations, *IEEE Trans. Auto. Control* **AC-19**, No. 6 (1974).

50. Goodwin, G. C., Application of curvature methods to parameter and state estimation, *Proc. IEE* **116**, No. 6 (1969), 1107–1110.

51. Clarke, D. W., Generalized least-squares estimation of the parameters of a dynamic model, *Proc. IFAC Symp. Identification Auto. Control Syst.*, Prague, Czechoslovakia, Paper 3.17 (1967).

52. Söderström, T., Convergence properties of the generalized least-squares identification method, *Proc. IFAC Symp. Identification Syst. Param. Identification*, Delft/The Hague (1973).

53. Aström, K. J., and Söderström, T., Uniqueness of the maximum likelihood estimates of the parameters of an ARMA model, *IEEE Trans. Auto. Control* **AC-19**, No. 6 (1974).

54. Bohlin, T., On the problem of ambiguities in maximum likelihood identification, *Proc. IFAC Symp. Identification Process Param. Estimation*, Prague, Czechoslovakia, Paper 3.2 (1970). (Appeared also in *Automatica* **7**, 1971.)

55. Hastings-James, R., and Sage, M. W., Recursive generalized least-squares procedure for on-line identification of process parameters, *Proc. IEE* **116**, No. 12 (1969).

56. Gertler, J., and Bányász, Cs., A recursive (on-line) maximum likelihood identification method, *IEEE Trans. Auto. Control* **AC-19**, No. 6 (1974).

57. Dvoretzky, A., On stochastic approximation, *Proc. 3rd Berkeley Symp. Math. Stat. Prob.* **1** (1956), 39–55.

58. Saridis, G. N., Stochastic approximation methods for identification and control—a survey, *IEEE Trans. Auto. Control* **AC-19**, No. 6 (1974).

59. Jazwinski, A. H., "Stochastic Processes and Filtering Theory." Academic Press, New York, 1970.

60. Aström, K. J., and Wittenmark, B., On self-tuning regulators, *Automatica* **9** (1973), 185–199.

61. Wittenmark, B., A self-tuning predictor, *IEEE Trans. Auto. Control* **AC-19**, No. 6 (1974).

62. Jones, R. H., Exponential smoothing for multivariate time series, *J. R. Stat. Soc. B* **28** (1966), 241–251.

63. Neyman, J., and Pearson, E. S., On the use and interpretation of certain test criteria for the purposes of statistical inference, *Biometrika* **20A** (1928), 175–263.

64. Akaike, H., On a decision procedure for system identification, *Proc. IFAC Symp. System Eng. Approach Computer Control*, Kyoto, Japan, Paper 30.1 (1970).

65. Payne, R. L., and Goodwin, G. C., A Bayesian criterion for structure determination based on ultimate model use, *Publication 74/5*, Dept. Computing and Control, Imperial College, London (1974).

66. Caines, P. E., and Chan, C. W., Feedback between stationary stochastic processes, *Control Syst. Rep.*, No. 7421, Dept. Elec. Eng., University of Toronto, Canada (1974).

67. Gauss, K. G., "Theory of Motion of Heavenly Bodies." Dover, New York, 1963.

68. Halmos, P. R., and Savage, L. J., Application of Radon–Nikodym theorem to the theory of sufficient statistics, *Ann. Math. Stat.* **20** (1949), 225–241.

69. Lehmann. E. L., "Testing Statistical Hypotheses." Wiley, New York, 1959.

70. Athans, M., The matrix minimum principle, *Information Control*, No. 11 (1968), 500–606.

71. De Groot, M. H. "Optimal Statistical Decisions." McGraw-Hill, New York, 1970.

72. Hannan, E. J., The identification problem for multiple equation systems with moving average errors *Econometrics* **39**, No. 5 (1971).

73. Rozanov, Y. A., "Stationary Random Processes." Holden–Day, San Francisco, 1967.

74. Fisher, F., "The Identification Problem in Econometrics." McGraw-Hill, New York, 1966.
75. Hannan, E. J., Regression for time series *in* "Time Series Analysis" (M. Rosenblatt, ed.), pp. 17–37. Wiley, New York, 1963.
76. Kalman, R. E., When is a linear control system optimal?, *Trans. ASME J. Bas. Eng.* **86-D** (1964), 51–60.
77. Desoer, C. A., "Notes for a Second Course on Linear Systems." Van Nostrand–Reinhold, New York, 1970.
78. Kwakernak and Sivan, "Linear Optimal Control Systems." Wiley, New York, 1972.
79. Itô, K., On stochastic differential equations, *Mem. Amer. Math. Soc.*, No. 4 (1951).
80. Mayne, D. Q., Canonical forms for identifications, Publication No. 74/49 (1974), Imperial College CCD.
81. Wong, K. Y., and E. Polak, Identification of linear discrete time systems using the instrumental variable method, *IEEE Trans.* **AC-12**, No. 6 (1967), 707–718.
82. Wald, A., Note on the consistency of maximum likelihood estimate, *Ann. Math. Stat.* **20** (1949), 595–601.
83. Roussas, G. C., Extension to Markov processes of a result of A. Wald about the consistency of the maximum likelihood estimate, *Z. Wahrscheinlichkeitsteorie* **4** (1967), 69–73.
84. Ralston, A., and Wilf, H. S., (eds.), "Mathematical Methods for Digital Computers." Wiley, New York, 1960–1967 (2 vols.).
85. Söderström, T., On the uniqueness of maximum likelihood identification, *Automatica* **11**, No. 2 (1975), 193–198.
86. Golub, C., Numerical method for solving linear least squares problems, *Numerische Mathematik* **7**, No. 3 (1965), 206–216.
87. MacDuffee, C. G., "The Theory of Matrices." Chelsea, New York, 1956.
88. Rosenbrock, H. H., "State Space and Multivariable Theory." Wiley, New York, 1970.
89. Popov, V. M., Some properties of the control systems with irreducible matrix-transfer functions, *in* "Seminar on Differential Equations and Dynamical Systems, Lecture Notes in Mathematics," No. 144. Springer-Verlag, Berlin, 1969.
90. Wolovich, W. A., The determination of state space representations for linear multivariable systems, *Automatica* **9** (1973), 97–106.
91. Forney, G. D., Jr., Minimal bases of rational vector spaces with applications to multivariable linear systems, *SIAM J. Control* **13**, No. 3 (1975), 493–520.
92. Gantmacher, F. R., "The Theory of Matrices." Chelsea, New York, 1959.
93. Luenberger, D. G., Canonical forms for linear multivariable systems, *IEEE Trans.* **AC-12** (1967), 290–293.
94. Kalman, R. E., Kronecker invariants and feedback, *Proc. Conf. Ordinary Differential Equations*, NRL Mathematics Research Center (1971).
95. Mayne, D. Q., A canonical form for identification of multivariable linear systems, *IEEE Trans.* **AC-17** (1972), 728–729.
96. Popov, V. M., Invariant description of linear time invariant controllable systems, *SIAM J. Control* **10**, No. 2 (1972), 252–264.
97. Bourbaki, N., "Theory of Sets." Herman, Paris, 1968.
98. Ljung, L., On consistency for prediction error identification methods, *Report* 7405, Div. Auto. Control, Lund Inst. Tech. (1974).
99. Anderson, B. D. O., An algebraic solution to the spectral factorization problem, *IEEE* **AC-12**, No. 4 (1967), 410–414,
100. Anderson, B. D. O., Algebraic properties of minimal degree spectral factors, *Automatica* **9** (1973), 491–500.
101. Deyst, J. J., and Price, C. F., Conditions for the asymptotic stability of the discrete minimum variance linear estimator, *IEEE* **AC-13**, No. 6 (1968), 702–705.

102. Anderson, B. D. O., and Moore, J. B., A matrix Kronecker lemma, *Tech. Rep.*, Department of Electrical Engineering, University of Newcastle (1975).
103. Doob, J. L., "Stochastic Processes." Wiley, 1953.
104. Meyer, P. A., "Probabilities and Potentials." Ginn (Blaisdell), Boston, 1966.
105. Neveu, J., "Mathematical Foundations of the Calculus of Probability." Holden-Day San Francisco, 1965.
106. Renyi, A., "Foundations of Probability." Holden-Day, San Francisco, 1970.
107. Finigan, B. M., and Rowe, I. H., Strongly consistent parameter estimation by the introduction of strong instrumental variables, *IEEE Trans.* **AC-19**, No. 6 (1974).
108. Young, P. C., An instrumental variable method for real-time identification of a noisy process, *Automatica* **6** (1970), 271–287.
109. Young, P. C., Some observations on instrumental variable methods of time series analysis, *Centre Resource Environ. Studies, Rep.* CRES/SS/RR1, Australian National Univ. (1975).
110. Clarke, D. W., Dyer, D. A. J., Hastings-James, R., Ashton, R. P., and Emery, J. B., Identification and control of a pilot-scale boiling rig, *Proc. 3rd IFAC Symp. Ident. Syst. Param. Ident.*, The Hague/Delft (1973).
111. Rissanen, J., and Caines, P. E., Consistency of maximum likelihood estimators for multivariate Gaussian processes with rational spectrum, *Tech. Rep.*, Dept. Elec. Eng., Linköping Univ. (1974).
112. Fagin, S. L., Recursive linear regression theory, optimal filter theory, and error analysis of optimal systems, *IEEE Intern. Conv. Record* **12** (1964), 216–240.
113. Reiersøl, O., Confluence analysis by means of instrumental sets of variables, *Ark. Mat. Astron. Fys.* **32** (1945).
114. Swiney, P., Multiple time series analysis using the generalized least squares principle, Final Year Undergraduate Project, University of Newcastle (1975).
115. Goodwin, G. C., Payne, R. L., and Kabaila, P. V., On canonical forms and algorithms for multivariable time series analysis, *4th IFAC Symp. Ident. Syst. Param. Estimation*, Tbilisi USSR (1976).
116. Roussas, G. C., Asymptotic normality of the maximum likelihood estimate in Markov processes, *Metrika*, **14** (1968), 62–70.
117. Caines, P. E., Prediction error identification methods for stationary stochastic processes, *Control Syst. Rep.* 7516, Department of Elec. Eng., Univ. of Toronto (1975).
118. Ljung, L., On consistency and identifiability, Mathematical Programming Studies No. 5, North Holland (1976).
119. Rissanen, J., and Caines, P. E., Consistency of maximum likelihood estimators for ARMA processes, *Control Syst. Rep.* 7424, Dept. of Elec. Eng., Univ. of Toronto (1974).
120. Ljung, L., On the consistency of prediction error identification methods, *in* "System Identification, Advances, and Case Studies" (R. K. Mehra and D. G. Lainiotis, eds.). Academic Press, New York, 1976.
121. Ng. T. S., Ph.D. Dissertation, University of Newcastle (1977).
122. Söderström, T., Gustavsson, I., and Ljung, L., Identifiability conditions for linear systems operating in closed loop, *Int. J. Control* **21**, No. 2 (1975), 243–255.
123. Wellstead, P. E., and Edmunds, J. M., Least squares identification of closed loop systems, *Int. J. Control* **21**, No. 4 (1975), 689–699.
124. Ljung, L., Gustavsson, I., and Söderström, T., Identification of linear multivariable systems operating under linear feedback control, *IEEE* **AC-19**, No. 6 (1974), 836–840.
125. Söderström, I., Ljung, L., and Gustavsson, I., A comparative study of recursive identification methods, *Report* 7427, Lund Inst. Tech., Dept. Aut. Control (1974).
126. Söderström, T., An on-line algorithm for approximate maximum likelihood identification

of linear dynamic systems, *Report* 7308, Lund Inst. Tech., Dept. Auto. Control (1973).

127. Ljung, L., Convergence of recursive stochastic algorithms, *Report* 7403, Lund Inst. Tech., Dept. Auto. Control (1974).

128. Heyde, C. C., On martingale limit theory and strong convergence results for stochastic approximation procedures, *Stoc. Proc. Their Applic.* **2** (1974), 359–370.

129. Bellman, R., "Adaptive Processes—A Guided Tour." Princeton Univ. Press, Princeton, New Jersey, 1961.

130. Bar Shalom, Y., and Tse, E., Dual effect certainty equivalence, and separation in stochastic control, *IEEE Trans.* **AC-19**, No. 5 (1974).

131. Peterka, V., Adaptive digital regulation of noisy systems, *IFAC Symp. Ident. Process. Param. Estimation*, Prague (1970).

132. Ljung, L., and Wittenmark, B., Asymptotic properties of self-tuning regulators, *Report* 7404, Lund, Inst. Tech., Div. Auto. Control, (1974).

133. Wieslander, J., and Wittenmark, B., An approach to adaptive control using real-time identification, *Automatica* **7** (1971), 211.

134. Goodwin, G. C., and Payne, R. L., An approximate Bayesian one-step-ahead regulator, *Publication* 73/27, Dept. Computing and Control, Imperial College (1973).

135. Åström, K. J., Borisson, U., Ljung, L., and Wittenmark B., Theory and applications of adaptive regulators based on recursive parameter estimation, *IFAC 6th World Conference, Boston* (1975).

136. Alster, J., and Bélanger, P. R., A technique for dual adaptive control, *Automatica* **10**, No. 6 (1974), 627–634.

137. Wittenmark B., An active suboptimal dual controller for systems with stochastic parameters, *Auto. Control Theory Appl.* **3**, No. 1 (1975).

138. Koopmans, T. C., Identification problems in economic model construction, *Econometrica* **17** (1949), 125–144.

139. Shannon, C. E., A mathematical theory of communication, *Bell Syst. Tech. J.* **27** (1948), 379–423, 623–656.

140. Lindley, A. B., On a measure of the information provided by an experiment, *Ann. Math. Stats.* **27**, (1956), 986–1005.

141. Kullback, S., "Information Theory and Statistics." Dover, New York, 1968 (Wiley, New York, 1959).

142. Jaynes, E. T., Information theory and statistical mechanics, *The Physical Review*, **106**, (1957), 620–630; **108** (1957), 171–190,

143. Weidemann, H. L., and Stear, E. B., Entropy analysis of parameter estimation, *Information Control* **14** (1969), 493–506.

144. Rissanen, J., and Ljung, L., Estimation of optimum structures and parameters for linear systems, *Conf. Alg. Syst. Theory*, Udine, Italy (1975).

145. Akaike, H., A new look at the statistical model identification, *IEEE Trans.* **AC-19**, No. 6 (1974), 716–722.

146. Parzen, E., Some recent advances in time series modeling, *IEEE Trans.* **AC-19**, No. 6 (1974), 723–729.

147. Tukey, J. W., The future of data analysis, *Ann. Math. Stat.* **33** (1962), 1–67.

148. Caines, P. E., and Chan, C. W., Feedback between stationary stochastic processes, *IEEE Trans. Auto. Control* **AC-20**, No. 4 (1975), 498–509.

149. Payne, R. L., and Goodwin, G. C., A Bayesian approach to experiment design with applications to linear multivariable dynamic systems, *IMA Conf. Comput. Problems Stat.*, Univ. of Essex (1973).

150. Wald, A., On the efficient design of statistical investigations, *Ann. Math. Stat.* **14** (1943), 134–140.

151. Cox, D. R., "Planning of Experiments." Wiley, New York, 1958.
152. Davies, O. L., "Design and Analysis of Industrial Experiments." Oliver and Boyd, Edinburgh, 1954.
153. Kempthorne, O., "Design and Analysis of Experiments." Wiley, New York, 1952.
154. Kiefer, J., and Wolfowitz, J., The equivalence of two extremum Problems, *Can. J. Math.* **12** (1960), 363–366.
155. Karlin, S., and Studden, W. J., Optimal experimental designs, *Annals Math. Stat.* **37** (1966), 783–815.
156. Federov, V., "Theory of Optimal Experiments." Academic Press, New York and London, 1971.
157. Whittle, P., Some general points in the theory of optimal experimental design, *J.R. Stat. Soc. B* **35**, No. 1 (1973), 123–130.
158. Wynn, H. P., Results in the theory and construction of D-optimum experimental designs, *J.R. Stat. Soc. B* **34**, No. 2 (1972), 133–147.
159. Levin, M. J., Optimal estimation of impulse response in the presence of noise, *IRE Trans. Circuit Theory* **CT-7** (1960), 50–56.
160. Litman, S., and Huggins, W. H., Growing exponentials as a probing signal for system identification, *Proc. IEEE* **51** (1963), 917–923.
161. Levadi, V. S., Design of input signals for parameter estimation, *IEEE Trans. Auto. Control* **AC-11**, No. 2 (1966), 205–211.
162. Gagliardi, R. M., Input selection for parameter identification in discrete systems, *IEEE Trans. Auto. Control* **AC-12**, No. 5 (1967).
163. Goodwin, G. C., and Payne, R. L., Design and characterisation of optimal test signals for linear single input–single output parameter estimation, *Paper TT-1, 3rd IFAC Symp.*, The Hague/Delft (1973).
164. Goodwin, G. C., Payne, R. L. and Murdoch, J. C., Optimal test signal design for linear single input–single output closed loop identification, *CACSD Conf., Cambridge* (1973).
165. Goodwin, G. C., Murdoch, J. C., and Payne, R. L., Optimal test signal design for linear single input–single output system identification, *Int. J. Control* **17**, No. 1 (1973), 45–55.
166. Arimoto, S., and Kimura, H., Optimal input test signals for system identification—an information theoretic approach, *Int. J. Syst. Sci.* **1**, No. 3 (1971), 279–290.
167. Mehra, R. K., Optimal inputs for system identification, *IEEE Trans. Auto. Control* **AC-19** (1974), 192–200.
168. Smith, P. L., Test input evaluation for optimal adaptive filtering, *Preprints 3rd IFAC Sym.*, The Hague/Delft, Paper TT-5 (1973).
169. Atkinson, A. C., and Cox, D. R., Planning experiments for discriminating between models, *J.R. Stat. Soc.* **B36**, No. 3 (1974), 321–348.
170. Nahi, N. E., and Wallis, D. R., Jr., Optimal inputs for parameter estimation in dynamic systems with white observation noise, Paper IV-A5, *Proc. JACC*, Boulder, Colorado (1969), 506–512.
171. Inoue, K., Ogino, K., and Sawaragi, Y., Sensitivity synthesis of optimal inputs for parameter identification, Paper 9-7, *IFAC Symp., Prague* (1970).
172. Rault, A., Pouliquen, R., and Richalet, J., Sensitivizing inputs and identification, *Preprints 4th IFAC Congr.*, Warsaw, Paper 26.2 (1969).
173. Van den Bos, A., Selection of periodic test signals for estimation of linear system dynamics, Paper TT-3, *Preprints 3rd IFAC Symp.*, The Hague/Delft (1973).
174. Goodwin, G. C., Optimal input signals for nonlinear system identification, *Proc. IEE* **118**, No. 7 (1971), 922–926.
175. Goodwin, G. C., Input synthesis for minimum covariance state and parameter estimation, *Electronics Letters* **5**, No. 21 (1969).
176. Reid, D. B., Optimal inputs for system identification, Stanford University Rep., No.

SUDAAR 440 (1972).

177. Sawaragi, Y., and Ogino, K., Game theoretic design of input signals for parameter identification, Paper TT-6, *Preprints 3rd IFAC Symp.*, The Hague/Delft (1973).

178. Keviczky, L., and Banyasz, C. S., On input signal synthesis for linear discrete-time systems, Paper TT-2, *Preprints 3rd IFAC Symp.*, The Hague/Delft (1973).

179. Keviczky, L., On some questions of input signal synthesis, *Report 7226(B)*, Lund Inst. Tech., Div. Aut. Control (1972).

180. Aoki, M., and Staley, R. M., On approximate input signal synthesis in plant parameter identification, presented at *1st Hawaii Internat. Conf. Syst. Sci.*, Univ. of Hawaii (1968).

181. Aoki, M., and Staley, R. M., Some computational considerations in input signal synthesis problems, *2nd Internat. Conf. Computing Methods Optimization Problems*, sponsored by SIAM, San Remo, Italy (1968).

182. Aoki, M., and Staley, R. M., On input signal synthesis in parameter identification, *Automatica* 6 (1970), 431–440; originally presented at *4th IFAC Congress, Warsaw* (1969).

183. Nahi, N. E., and Napjus, G. A., Design of optimal probing signals for vector parameter estimation, Paper W9-5, *Preprints IEEE Conf. Decision and Control, Miami* (1971).

184. Napjus, G. A., Design of optimal inputs for parameter estimation, Ph.D. Dissertation, Univ. of Southern California (1971).

185. Tse, E., Information matrix and local identifiability of parameters, Paper 20-3, *JACC*, Columbus, Ohio (1973).

186. Minnich, G. M., Some considerations in the choice of experimental designs for dynamic models, Ph.D. Dissertation, Univ. of Wisconsin (1972).

187. Kalaba, R. E., and Spingarn, I., Optimal inputs and sensitivities for parameter estimation, *JOTA* 11, No. 1 (1973), 56–67.

188. Atkinson, A. C., and Federov, V. V., The design of experiments for discriminating between two rival models, *Biometrika* 62, No. 1 (1975), 57–69.

189. Mehra, R. K., Frequency domain synthesis of optimal inputs for parameter estimation, *Tech. Rep.* No. 645, Div. Eng. Appl. Physics, Harvard Univ. (1973).

190. Mehra, R. K., Synthesis of optimal inputs for multiinput–multioutput (MIMO) systems with process noise, *Tech. Rep.* No. 649, Div. Eng. Appl. Physics, Harvard Univ. (1974).

191. Van den Bos, A., Construction of binary multifrequency test signals, Paper 4-6, *IFAC Symp. Identification Prague* (1967).

192. Mehra, R. K., Optimal measurement policies and sensor designs for state and parameter estimation, *Milwaukee Symp. Auto. Control* (1974).

193. Zarrop, M. B., Payne, R. L., and Goodwin, G. C., Experiment design for time series analysis—the multivariate case, *Tech. Rep.* EE7506, Univ. of Newcastle (1975).

194. Payne, R. L., and Goodwin, G. C., Simplification of frequency domain experiment design for single input–single output systems, *Report* No. 74/3, Department of Computing and Control, Imperial College, London (1974).

195. Viort, B., D-optimal designs for dynamic models—Part I. Theory, *Tech. Rep.* 314, Department of Statistics, Univ. of Wisconsin (1972).

196. Goodwin, G. C., Zarrop, M. B., and Payne, R. L., Coupled design of test signal, sampling intervals and filters for system identification, *IEEE Trans. Auto. Control* AC-19, No. 6 (1974), 748–753.

197. Payne, R. L., Goodwin, G. C., and Zarrop, M. B., Frequency domain approach for designing sampling rates for system identification, *Automatica* 11, No. 2 (1975), 189–193.

198. Åström, K. J., On the choice of sampling rates in parameter identification of time series, *Inform. Sci.* 1 (1969), 273–287.

199. Gustavsson, I., Choice of Sampling intervals for parametric identification, *Report* 7103, Lund Inst. Tech. (1971).

200. Zarrop, M. B., Experimental design for system identification, *M.Sc. Dissertation*, Imperial College of Science and Technology (1973)

201. Butler, P., and Cantoni, A., Noniterative automatic equalization, *IEEE Trans on Comm.* **COM-23**, No. 6 (1975), 621–633.

202. Brown, R. F., and Goodwin, G. C., New class of pseudorandom binary sequences, *Electronics Letters* **3**, No. 5 (1967).

203. Speedy, C. B., Brown, R. F., and Goodwin, G. C., "Control Theory: Identification and Optimal Control," p. 72. Oliver and Boyd, Edinburgh, 1970.

204. Ng, T. S., and Goodwin, G. C., On optimal choice of sampling strategies for linear system identification, *Int. J. Control* **23**, No. 4 (1976), 459–475.

205. Dvoretsky, A. On stochastic approximation, *Proc. 3rd Berkeley Symp. on Math. Stat. and Problem* **1** (1956), 39–56.

206. Wong, E., "Stochastic Processes in Information and Dynamical Systems." McGraw-Hill, New York, 1970.

207. Yen, J. L., On nonuniform sampling of bandwidth limited signals, *IRE Trans. Circuit Theory* (1956), 251–257.

208. Robbins, M., and Monro, S., A stochastic approximation method, *Ann. Math. Stat.* **22** (1951), 400–407.

209. Anderson, B. D. O., Hitz, K. L., and Diem, N. D., Recursive algorithm for spectral factorization, *IEEE Trans. Circuits and Syst.* **CAS-21**, No. 6 (1974).

210. Rockafellar, R., "Convex Analysis." Princeton, Univ. Press, Princeton, New Jersey, 1970.

211. Zarrop, M. B., *Ph.D. Dissertation*, Imperial College (1977).

212. Wolovich, W. A., "Linear Multivariable Systems." Springer-Verlag, Applied Mathematical Sciences, No. 11, New York, 1974.

213. Blum, J., Multidimensional stochastic approximation methods, *Ann. Math. Stat.* **25** (1954), 737–744.

214. Wieslander, J., Real time identification—Part I, *Report* 6908, Div. Auto. Control., Lund Inst. Tech. (1969).

215. Kiefer, J., and Wolfowitz, J., Stochastic estimation of a regression function, *Ann. Math. Stat.* **23** (1952), 462–466.

216. Veltman, B., van den Bos, A., de Bruine, R., de Ruiter, R., and Verloren, P., Some remarks on the use of autocorrelation functions with the analysis and design of signals, *Proc. NATO Advanced Study Inst. Signal Proc.*, Loughborough, England (1973).

217. Coelho, D. P., The generation of binary sequences with prescribed autocorrelations, M.Sc. Dissertation, Imperial College (1973).

218. Evans, R. J., Optimal signal processing with constraints, Ph.D. Thesis, University of Newcastle (1975).

219. Rissanen, J. J., Minimax entropy estimation of models for vector processes, IBM Res. Lab., San Jose, *Tech. Report* (1975).

220. Brewer, M. J., Estimation of the state vector of a dynamic system using limited memory least squares, M.Sc. Thesis, Dept. Computing Control, Imperial College, London (1971).

221. Moore, J. B., and Anderson, B. D. O., "On martingales and least squares linear system identification, *Tech. Rep. No. EE7522*, Dept. Elect. Eng., Univ. of Newcastle (1975).

222. Anderson, B. D. O., and Moore, J. B., "Optimal Filtering." Prentice Hall, New Jersey, 1977.

223. Caines, P. E., and Ljung, L., Asymptotic normality and accuracy of prediction error estimators, *Res. Rep. No. 7602*, Univ. of Toronto, Dept. of Elect. Eng. (1976).

224. Caines, P. E., Prediction error identification methods for stationary stochastic processes, *Res. Rep. No. 7516*, Univ. of Toronto, Dept. Elect. Eng. (1975).

225. Vorchik, V. G., Plant identification in a stochastic closed loop system, *Automatica i*

Telemekhanika, No. 4 (1975), 32–48.

226. Ng, T. S., Goodwin, G. C., and Anderson, B. D. O., On the identifiability of multiple input–multiple output linear dynamic systems operating in closed loop, *Automatica* (1977).

227. Kabaila, P. V., Ph.D. Thesis, Univ. of Newcastle, 1978.

228. Atkinson, A. C., and Federov, V. V., Optimal design: Experiments for discriminating between several models, *Biometrika* **62**, 2 (1975), 289–303.

229. Atkinson, A. C., Planning experiments for model testing and discrimination, *Math. Operationsforsch, U. Statist.* **6** (1975), 253–267.

230. Hanson, R. J., and Lawson, C. L., Extensions and applications of the Householder algorithms for solving linear least squares problems, *Math. Comput.* **23** (1969), 787–812.

231. Billingsley, P., The Lindeberg–Lévy theorem for martingales, *Proc. Am. Math. Soc.* **12** (1961), 788–792.

232. Ljung, L., Consistency of the least squares identification method, *IEEE Trans. Auto. Control* **21** (1976).

233. Levinson, N., The Wiener RMS error criterion in filter design and prediction, *J. Math. Phys.* **25** (1947), 261–278.

234. Morf, M., and Ljung, L., Fast algorithms for recursive identification, *Tech. Rep.*, Inform. Syst. Lab., Stanford Univ. (1976).

235. Gevers, M. R., On the identification of feedback systems, presented at *IVth IFAC Symp. Identification, Tbilisi, U.S.S.R* (1976).

236. Ljung, L., Soderstrom, T., and Gustavsson, I., Counterexamples to general convergence of a commonly used recursive identification method, *IEEE Trans. Auto. Control* **AC-20**, No. 5 (1975), 643–652, October 1975.

237. Gustavsson, I., Ljung, L., and Soderstrom, T., Identification of processes in closed loop: Identifiability and accuracy aspects, *4th IFAC Symp. Identification, Tbilisi, U.S.S.R.* (1976).

238. Phadke, M. S., and Wu, S. M., Identification of multiinput–multioutput transfer function and noise model of a blast furnace from closed loop data, *IEEE Trans. Auto. Control* (1974), 944–951.

239. Dablemont, S. P., and Gevers, M. R., Identification and feedback control of an industrial glass furnace, *Tech. Rep.*, Louvain Univ., Belgium (1976).

240. Panuska, V., A stochastic approximation method for identification of linear systems using adaptive filtering, *Proc. JACC* (1968).

241. Panuska, V., An adaptive recursive least squares identification algorithm, *Proc. 8th IEEE Symp. Adaptive Proc.* (1969).

242. Young, P. C., The use of linear regression and related procedures for the identification of dynamic processes, *Proc. 7th IEEE Symp. Adaptive Proc.* (1968).

243. Young, P. C., An extension to the instrumental variable method for identification of a noisy dynamic process, *Tech. Note CN/70/1*, Dept. Eng. Univ. of Cambridge, England (1970).

244. Young, P. C., Shellswelt, S. H., and Nethling, C. G., A recursive approach to time series analysis, *Rep. CUED/B-Control/TR.12*, Dept. Eng., Univ. of Cambridge (1971).

245. Young, P. C., and Hastings-James, R., Identification and control of discrete dynamic systems subject to disturbances with natural spectral density, *Proc. 9th IEEE Symp. Adaptive Proc.* (1970).

246. Kashyap, R. L., Estimation of parameters in a partially whitened representation of a stochastic process, *IEEE Trans. Auto. Control* **AC-19** (1974), 13–21.

247. Talmon, J. L., and van den Boom, A. J. W., On the estimation of transfer function parameters of process and noise dynamics using a single stage estimator, *Proc. 3rd IFAC Symp. Identification, The Hague* (1973).

248. Valis, J., and Gustavsson, I., Some computational results obtained by Panuska's method of stochastic approximations for identification of discrete time systems, Div. Auto. Cont., Lund Inst. of Tech., Sweden, *Report 6915* (1969).
249. Soderstrom, T., Ljung, L., and Gustavsson, I., On the accuracy problem in identification, *IFAC Cong., Boston* (1975).
250. Cadzow, J. A. "Discrete Time Systems." Prentice Hall, 1973.

Index

A
B 7
C 8
D 9
E 0
F 1
G 2
H 3
I 4
J 5